Perovskite:

A Structure of Great Interest to Geophysics and Materials Science

Geophysical Monograph Series

Including
Maurice Ewing Volumes
Mineral Physics Volumes

GEOPHYSICAL MONOGRAPH SERIES

Geophysical Monograph Volumes

1 **Antarctica in the International Geophysical Year** *A. P. Crary, L. M. Gould, E. O. Hulburt, Hugh Odishaw, and Waldo E. Smith (Eds.)*
2 **Geophysics and the IGY** *Hugh Odishaw and Stanley Ruttenberg (Eds.)*
3 **Atmospheric Chemistry of Chlorine and Sulfur Compounds** *James P. Lodge, Jr. (Ed.)*
4 **Contemporary Geodesy** *Charles A. Whitten and Kenneth H. Drummond (Eds.)*
5 **Physics of Precipitation** *Helmut Weickmann (Ed.)*
6 **The Crust of the Pacific Basin** *Gordon A. Macdonald and Hisashi Kuno (Eds.)*
7 **Antarctica Research: The Matthew Fontaine Maury Memorial Symposium** *H. Wexler, M. J. Rubin, and J. E. Caskey, Jr. (Eds.)*
8 **Terrestrial Heat Flow** *William H. K. Lee (Ed.)*
9 **Gravity Anomalies: Unsurveyed Areas** *Hyman Orlin (Ed.)*
10 **The Earth Beneath the Continents: A Volume of Geophysical Studies in Honor of Merle A. Tuve** *John S. Steinhart and T. Jefferson Smith (Eds.)*
11 **Isotope Techniques in the Hydrologic Cycle** *Glenn E. Stout (Ed.)*
12 **The Crust and Upper Mantle of the Pacific Area** *Leon Knopoff, Charles L. Drake, and Pembroke J. Hart (Eds.)*
13 **The Earth's Crust and Upper Mantle** *Pembroke J. Hart (Ed.)*
14 **The Structure and Physical Properties of the Earth's Crust** *John G. Heacock (Ed.)*
15 **The Use of Artificial Satellites for Geodesy** *Soren W. Henricksen, Armando Mancini, and Bernard H. Chovitz (Eds.)*
16 **Flow and Fracture of Rocks** *H. C. Heard, I. Y. Borg, N. L. Carter, and C. B. Raleigh (Eds.)*
17 **Man-Made Lakes: Their Problems and Environmental Effects** *William C. Ackermann, Gilbert F. White, and E. B. Worthington (Eds.)*
18 **The Upper Atmosphere in Motion: A Selection of Papers With Annotation** *C. O. Hines and Colleagues*
19 **The Geophysics of the Pacific Ocean Basin and Its Margin: A Volume in Honor of George P. Woollard** *George H. Sutton, Murli H. Manghnani, and Ralph Moberly (Eds.)*
20 **The Earth's Crust: Its Nature and Physical Properties** *John G. Heacock (Ed.)*
21 **Quantitative Modeling of Magnetospheric Processes** *W. P. Olson (Ed.)*
22 **Derivation, Meaning, and Use of Geomagnetic Indices** *P. N. Mayaud*
23 **The Tectonic and Geologic Evolution of Southeast Asian Seas and Islands** *Dennis E. Hayes (Ed.)*
24 **Mechanical Behavior of Crustal Rocks: The Handin Volume** *N. L. Carter, M. Friedman, J. M. Logan, and D. W. Stearns (Eds.)*
25 **Physics of Auroral Arc Formation** *S.-I. Akasofu and J. R. Kan (Eds.)*
26 **Heterogeneous Atmospheric Chemistry** *David R. Schryer (Ed.)*
27 **The Tectonic and Geologic Evolution of Southeast Asian Seas and Islands: Part 2** *Dennis E. Hayes (Ed.)*
28 **Magnetospheric Currents** *Thomas A. Potemra (Ed.)*
29 **Climate Processes and Climate Sensitivity (Maurice Ewing Volume 5)** *James E. Hansen and Taro Takahashi (Eds.)*
30 **Magnetic Reconnection in Space and Laboratory Plasmas** *Edward W. Hones, Jr. (Ed.)*
31 **Point Defects in Minerals (Mineral Physics Volume 1)** *Robert N. Schock (Ed.)*
32 **The Carbon Cycle and Atmospheric CO_2: Natural Variations Archean to Present** *E. T. Sundquist and W. S. Broecker (Eds.)*
33 **Greenland Ice Core: Geophysics, Geochemistry, and the Environment** *C. C. Langway, Jr., H. Oeschger, and W. Dansgaard (Eds.)*

34 **Collisionless Shocks in the Heliosphere: A Tutorial Review** *Robert G. Stone and Bruce T. Tsurutani (Eds.)*

35 **Collisionless Shocks in the Heliosphere: Reviews of Current Research** *Bruce T. Tsurutani and Robert G. Stone (Eds.)*

36 **Mineral and Rock Deformation: Laboratory Studies—The Paterson Volume** *B. E. Hobbs and H. C. Heard (Eds.)*

37 **Earthquake Source Mechanics (Maurice Ewing Volume 6)** *Shamita Das, John Boatwright, and Christopher H. Scholz (Eds.)*

38 **Ion Acceleration in the Magnetosphere and Ionosphere** *Tom Chang (Ed.)*

39 **High Pressure Research in Mineral Physics (Mineral Physics Volume 2)** *Murli H. Manghnani and Yasuhiko Syono (Eds.)*

40 **Gondwana Six: Structure, Tectonics, and Geophysics** *Gary D. McKenzie (Ed.)*

41 **Gondwana Six: Stratigraphy, Sedimentology, and Paleontoloty** *Garry D. McKenzie (Ed.)*

42 **Flow and Transport Through Unsaturated Fractured Rock** *Daniel D. Evans and Thomas J. Nicholson (Eds.)*

43 **Seamounts, Islands, and Atolls** *Barbara H. Keating, Patricia Fryer, Rodey Batiza, and George W. Boehlert (Eds.)*

44 **Modeling Magnetospheric Plasma** *T. E. Moore, J. H. Waite, Jr. (Eds.)*

Maurice Ewing Volumes

1 **Island Arcs, Deep Sea Trenches, and Back-Arc Basins** *Manik Talwani and Walter C. Pitman III (Eds.)*

2 **Deep Drilling Results in the Atlantic Ocean: Ocean Crust** *Manik Talwani, Christopher G. Harrison, and Dennis E. Hayes (Eds.)*

3 **Deep Drilling Results in the Atlantic Ocean: Continental Margins and Paleoenvironment** *Manik Talwani, William Hay, and William B. F. Ryan (Eds.)*

4 **Earthquake Prediction—An International Review** *David W. Simpson and Paul G. Richards (Eds.)*

5 **Climate Processes and Climate Sensitivity** *James E. Hansen and Taro Takahashi (Eds.)*

6 **Earthquake Source Mechanics** *Shamita Das, John Boatwright, and Christopher H. Scholz (Eds.)*

Mineral Physics Volumes

1 **Point Defects in Minerals** *Robert N. Schock (Ed.)*

2 **High Pressure Research in Mineral Physics** *Murli H. Manghnani and Yasuhiko Syono (Eds.)*

Geophysical Monograph 45

Perovskite:

A Structure of Great Interest to Geophysics and Materials Science

Alexandra Navrotsky
Donald J. Weidner
Editors

American Geophysical Union, Washington, D.C.

Published under the aegis of AGU Books Board.

Library of Congress Cataloging-in-Publication Data

Perovskite: a structure of great interest to geophysics
and materials science.

(Geophysical monograph; 45)
Consists of proceedings and papers related to the Chapman Conference on Perovskite—a Structure of Great Interest to Geophysics and Materials Science; held in Bisbee, Arizona, Oct. 29–Nov. 2, 1987.

1. Perovskite—Congresses. I. Navrotsky, Alexandra II. Weidner, Donald J. 1945— . III. Chapman Conference on Perovskite—a Structure of Great Interest to Geophysics and Materials Science (1987: Bisbee, Ariz.) IV. Series.

QE391.P47P47 1989 549′.6 88-34984
ISBN 1-887590-071-2

Printed in the United States of America.

CONTENTS

PREFACE

Perovskite, $CaTiO_3$, was discovered and named in 1839 by Gustav Rose, German chemist and mineralogist (1798–1873), the year he was appointed professor at Berlin University. To Rose we owe sanidine (1808), anorthite (1823), and cancrinite (1859) as well. Alexander von Humboldt whom the Tsar of Russia had asked to explore the far reaches of his empire chose Rose as a fellow traveller. Rose's report "Reise nach dem Ural, Altai und dem Kaspischen Meer", was published in Berlin between 1837 and 1842. It is presumably there that Rose first mentioned perovskite.

Lev Alexeievitch Perovsky was born in 1792 at Kharkov, the son of a Ukrainian nobleman of Tartar stock. He pursued a military career and fought in the Napoleonic wars, until he was wounded in the campaign of 1814. He was promoted to colonel in 1818. He left the army in 1823. In 1841, he was appointed Secretary of the Interior and in 1852, Director of the Imperial Cabinet; he also was chairman of the committee set up to supervise the building of St. Isaac cathedral in St. Petersburg. In 1855, he devoted much effort to promoting the exploitation of the mines of the Altai district. During the Crimean war he mustered a regiment of riflemen and he was promoted to general and adjutant general to the Emperor. He died in 1856 in St. Petersburg.

This minor accessory mineral, named after a minor dignitary, has lent its name to all materials possessing similar structures. Because the perovskite structure uniquely accomodates both large and small cations, because distortions of the ideal cubic structure provide further flexibility for incorporating cations of different sizes, and because the structure is remarkably tolerant of vacancy formation and atomic-scale intergrowths with other structural motifs, perovskite-related compounds can be synthesized for an extremely wide variety of combinations of chemical elements. The resulting materials can be insulators, semiconductors, metals, and, as the past two years have shown, superconductors. They find technical application today in ceramics, refractories, and electronics, as possible hosts for nuclear waste. When the high T_c superconductors are commercialized, perovskite-related materials will presumably pioneer a host of new technologies.

Nature has put the perovskite structure to use in the dense hot ceramic interior of our planet. From the 670 km seismic discontinuity to the core-mantle boundary, an $MgSiO_3$-rich perovskite phase probably accounts for 50-90% of the volume of that region. Thus this material, accessible to us on the surface in microgram amounts in diamond-cell experiments and in milligram amounts by laborious "large volume" synthesis, may in fact be the most abundant single mineral in the Earth.

Even before the excitement of high-T_c made front-page news in late 1986, a group of mineral physicists had started planning a small interdisciplinary conference on the perovskite structure. Its purpose was to bring together the geophysics and materials science community to discuss fundamental questions of structure, stability, and properties in perovskite-related materials. We felt that the geophysicists, trying to model the lower mantle based on scanty data for the properties of $MgSiO_3$ perovskite had much to learn from the systematic study of perovskite structures done in the materials community. At the same time, the materials scientists could broaden their pressure and temperature horizons by considering $MgSiO_3$ perovskite. The variety of papers and of fields represented, the interest with which these were received, and the spirited discussions which ensued, indeed confirmed this impression and contributed to the success of the Chapman Conference on "Perovskite—A Structure of Great Interest to Geophysics and Materials Science" held in Bisbee, Arizona, Oct. 29–Nov. 2, 1987. This book is more a collection of papers related to the talks presented than a strict proceedings. It intermingles review material and new research results. We hope that it finds readers in both the earth science and materials science communities.

The organizers and editors owe thanks to our colleagues for their enthusiastic participation and to the National Science Foundation for financial assistance. Many people associated with the Bisbee Convention Center helped make the meeting easy and pleasant, and the town and its people provided a unique friendly atmosphere, splendid scenery, and a Halloween celebration to complement our serious deliberations. We thank you all.

Alexandra Navrotsky
Department of Geological and Geophysical Sciences
Princeton University

Donald J. Weidner
Department of Earth and Space Sciences
State University of New York at Stony Brook

Editors

Geophysical and Crystal Chemical Significance of (Mg, Fe)SiO3 Perovskite

Q. WILLIAMS,[1] E. KNITTLE[1] AND R. JEANLOZ

Department of Geology and Geophysics, University of California, Berkeley, CA 94720

Introduction

The importance of magnesium silicate perovskite in determining the thermal and chemical properties of the Earth's lower mantle is unquestionable. Available phase equilibrium data document that silicate perovskite is not only the most abundant mineral in the lower mantle, but also within the Earth itself [*Knittle and Jeanloz*, 1987a]. Both the density distribution and the thermal state of the lower mantle are critically dependent on the physical properties of silicate perovskite. Thus, an understanding of silicate perovskite's material properties is vital in the construction of all compositional and geodynamic models of the Earth.

This importance of silicate perovskite for the Earth sciences is a fortunate circumstance. The perovskite structure, or its close variants, are adopted by an extraordinarily broad range of chemical compositions (at least 700 distinct compositions as of 1970 [*Goodenough and Longo*, 1970]). As such, the material properties of substances with this structure have been extensively studied at zero pressure. Despite the interest in these materials from both the earth science and materials science communities, however, it is only in the last three years that the behavior of perovskites in general, and silicate perovskite in particular, has been probed at lower mantle pressures (P>24 GPa). Additionally, the effect of compositional variations on the internal structure of $MgSiO_3$ silicate perovskite has previously been uncharacterized. Again, only within the last two years has direct experimental evidence been brought to bear on the question of where iron, an element that is inferred to comprise between 5 and 10 weight per cent of the lower mantle, resides within the silicate perovskite structure.

In this paper, we review a series of measurements performed at both high pressures and zero pressure designed to illuminate the thermoelastic, structural, kinetic and electronic properties of $(Mg,Fe)SiO_3$ perovskites. These measurements include the synthesis pressures, thermal expansion at zero pressure, and equation of state across a wide range of pressures [*Knittle et al.*, 1986; *Knittle and Jeanloz*, 1987a]. Raman spectroscopy and pressure-dependent infrared spectra characterize the vibrational frequencies of $MgSiO_3$-perovskite [*Williams et al.*, 1987]. From these data, we derive values not only for the thermal expansion, bulk modulus and Grüneisen parameter of this important lower mantle phase, but we also document the extraordinary structural stability of magnesium silicate perovskite with respect to extremes of temperature and pressure. Our thermoelastic results are compared with the seismologically observed elastic properties of the lower mantle; such comparisons are required to infer possible constraints on the composition of the mantle, and thus on the style of mantle convection.

The Thermoelastic Properties of Silicate Perovskite

Constructing compositional models of the Earth's deep interior requires constraints on the density of silicate perovskite at high pressures and temperatures. Therefore, we have measured the isothermal bulk modulus ($K_T = -V(dP/dV)_T$), its pressure derivative ($K_T' = dK_T/dP$) and the zero-pressure thermal expansion coefficient ($\alpha = (1/V)(dV/dT)_P$) of $(Mg_{0.88}Fe_{0.12})SiO_3$ in the high-pressure perovskite structure [*Knittle et al.*, 1986; *Knittle and Jeanloz*, 1987a].

The room temperature equation of state of $(Mg_{0.88}Fe_{0.12}SiO_3$ perovskite measured to pressures of 112 GPa is reproduced in Figure 1. The experiments were done on perovskite synthesized at pressures between 25 and 127 GPa, and temperatures of about 2000 K, using the laser-heated diamond cell. The samples were x-rayed *in situ* without being quenched in pressure (further details are given in *Knittle and Jeanloz* [1987a]). Our data, and the earlier measurements of Yagi et al. [1982] to pressures of 10 GPa, were analyzed using a Birch-Murnaghan (Eulerian) third-order finite-strain equation of state formalism [*Birch*, 1978]. These yield zero-

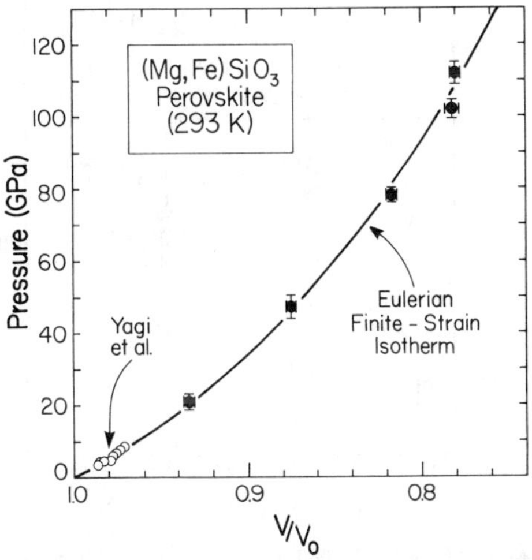

Fig. 1. Room-temperature isotherm of $(Mg_{0.9}Fe_{0.1})SiO_3$ perovskite measured to 112 GPa [*Knittle and Jeanloz*, 1987a]. The data were obtained by measuring the change in the unit cell dimensions of perovskite using x-ray diffraction through the diamond cell. The study demonstrates that perovskite is stable throughout the pressure range of the Earth's lower mantle.

---

[1]Present address: Institute of Tectonics, Board of Earth Sciences, University of California, Santa Cruz, CA 95064, U.S.A.

pressure values of K_T= 266 (+/- 6) GPa and K_T' = 3.9 (+/- 0.4) for this perovskite. The ten-fold increase in the pressure range of our data over that of Yagi et al., results in a robust constraint not only on the bulk modulus but also on its pressure derivative.

Notably, $MgSiO_3$-perovskite has the highest bulk modulus of any oxide perovskite known; while most titanate and aluminate perovskites have bulk moduli between 175 and 215 GPa, only $ScAlO_3$ (K_T= 249 (+/- 13) GPa) approaches $MgSiO_3$ in its incompressibility [*Bass*, 1984]. Also, this study demonstrates that the "external" structural distortion of perovskite, which is manifested by the ratios of the lattice parameters, remains constant over the 25 percent range of compression within the resolution of these measurements.

Our results were measured for a perovskite composition with a model lower mantle composition, $(Mg_{0.88}Fe_{0.12})SiO_3$, whereas the data of Yagi et al. were collected on the $MgSiO_3$ endmember. The difference between the bulk modulus derived from the lower pressure data of Yagi et al. (260 +/- 20 GPa), and the value based uniquely on our five static compression data points above 20 GPa (263 +/- 8 GPa) is not significant, however. The effect of iron on the bulk modulus of magnesium-silicate perovskite is therefore too small to resolve with the present data set.

The value for the bulk modulus of $MgSiO_3$-perovskite derived here is in accord with that expected from empirical correlations between bulk modulus and molar volume, and between sound speed and mean atomic weight, for a wide variety of oxide perovskites (K_T= 260 (+/-30) GPa) [*Bass*, 1984; *Liebermann et al.*, 1977]. Because of the small change in volume associated with iron substitution into silicate perovskite (as discussed below), these elasticity systematics also indicate that the effect on the bulk modulus of substituting 12 percent iron for magnesium on the bulk modulus of this material would be less than 1 percent [*Bass*, 1984; *Liebermann et al.*, 1977].

The only information to date on the effect of temperature on the volume of perovskite is a measurement of the zero-pressure thermal expansion of $(Mg_{0.88}Fe_{0.12})SiO_3$ [*Knittle et al.*, 1986]. The increase in volume as a function of temperature is shown in Figure 2. For these experiments, thirty to fifty samples of $(Mg_{0.88}Fe_{0.12})SiO_3$ perovskite (~1 microgram/sample) were quenched to ambient conditions after being synthesized at pressures of about 40 GPa and temperatures of about 2000 K using the laser-heated diamond cell [*Knittle et al.*, 1986]. It was only possible to collect data from room temperature to 850 K because above this temperature the metastable perovskite structure reverts to the low-pressure enstatite polymorph. The average thermal expansion coefficient over the 550 K temperature range of the data is 3.3 (+/- 0.5) x 10^{-5} K^{-1}. This value is comparable to the thermal expansion coefficients measured for other oxide perovskites over similar temperature ranges [Table 1: *Touloukian et al.*, 1977].

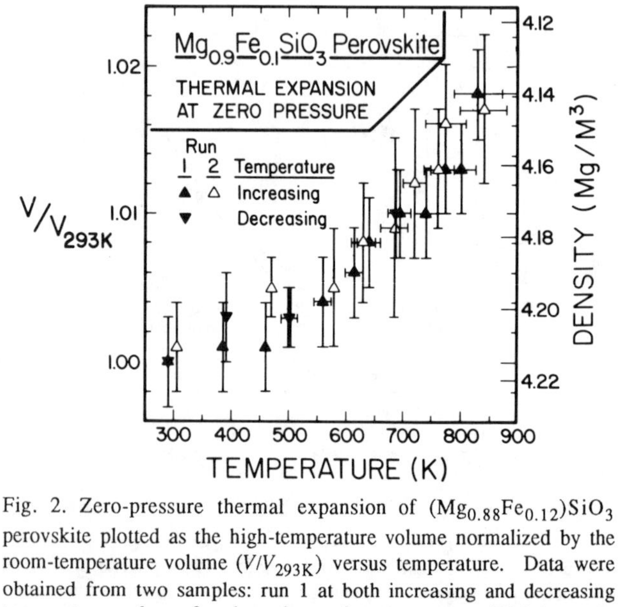

Fig. 2. Zero-pressure thermal expansion of $(Mg_{0.88}Fe_{0.12})SiO_3$ perovskite plotted as the high-temperature volume normalized by the room-temperature volume (V/V_{293K}) versus temperature. Data were obtained from two samples: run 1 at both increasing and decreasing temperature and run 2 only at increasing temperature [*Knittle et al.*, 1986].

TABLE 1. Thermal Expansion Coefficients of Perovskites[1]

Compound	Comments	$<\alpha>_{1000\ K}$ (10^{-5} K^{-1})
$BaTiO_3$	cubic phase	4.4 (above 393 K)
$SrTiO_3$	cubic phase	3.6
$KTaO_3$	cubic phase	2.7
$CaZrO_3$	cubic phase	3.6
$BaZrO_3$	cubic phase	3.0
$PbZrO_3$	cubic phase	3.2
$SrZrO_3$	orthorhombic phase	3.3
$Mg_{0.88}Fe_{0.12}SiO_3$	orthorhombic phase	3.9 ($\pm$0.5)[2]

[1]All data from Touloukian et al. [1977], except for the silicate perovskite value which is from Knittle et al. [1986].

[2] This value is extrapolated to 1000 K. The value of the thermal expansion coefficient at the highest temperature at which data was obtained (850 K) is 3.7 ($\pm$0.5) X 10^{-5} K^{-1}.

In order to be directly applicable to the Earth's lower mantle, the thermal expansion coefficient for perovskite must be extended to temperatures considerably beyond 850 K. We have therefore analyzed our high-temperature data using Grüneisen's theory of thermal expansion [*Knittle and Jeanloz*, in preparation; *Knittle et al.*, 1986; *Suzuki*, 1975; *Suzuki et al.*, 1979]. In this formalism, the high temperature volumes are fit to obtain values for the Grüneisen parameter (γ) and the Debye temperature (Θ) which can be used to calculate the thermal expansion coefficients at temperatures beyond the range of the available data. For $(Mg_{0.88}Fe_{0.12})SiO_3$ perovskite, values derived from the best fit to our data are tabulated in Table 2. These values differ slightly from those given by Knittle et al. [1986] because they are based on an improved theory of thermal expansion [*Jeanloz and Knittle*, in preparation]. Within the uncertainties of each fit however, they are indistinguishable.

The Grüneisen Parameter of Silicate Perovskite

The Grüneisen parameter of $MgSiO_3$-perovskite is of critical importance in determining the adiabatic gradient through the convecting lower mantle. In general, the value of this dimensionless parameter lies between 0 and 2, with higher values implying a higher thermal gradient. Two independent methods are available for evaluating this parameter, one based on the macroscopic thermodynamic properties of the crystal, and the

TABLE 2. Thermodynamic Properties of $(Mg,Fe)SiO_3$ Perovskite

	K_T (GPa)	K_T'	$<\alpha>_{850\ K}$ (10^{-5} K^{-1})	γ	Θ_{Debye} (K)
Knittle and Jeanloz, 1987a	266 ($\pm$6)	3.9 ($\pm$0.4)	---	---	---
Yagi et al., 1982	260 ($\pm$20)	4.0 (assumed)	---	---	---
Knittle et al., 1986					
Knittle and Jeanloz, 1988	---	---	3.7 ($\pm$0.5)	1.7 ($\pm$0.05)	700 ($\pm$50)
Williams et al., 1987	---	---	---	1.9 ($\pm$0.2)	620 ($\pm$50)[1]

[1]Value estimated from the observed vibrational spectrum of the crystal.

other on the behavior of the vibrational frequencies of the crystal under pressure. Thus, we may express the Grüneisen parameter as either:

$$\gamma = \alpha K_T V / C_V \qquad (1)$$

or

$$-\sum_i \frac{C_i V}{\omega_i} \frac{d\omega_i}{dV} = \sum_i c_i \gamma_i \qquad (2)$$

where α is the thermal expansion coefficient, K_T the isothermal bulk modulus, V the crystal volume and C_V the heat capacity at constant volume. Within the summation over the i vibrational modes of the crystal in (2), C_i is the contribution of the ith mode to the crystal heat capacity,

ω_i is the frequency of that mode, and γ_i is the mode Grüneisen parameter.

Using the thermodynamic definition of the Grüneisen parameter and a fit to the thermal expansion data of Knittle et al. [1986], a value of about 1.7 (+/- 0.05) is arrived at for the 300 K Grüneisen parameter [*Suzuki et al.*, 1979; *Knittle and Jeanloz*, 1988]. As an independent check on this number, we have undertaken a study of the shift in the mid-infrared vibrational bands of $MgSiO_3$-perovskite as a function of pressure [*Williams et al.*, 1987]. Our zero-pressure spectrum is shown in Figure 3, and the shifts of the four bands with pressure are shown in Fig. 4. The

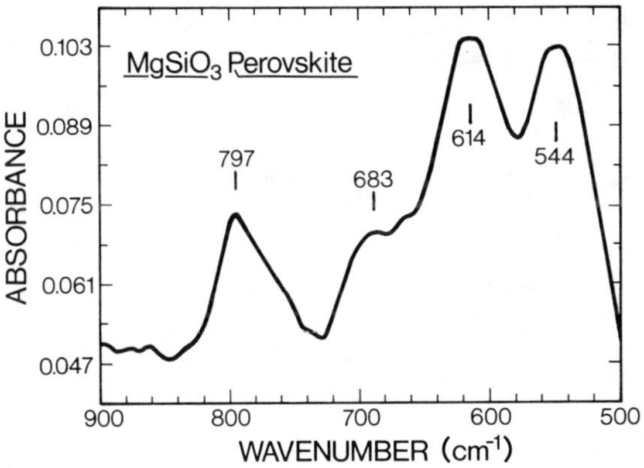

Fig. 3. Mid-infrared spectrum of $MgSiO_3$ perovskite measured at zero pressure and 300 K [*Williams et al.*, 1987].

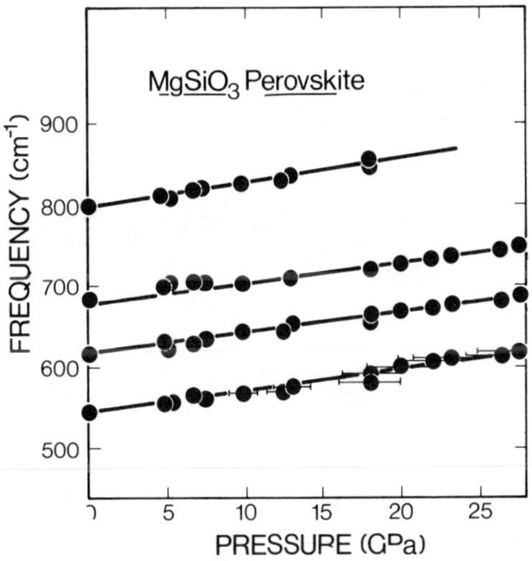

Fig. 4. Shift in the mid-infrared peak positions of $MgSiO_3$ perovskite as a function of pressure. The error bars represent the pressure uncertainty in each measurement [*Williams et al.*, 1987].

highest frequency band, with a zero- pressure frequency of 797 cm^{-1}, is assigned by analogy with the spectra of other perovskites to a stretching vibration of the SiO_6 structural unit. The three other bands are similarly associated with bending and twisting vibrations of this octahedron. The average mode Grüneisen parameter of these vibrations is 1.36 (+/-0.15), with the bulk of this error being due to the uncertainty in q, the logarithmic derivative of the Grüneisen parameter with respect to volume.

To derive an estimate of the thermodynamic Grüneisen parameter from mid-infrared data requires two assumptions: first, that the shifts of the Brillouin zone-center modes observed in infrared studies are representative of all vibrational modes of the crystal within the energy range of the measurement (that is, the effect of dispersion is small on average, and Raman-active and spectroscopically inactive modes at comparable energies behave similarly). Second, an assumption must be made concerning the behavior of lower frequency modes, particularly those involving motion of the magnesium cation. The probable location of these lower frequency bands may be derived from the Raman spectrum of $MgSiO_3$-perovskite (Figure 5). While the two higher frequency Raman bands are likely to be due to vibrations of the SiO_6 octahedral group, the two lower frequency modes are likely to be associated with motion of the magnesium ion (for

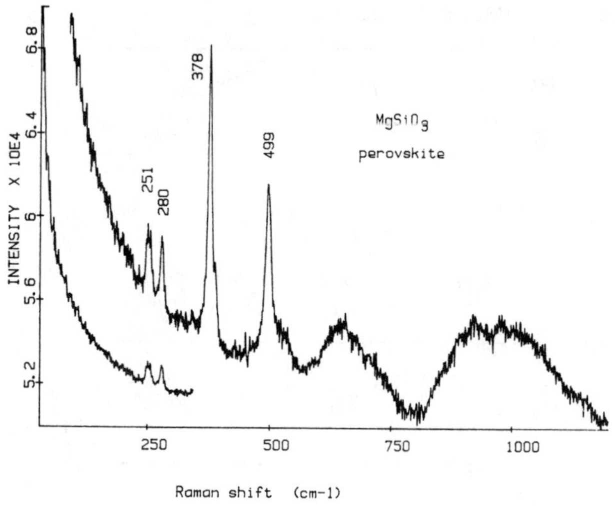

Fig. 5. The Raman spectrum of $MgSiO_3$ perovskite measured at zero pressure and 300 K [*Williams et al.*, 1987].

a summary of the vibrational normal modes of perovskite, see Figure 6). We estimate the thermodynamic Grüneisen parameter by deriving a model zero-pressure density of states for perovskite based upon the observed (Brillouin zone-center) vibrational spectra. Noting that the absolute spectral shift with pressure of our higher frequency bands is essentially constant, we obtain an estimate for the thermodynamic Grüneisen parameter of 1.9 (+/-0.2). More recent estimates of the Grüneisen parameter, based upon far-infrared spectra of perovskite under pressure [*Hofmeister et al.*, 1987], indicate that our assumption of constant frequency shift may slightly overestimate γ. Also, the Grüneisen parameter derived from a larger spectroscopic data set may be in better accord with the value derived from the thermal expansion data (Table 2).

Effect of Pressure and Temperature on Perovskite Distortions

Considerable mineralogical significance has been attached to the possibility that the distortion of the silicate perovskite structure may be sensitive to external conditions of temperature and pressure [*O'Keeffe et al.*, 1979; *Price*, 1986; *Wall et al.*, 1986]. However, the problem of the stability of the orthorhombic silicate perovskite structure has now been directly studied both experimentally and theoretically. X-ray diffraction shows that the lattice parameter ratios b/a and c/a of

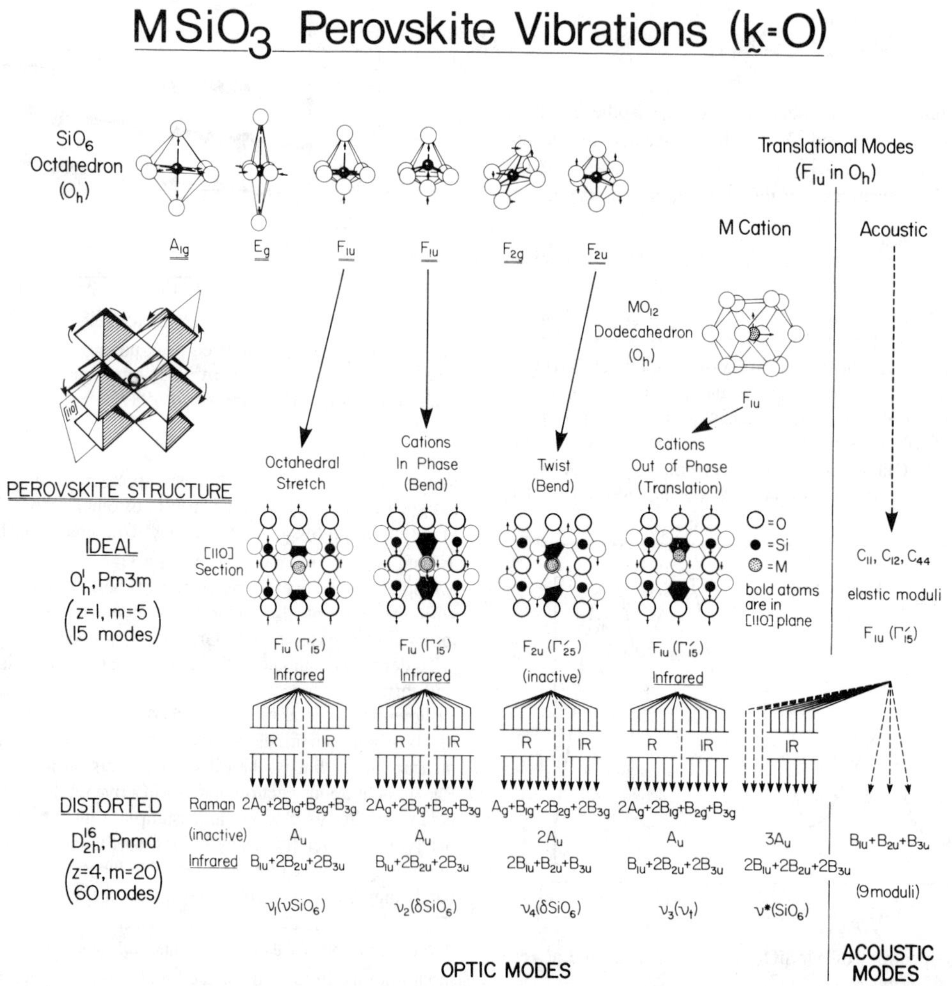

Fig. 6. Summary of the long-wavelength (k=0) vibrational modes of $MSiO_3$ (M=Mg, Ca) [see *Williams et al.*, 1987].

$(Mg_{0.88}Fe_{0.12})SiO_3$ perovskite, which are determined by the rotation and tilting of the silicon octahedra, do not change over a broad range of either pressure or temperature [*Knittle et al.*, 1986; *Knittle and Jeanloz*, 1987a; *Kudoh et al.*, 1987]. Although Kudoh et al. [1987] claim that they observe a change in the lattice parameter ratios with pressure, an examination of their data shows that these changes are not present in all of their high pressure measurements and are, if present at all, smaller in magnitude than their experimental uncertainties. In fact, their error bars on b/a and c/a are of the same magnitude as those measured by Knittle and Jeanloz [1987a] over a pressure range approximately 12 times larger. Thus, Kudoh et al.'s interpretation that their data show an increase in b/a

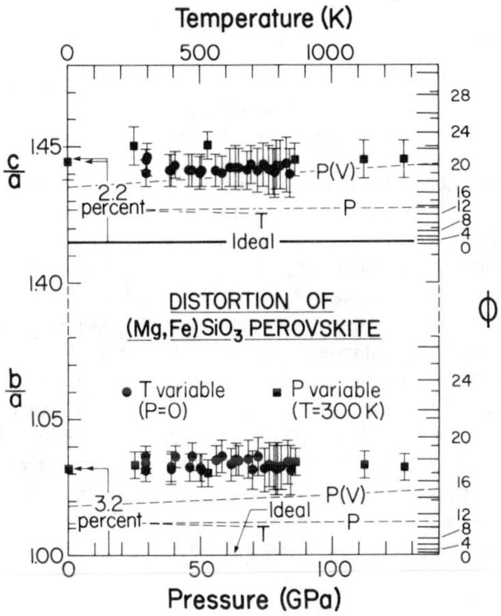

Fig. 7. Effect of pressure and temperature on the orthorhombic perovskite structure of $(Mg,Fe)SiO_3$. The degree of external distortion is characterized by either the lattice parameter ratios b/a and c/a (left-hand scale) or by Φ (right-hand scale): the angle of rotation of rigid SiO_6 octahedra about trigonal axes [see *O'Keeffe et al.*, 1979]. The unit cell of the perovskite is defined such that $b/a = 1$ and $c/a = \sqrt{2}$. The effect of temperature (top axis) on the perovskite structure is shown between room temperature and 840 K at zero pressure (circles with error bars: see *Knittle et al.* [1986] for details). The effect of pressure (bottom axis) on the lattice parameter ratios is shown to over 100 GPa at room temperature (square symbols: see *Knittle and Jeanloz* [1987a]). For both sets of experiments, the lattice parameters are constant within the error bars and are the same as the values at ambient conditions: b/a = 1.032 +/- 0.004 and c/a = 1.444 +/- 0.006. Arrows at left indicate the values of b/a and c/a reported by Ito and Matsui [1978] and Horiuchi et al. [1987]. The theoretical predictions of the changes in structural distortion with pressure and temperature are included as dashed lines. $P(V)$ labels the change in b/a and c/a with decreasing volume (increasing pressure) calculated by Hemley et al. [1987]. The results of Wolf and Bukowinski [1987] for high pressure and temperature are designated P and T, respectively. Although the calculations do not always reproduce the absolute values of b/a and c/a measured experimentally, they do show that the perovskite structure is not expected to undergo external distortions (octahedral tilting) that are resolvable by our experiments at elevated pressures and temperatures. That the angle Φ differs slightly for the observed (and calculated) values of b/a and c/a indicates that the perovskite structure involves internal distortions as well as tilting of the octahedra.

and a decrease in c/a is equivocal, and their data are equally interpretable as showing that both b/a and c/a are constant.

Combining the thermal expansion and compression measurements, the experimental values span a 25 percent range in the volume of the perovskite, with the volume ratio V/V_0 being between 1.02 and 0.78. Recent theoretical calculations of the structure of silicate perovskite are in accord with the experimental data [*Hemley et al.*, 1987; *Wolf and Bukowinski*, 1987]. Hence, both experiment and theory demonstrate that the external distortion, that is the tilt of the octahedra which is reflected in the b/a and c/a ratios, is constant within current limits of resolution (Figure 7). Rather, changes in the Si-O, Fe-O and Mg-O bond lengths are likely to be accomodated by internal distortions of the polyhedra.

The lack of any measureable effect of pressure and temperature on the external distortion of silicate perovskite contrasts with the behavior of most other perovskite-structured compounds. Among alkaline earth perovskites, only the titanates ($CaTiO_3$, $SrTiO_3$, $BaTiO_3$) have been examined in situ at pressures above 10 GPa, with $CaTiO_3$ having been heated and examined while held at high pressures [*Xiong et al.*, 1986]. Of the three, cubic $SrTiO_3$ alone has no documented room temperature phase transitions to a pressure of 20 GPa [*Edwards and Lynch*, 1970], despite the expectation that it is likely to undergo a tetragonal distortion at a pressure of 6-7 GPa [*Okai and Yoshimoto*, 1975]. $CaTiO_3$ undergoes a phase transition to a hexagonal form at about 10 GPa and room temperature. When it is quenched from high temperature at high pressures, $CaTiO_3$ is found to have transformed to a tetragonal form at 8.5 GPa, and to the hexagonal polymorph at 16 GPa [*Xiong et al.*, 1986]. $BaTiO_3$ transforms from a tetragonal to a cubic structure at 1.9 GPa, and remains stable in the cubic form to at least 11 GPa at ambient temperatures [*Samara*, 1966; *Fischer and Polian*, 1987].

As temperature is varied at zero pressure, both $SrTiO_3$ and $BaTiO_3$ undergo thermally induced phase transitions. At temperatures below 105 K, $SrTiO_3$ becomes tetragonally distorted. Similarly, $BaTiO_3$ occurs in the ideal cubic structure above 398 K, while between 278 K and 398 K it is in a tetragonal structure. At lower temperatures, this compound is orthorhombically distorted. By contrast, orthorhombic $CaTiO_3$ appears to be stable between 148 and 1073 K, although its distortion appears to lessen with increasing temperature [*Kay and Bailey*, 1957]. Thus, the effect of increased temperature on these perovskites is, in general, to increase their symmetry. The effect of pressure on the temperatures of the thermally-induced transitions is more ambiguous. While in $BaTiO_3$, pressure enlarges the stability field of the cubic structure [*Decker and Zhao*, 1984], the effect of pressure is the opposite in $SrTiO_3$: the lower symmetry tetragonal phase is stabilized relative to the cubic structure with increased pressure [*Okai and Yoshimoto*, 1975]. Thus, the remarkable stability of $MgSiO_3$ perovskite in a structure with a specific amount of (orthorhombic) distortion (Figure 7), is one of the most interesting aspects of the crystal chemistry of this phase. Silicate perovskite exhibits not only the largest pressure-temperature stability field known for any perovskite, but one of the largest that is known for any mineral.

Role of Iron in Perovskite

Figure 8 shows the wide range of bond lengths and valences which iron takes on in the octahedral sites of perovskite structures. Within the $MgSiO_3$ framework, EXAFS (extended x-ray absorption fine structure) results for perovskite synthesized from Bamble enstatite, $(Mg_{0.88}Fe_{0.12})SiO_3$, indicate that Fe^{2+} enters into the octahedral site, displacing silicon. Modelling of the EXAFS yields an Fe-O bond length of 214 (+/- 4) pm. The pattern of Fe substituting for Si, although unexpected from the behavior of Fe in magnesium silicates at low pressure, is well-precedented for iron in other perovskites (Figure 8). Indeed, divalent iron doped into $SrTiO_3$ enters into octahedral coordination, although such substitution is often associated with an

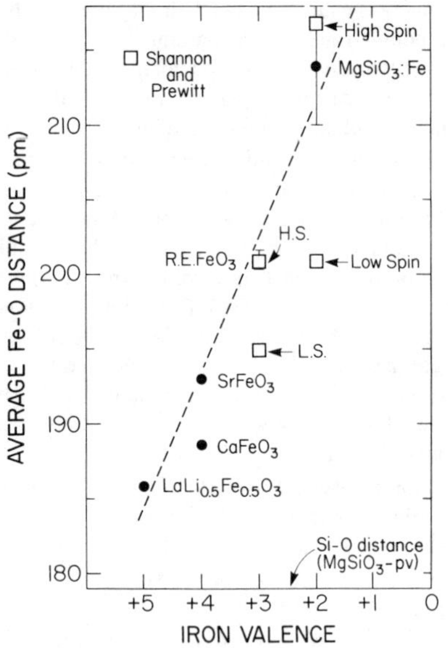

Fig. 8. Iron valence in five iron-bearing perovskites as a function of measured iron-oxygen distance. The square symbols are the Fe-O distances for high-spin and low-spin iron with iron valences of +2, +3 and +4 as tabulated by Shannon and Prewitt [1969] and Shannon [1976]. In all of the perovskites shown here, iron is six-fold coordinated (i.e. it resides in the octahedral site) and is in the high-spin state. Also plotted is the Si-O distance (178 pm) in the octahedral site of silicate perovskite ($MgSiO_3$). This illustrates the expansion that is required of the octahedron when iron is substituted for silicon into that site. The data are from Jackson et al. [1987], Buffat et al. [1986], and Takeda et al. [1978, 1986].

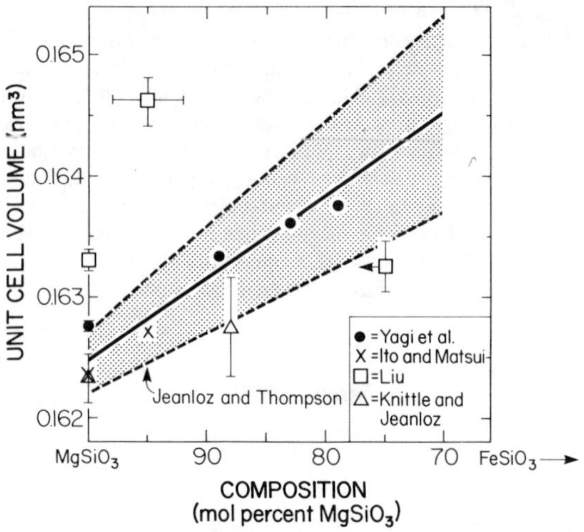

Fig. 9. Expansion of the unit cell volume of Mg-silicate perovskite with increasing iron content [*Yagi et al.*, 1978b; *Ito and Matsui*, 1978; *Liu*, 1975, 1976; *Knittle et al.*, 1986]. The stippled region indicates the unit cell volume and its uncertainty, as given by Jeanloz and Thompson [1983]. The $Mg_{0.95}Fe_{0.05}SiO_3$ point of Liu [1976] is the calculated composition of a perovskite run-product synthesized from $(Mg_{0.7}Fe_{0.3})_2SiO_4$ starting material. The calculation assumes stoichiometric decomposition to an iron-bearing perovskite of this volume and a magnesiowüstite whose 1-bar lattice parameter indicates a stoichiometry of $Fe_{0.55}Mg_{0.45}O$ [*Rosenhauer et al.*, 1975]. The arrow pointing from the higher iron concentration point of Liu [1976] indicates that this perovskite may be of lower iron content than his pyroxene stoichiometry starting material: (Mg,Fe)O magnesiowüstite of unknown composition was present within his samples.

oxygen vacancy [*Berney and Cowan*, 1981]. Also, there are a number of fluoride perovskites based upon Fe^{2+} in octahedral coordination [*Goodenough and Longo*, 1970].

In the case of $MgSiO_3$ perovskite, the difference between the Fe-O octahedral bond and the Si-O octahedral bond is approximately 35 (+/- 4) pm, reflecting the large size variations permitted by the perovskite structure on substitution. The tendency of the iron to partition into the octahedral site, in preference to the 8-12 fold coordinated site, is likely to be due to the highly directional d-orbitals in the valence shell of iron. The Fe^{2+} ion in the octahedral site has oxygen ions on-axis with each of its d-orbitals, allowing strong σ-type bonding within the octahedron. In fact, it appears that this affinity of iron for the octahedral site in silicate perovskite is large enough to actually force some silicon into the quasi-eightfold coordinate site, in a reaction which may be written:

$$(Mg^{VI}_{1-x}\,Fe^{VI}_{x})SiO_3\text{-pyroxene} \dashrightarrow (Mg_{1-x}Si_x)^{VIII-XII}(Fe_xSi_{1-x})^{VI}O_3\text{-perovskite}, \quad (3)$$

where the Roman numerals represent the number of nearest-neighbor oxygens surrounding each cation site. The EXAFS data therefore represent the first experimental indication for any oxide phase that silicon may occur in higher than 6-fold coordination.

The effect of iron content on the unit cell volume of $MgSiO_3$ perovskite is shown in Figure 9. The increase in volume of the crystal observed on substitution of 20 percent iron for magnesium (occupying the silicon site) is only about 0.6%. Similarly, the ratios of the lattice parameters, c/a and b/a, change by only a small amount, with the structure becoming slightly less distorted on iron substitution (Figure 10). Both of these observations are, we believe, reconcilable with the presence of comparatively large iron octahedra and with silicon ions being located in distorted sites of quasi-eight-fold coordination. That only small changes in unit cell volume accompany substitution of iron dictates that major internal distortions must occur within the unit cell of iron-bearing silicate perovskite.

It is notable that no silicate perovskite has been synthesized with an iron content greater than that corresponding to a formula of $Mg_{0.75}Fe_{0.25}SiO_3$ [*Yagi et al.*, 1978b]. For synthesis runs with somewhat higher iron content, mixtures of magnesiowüstite, stishovite and magnesium-rich ($\cong Mg_{0.95}$) perovskite are produced. It appears that the substitution of iron in quantities greater than 0.25 percent of the silicon content results in destabilization of the perovskite structure. That iron destabilizes the perovskite structure is also indicated by the strong tendency for iron to partition into any coexisting magnesiowustite: the distribution coefficient ($K_{pv/mw}$) is 0.08 (+/-0.03) [*Bell et al.*, 1979; *Ito and Yamada*, 1982].

These observations, combined with our earlier conclusion that there must be internal distortions associated with iron octahedra, suggests the presence of neighboring iron-bearing octahedra causes the silicate perovskite structure to break down. That is, the structure tolerates octahedral divalent iron as long as the Fe content is sufficiently low (x<0.25 in Equation 3) that two iron octahedra are not present within the endmember Pbnm (Z=4) unit cell. If correct, this model implies that iron is likely to be ordered within the silicate perovskite structure.

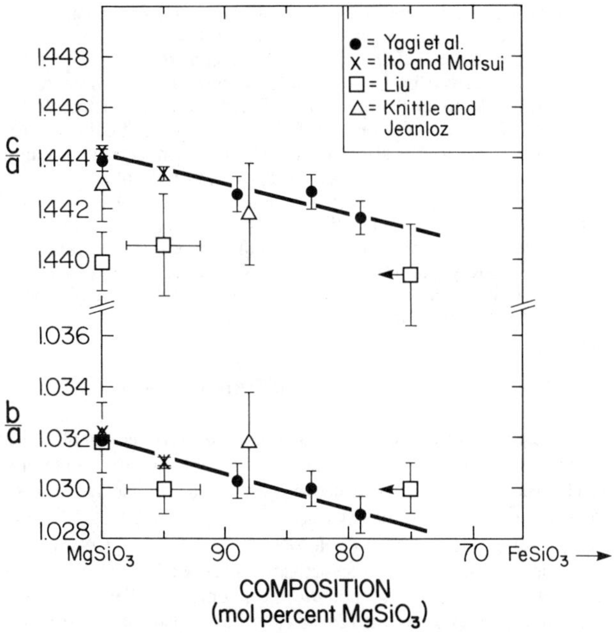

Fig. 10. Ratios of silicate-perovskite unit-cell parameters, *c/a* and *b/a*, as functions of iron content. The symbols and references are identical to those used in Figure 9. The addition of iron into the site normally occupied by silicon, increases all of the lattice parameters of perovskite and slightly decreases the lattice parameter ratios. Although the difference in the bond lengths of Si-O and Fe-O are large (178 pm versus 214 pm), there is only a 0.1 percent change in the *b/a* and *c/a* ratios. This indicates that the enlarged iron-bearing octahedron is accomodated by internal distortions within the unit cell, but these do not significantly change the external distortion of the unit cell as measured by the lattice parameter ratios.

The suggestion that there are no shared oxygens between iron octahedra is similar to a long-standing empirical rule concerning the structural role of tetrahedral aluminum in silicate glasses and minerals [*Lowenstein*, 1954]. We accordingly refer to this as "iron avoidance" in silicate perovskite. This iron avoidance resolves an apparent discrepancy between x-ray structural refinements performed on endmember $MgSiO_3$ perovskite [*Yagi et al.*, 1978a; *Ito and Matsui*, 1978; *Horiuchi et al.*, 1987], and the transmission electron microscopic observations of $Mg_{0.93}Fe_{0.07}SiO_3$ perovskite [*Madon et al.*, 1980]. The latter authors reported that the structure of their perovskite sample could only be indexed with a unit cell twice as large in every direction (Z=32) as that inferred from the x-ray diffraction measurements on endmember $MgSiO_3$-perovskite. Although Madon et al. [1988] have suggested that the extra reflections leading to this interpretation of their electron diffraction pattern may have been caused by twinning within the sample, we note that this new suggestion is based on observations of $MnGeO_3$, and not on new data on silicate perovskite. Thus, we attribute the apparent structural difference between $MgSiO_3$-perovskite and iron-bearing silicate perovskite to the ordering of iron defects within the sample containing iron. This would produce a unit cell that is twice as large in each direction due to the different cantings of alternately placed iron octahedra.

A possible structural model for the substitution of iron into the silicon site in $MgSiO_3$-perovskite is shown in Figure 11. We have expanded the central octahedron within the unit cell of orthorhombic (Z=4) perovskite to accomodate the larger radius of the Fe^{2+} ion. This expansion is performed in accord with a number of crystal-chemical constraints on the nearest-neighbor configurations of ions. First, all Si^{VI}-O bonds are held between 0.178 and 0.180 nm, the range of octahedral silicon-oxygen bond distances in $MgSiO_3$-perovskite [*Horiuchi et al.*, 1987]. The Fe^{VI}-O bonds are similarly constrained to lie between 0.210 and 0.218 nm, as dictated by the EXAFS results [*Jackson et al.*, 1987], and all Mg-O bonds are 0.200 nm or longer (the shortest Mg-O distance in endmember $MgSiO_3$-perovskite is 0.2014 nm: *Horiuchi et al.* [1987]). Furthermore, all oxygen-oxygen distances are longer than 0.216 nm., the shortest O-O separation in SiO_2-stishovite. We have also allowed oxygens to approach to within 0.185 nm. of the eight- to twelve-fold coordinated silicon displaced by the octahedral iron, in accord with the radius expected for silicon in this site based on percentage increases in radii observed in other ions for a comparable change in coordination [*Shannon and Prewitt*, 1969]. Finally, we have fixed the iron at the center of the orthorhombic unit cell. In actual fact, it may be energetically favorable for the iron to be off-center within the unit cell (particularly since the octahedron in which it sits is in an inherently asymmetric environment, with an eight- to twelve-fold coordinated silicon in an adjoining site). In this sense, our

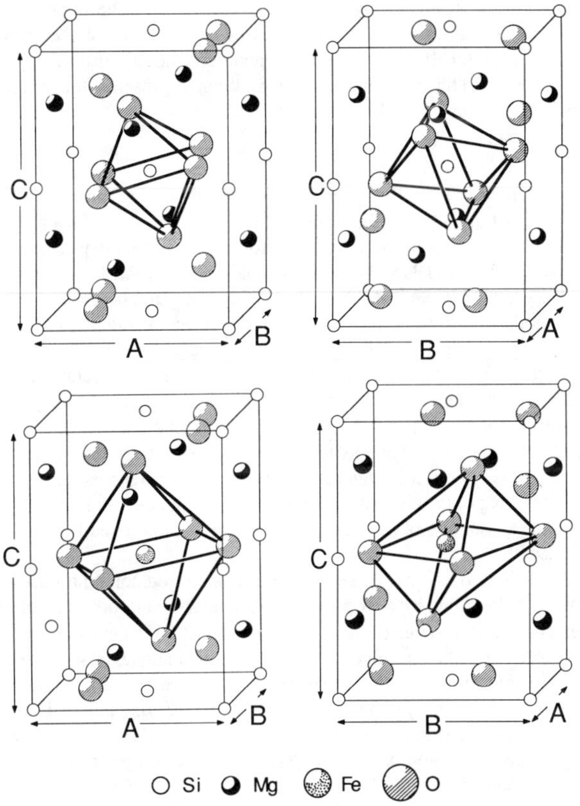

Fig. 11a-d. Perspective view of the orthorhombic (Z=4) unit cell of $MgSiO_3$-perovskite, viewed perpendicular to the 101 and 011 planes of the perovskite structure (11a and b, respectively), and the same crystallographic perspectives of our hypothetical structure for an $Mg_{0.9}Fe_{0.1}SiO_3$-perovskite with the larger Fe^{2+} ion substituting into the central octahedral site (11c and d). Note that a silicon ion has been substituted into a magnesium site on the left hand side of these cells. For more details, see text.

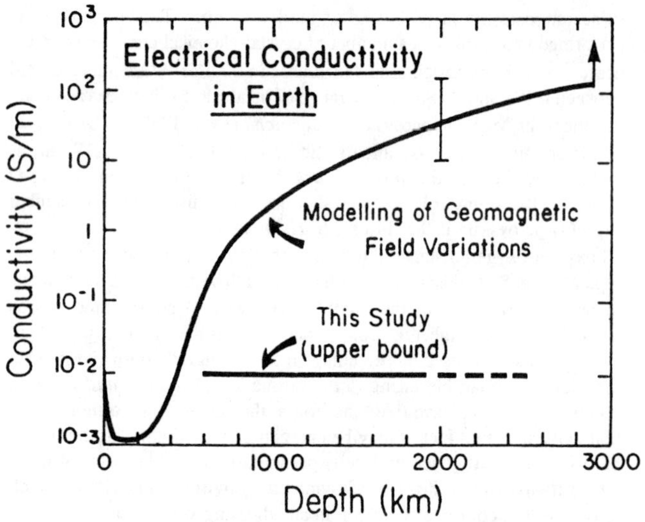

Fig. 12. Comparison of the electrical conductivity inferred for the mantle as a function of depth with the upper bound on the conductivity of perovskite as measured by Li and Jeanloz [1987]. The upper mantle conductivity values are derived from Banks' interpretation of geomagnetic variations [1969, 1972] and the lower mantle values are from Ducruix et al. [1980]. The error bar on the diagram represents the range of published values for the lower mantle.

model almost certainly represents an oversimplification of the actual atomic configuration in silicate perovskite.

Additionally, our model for the location of iron is not unique; there are probably other models which account equally well for the data. Its primary purpose is to indicate that the comparatively large iron octahedron can be accomodated within the silicate perovskite structure. Perhaps the most remarkable aspect of this simulation is that, for the Pbnm (Z=4) orthorhombic unit cell, the substitution of iron into the octahedral site with silicon entering the dodecahedral site can be accomplished without changes in the lattice parameters of the orthorhombic unit cell. Still, our model suggests that perovskite is likely to exhibit internal distortions which probably differ in sign between nearest-iron neighbors (particularly in the canting of the alternately-placed iron-bearing octahedra). This would produce a doubling of the unit cell similar to that discussed by Madon et al. [1980].

According to our model, six-fold coordinated iron could produce such unit-cell doubling by two possible mechanisms: 1) ordering of iron defects, due to variable cantings of the expanded Fe^{2+} octahedra, as just described; and 2) inter-unit cell distortions. For example, relaxation of the oxygen ions around silicon in the eight- to twelve-fold coordinated site could change the oxygen arrangement in the neighboring unit cell because this silicon polyhedron overlaps between two unit cells. Additionally, the expansion of the iron octahedron produces some oxygen rearrangement within neighboring unit cells, which could also generate the observed doubling of each unit cell dimension on introduction of iron into the lattice.

Finally, we note that as a consequence of our iron avoidance rule, it is likely that electron transfer between iron ions, one of the dominant conduction mechanisms in iron-rich silicates [*Schock and Duba*, 1984; *Bradley et al.*, 1964], is negligible even at the high temperatures of the Earth's lower mantle. In the case of iron-bearing magnesium silicate perovskite, we expect the iron ions to be isolated from one another by at least one silicon octahedron. This dictates that each iron center is separated from its nearest counterpart by more than 750 pm, thus inhibiting electron hopping between iron ions. Our expectation of low conductivity in iron-bearing silicate perovskites is in accord with recent measurements in perovskite-dominated assemblages under lower mantle conditions (Figure 12, [*Li and Jeanloz*, 1987]). Indeed, the upper bound for conductivity derived by Li and Jeanloz is substantially lower than the electrical conductivity that has been inferred for the lower mantle from modelling of the temporal variations of the geomagnetic field [*Ducruix et al.*, 1980]. These results also indicate that ionic conductivity within the anion sublattice, which has been proposed to take place in several analogue fluoride perovskites [*O'Keeffe and Bovin*, 1979; *Poirier et al.*, 1983], does not take place in silicate perovskite at mantle conditions. As the observations of ionic conductivity within these analogue perovskites has been plausibly attributed to non-intrinsic surface effects [*Andersen et al.*, 1985], it appears that ionic conductivity is not significant for perovskite within the Earth.

Kinetics of Perovskite Decomposition

We next address whether any natural occurances of metastable silicate perovskite might occur at the surface of the Earth, and the nature of the mechanism by which it could decompose. Figure 13 displays results on the back-transformation of $(Mg_{0.88}Fe_{0.12}SiO_3$-perovskite at 1 bar to its low pressure polymorph, enstatite [*Knittle and Jeanloz*, 1987b]. From these kinetic data, we obtain an activation energy of 70 (+/-20) kJ/mole. This value is a factor of two to seven times smaller than typical activation energies determined for creep, diffusion and phase transitions in low-pressure silicates. One consequence of this relatively small activation energy is that there should be essentially no kinetic hindrance to the perovskite phase transition occurring in the Earth's mantle.

The reversion can be modelled as a conventional nucleation and growth reaction, which may be associated with motion of either the Mg or Si cations through the faces of their respective polyhedra. Modelling this process suggests that the temperature dependence of this mobility is small

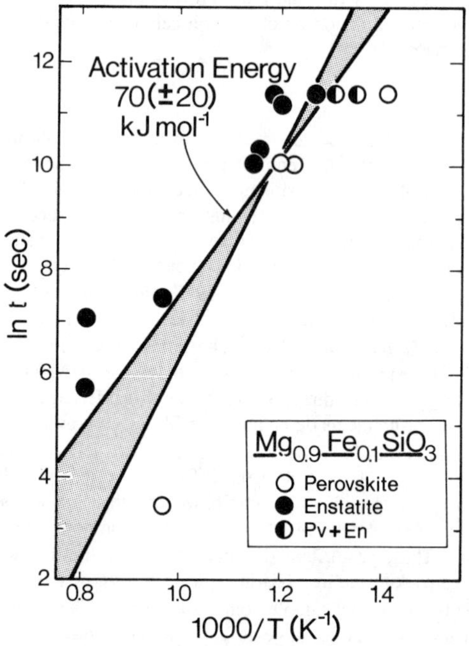

Fig. 13. Data on the back transformation of $(Mg,Fe)SiO_3$ perovskite to enstatite at zero pressure. The logarithm of the time of transformation (t) versus the reciprocal temperature (T) gives the activation energy of transformation as being 70 (+/- 20) kJ/mole [from *Knittle and Jeanloz*, 1987b].

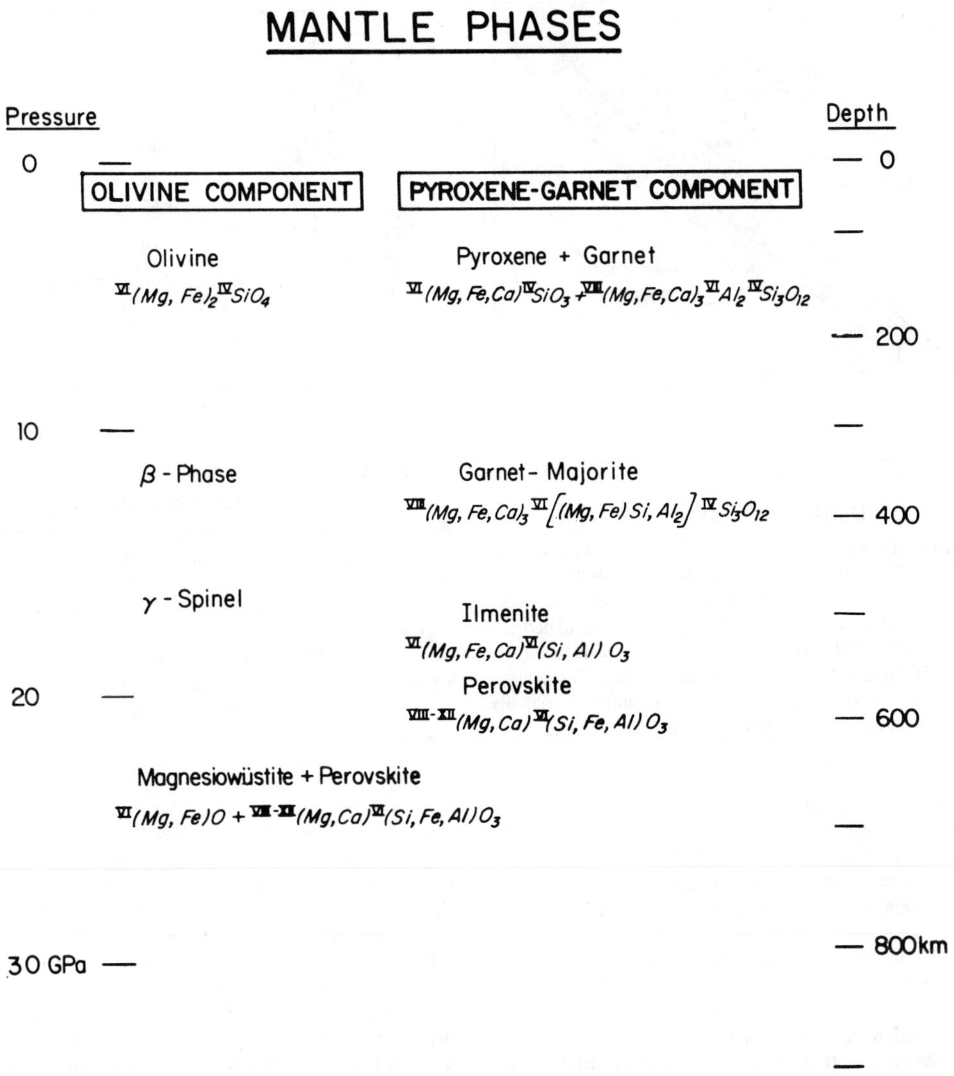

Fig. 14. Summary of the high-pressure mineral phases in the olivine and pyroxene-garnet components of an upper-mantle mineral assemblage as a function of pressure (left-hand scale) or depth in the Earth (right-hand scale). The coordination number for the cations (Roman numerals) is included with the chemical formula for each phase, and demonstrates the increase in cation coordination with pressure. The chemical formulae for β-phase and γ-spinel are identical to that of olivine [after *Jeanloz and Thompson*, 1983; see also *Jackson et al.*, 1987].

[*Knittle and Jeanloz*, 1987b]. This approach assumes that the cation motion is the rate determining step in the decomposition, and we note that the destabilization of the structure may be accomplished by simply moving each cation in the unit cell by about 0.4 nm over the time range necessary to decompose perovskite at a given temperature (Figure 13). Our model can apply to decomposition of perovskite at low pressure, or to chemical diffusion in perovskite at high pressures. Ionic diffusivities derived using these assumptions are less than 10^{-20} m^2/sec, indicating that this is unimportant. Also, the Nernst-Einstein relation implies that the electrical conductivity due to ionic mobility is orders of magnitude less than the bound derived in Figure 12. Therefore, both the absolute magnitude and temperature dependence of ionic motion within silicate perovskite may be small in comparison with low pressure silicates.

In the past, considerable excitement has accompanied observations of naturally occurring high-pressure phases, such as stishovite within meteorite craters and olivine composition spinels within meteorites [*Chao et al.*, 1962; *Binns et al.*, 1969]. However, because of the small activation energy associated with the back transformation of silicate perovskite, neither xenoliths containing silicate perovskite nor impact-generated perovskite would survive at 1 bar and 300 K for more than about 10^2 years before reverting to enstatite.

Silicate Perovskite and the Earth's Lower Mantle

High-pressure, high-temperature experiments show that the major upper mantle mineral phases, olivine ($(Mg,Fe)_2SiO_4$), pyroxene ($(Mg,Fe)SiO_3$) and garnet ($(Mg,Fe,Ca)_3Al_2Si_3O_{12}$), all transform to assemblages dominated by perovskite-structured silicates ($(Mg,Ca,Si)(Si,Al,Fe)O_3$) at the conditions of the Earth's lower mantle as shown in Figure 14 [*Liu*, 1979; *Akimoto et al.*, 1977; *Yagi et al.*, 1978b]. To determine the

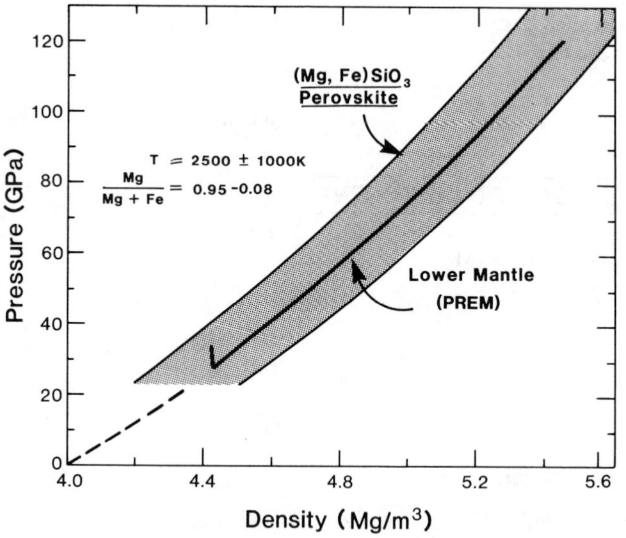

Fig. 15. A comparison of the density of $(Mg,Fe)SiO_3$ perovskite at high pressures and temperatures with the seismologically determined density in the lower mantle as given by PREM [*Dziewonski and Anderson*, 1981]. The stippled band represents perovskite densities calculated for Mg/(Mg+Fe)=0.80 to 0.95 and temperatures of 2500 +/- 1000 K using the data of Knittle et al. [1986] and Knittle and Jeanloz [1987a]. The density of perovskite calculated for a wide range of temperatures and iron contents is broadly compatible with the observed density of the lower mantle.

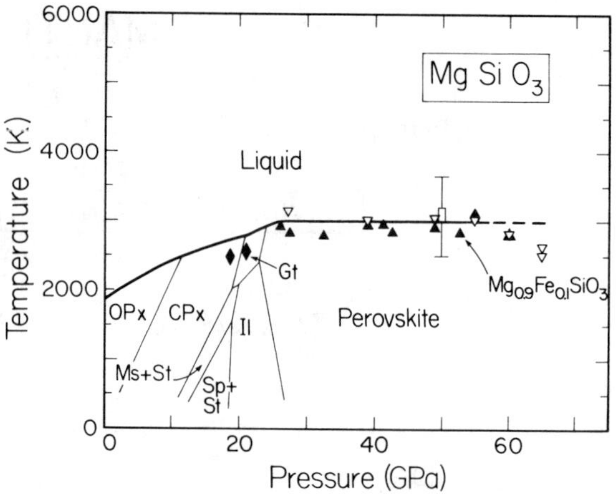

Fig. 16. Melting curve of $Mg_{0.9}Fe_{0.1}SiO_3$ perovskite plotted along with the currently accepted lower pressure phase diagram of $MgSiO_3$ [*Kato and Kumazawa*, 1985]. Solid upward pointing triangles represent the highest temperatures observed in solid perovskite at a given pressure, while open downward pointing triangles represent the lowest temperatures observed with molten perovskite being present. The open square and error bar represents an independent determination of the liquid-solid coexistence temperature. The two diamonds at pressures of about 20 GPa are experimental determinations of the melting temperature of this composition at pressures below the perovskite stability field [from *Heinz and Jeanloz*, 1987].

effect of both pressure and temperature on the density of silicate perovskite, we calculate a thermal equation of state using an anharmonic finite-strain theory of thermal expansion that is fitted to the thermal expansion and compressibility data that are summarized above [*Knittle and Jeanloz*, in preparation; *Jeanloz and Knittle*, in preparation]. A range of possible mantle iron contents (Mg/(Mg+Fe) = 0.8 to 0.95: *Ringwood* [1975]) is considered, with Figure 15 summarizing the results. It shows that a broad range of silicate perovskite compositions have densities consistent with that observed seismically within the lower mantle. This observation is in accord with the conclusion drawn from the phase equilibrium results that the silicate perovskite phase predominates within the lower mantle.

The melting curve of perovskite measured by Heinz and Jeanloz [1987] using the laser-heated diamond cell is shown in Figure 16. Over the range of the data, 25-60 GPa, the melting temperature of perovskite is unchanged by pressure. Interestingly, the melting temperatures of both pyrope and diopside have also been found to be relatively constant at pressures above about 8 GPa [*Ohtani et al.*, 1982; *Scarfe and Takahashi*, 1986]. The shallowness of the perovskite melting curve indicates that the volume change on fusion of this compound is nearly zero, in accord with shock-wave measurements which show that the densities of solid and liquid silicates tend to approach one another at high pressures [*Rigden et al.*, 1984]. Based on recent spectroscopic measurements, we conclude that this similarity in densities is caused by a change of oxygen coordination around silicon from four to six, as the melt is taken to high pressure [*Williams and Jeanloz*, 1988].

The pressure at which the silicon coordination begins to change in the liquid may be significantly lower than that at which the crystalline polymorphs undergo this change. In effect, the liquid undergoes anomalous compression as the pressure-induced change in coordination proceeds, thus reducing the volume difference between liquid and solid. The low fusion slope of perovskite is then likely to be produced by the close approach in coordination of the solid and liquid phases over a broad interval in pressure. At higher pressures, above 60 GPa, we would expect the transition from four- to six-fold coordination of oxygen around silicon to be essentially complete within the liquid; consequently, the fusion slope deep in the lower mantle is likely to become dictated by the relative compressibilities of the octahedrally coordinated liquid and silicate perovskite, rather than by coordination changes in the melt.

Because of this change in the relative densities of magmas and their coexisting crystals with increasing pressure, neutrally buoyant or even negatively buoyant melts may have been present within the deep mantle early in Earth's history. It is thus likely that the processes of chemical differentiation in the lower mantle and transition zone were markedly different than those presently observed at shallower depths, with incompatible elements that preferentially partition into a melt phase being enriched at depth.

Acknowledgments. We are grateful to T. Yagi and G.D. Price for helpful reviews, and to F. Guyot for preprints. The discussion of the unpublished work of Madon et al. was included at the request of one of the editors of this volume (D.W.). This work was supported by NSF and NASA. R.J. thanks the Fairchild Scholar Program at the California Institute of Technology for their support. Contribution no. 59 of the Institute of Tectonics at U.C. Santa Cruz.

References

Akimoto, S.I., T. Yagi and K. Inoue, High-temperature-pressure phase boundaries in silicate systems using in situ x-ray diffraction, in *High Pressure Research Applications in Geophysics*, edited by M.H. Manghnani and S.I. Akimoto, pp. 585-592, Academic, Orlando, Fla, 1977.

Andersen, N.H., J.K. Kjems and W. Hayes, Ionic conductivity of the perovskites $NaMgF_3$, $KMgF_3$ and $KZnF_3$ at high temperatures, *Solid State Ionics, 17*, 143-145, 1985.

Banks, R.J., Geomagnetic variations and the electrical conductivity of the upper mantle, *Geophys. J. R. Astr. Soc., 17*, 457-487, 1969.

Banks, R.J., The overall conductivity distribution of the Earth, *J. Geomagn. Geoelect., Kyoto, 24*, 337-351, 1972.

Bass, J., Elasticity of single-crystal $SmAlO_3$, $GdAlO_3$ and $ScAlO_3$ perovskites, *Phys. Earth Planet. Inter., 36*, 145-156, 1984.

Bell, P.M., T. Yagi and H.-K. Mao, Iron-magnesium distribution coefficients between spinel [$(Mg,Fe)_2SiO_4$], magnesiowüstite [(Mg,Fe)O], and perovskite [$(Mg,Fe)SiO_3$], *Carnegie Inst. Washington Year Book, 78*, 616, 1979.

Berney, R.L. and D.L. Cowan, Photochromism of three photosensitive Fe centers in $SrTiO_3$, *Phys. Rev. B 23*, 37-50, 1981.

Binns, R.A., R.J. Davis and S.B.J. Reed, Ringwoodite, Natural $(Mg,Fe)_2SiO_4$ spinel in the Tenham meteorite, *Nature, 221*, 943-944, 1969.

Birch, F., Finite strain isotherm and velocities for single-crystal and polycrystalline NaCl at high pressure and 300 K, *J. Geophys. Res., 83*, 1257-1267, 1978.

Bradley, R.S., A.K. Jamil and D.C. Munro, The electrical conductivity of olivine at high temperatures and pressures, *Geochim. Cosmochim. Acta, 28*, 1669-1678, 1964.

Buffat, B. , M.H. Tuilier, H. Dexpert, G. Demazeau and P. Hagenmuller, X-ray absorption investigation of some high oxidation states of six-coordinated iron in oxides of perovskite or K_2NiF_4-type structures, *J. Phys. Chem. Solids 47*, 491-496, 1986.

Chao, E.C.T., J.J. Fahey, J. Littler and D.J. Milton, Stishovite, a very high pressure new mineral from Meteor Crater, Arizona, *J. Geophys. Res. 67*, 419-421, 1962.

Decker, D.L. and Y.X. Zhao, Change in the $BaTiO_3$ phase transition to 40 kbar, in *High Pressure in Science and Technology, Part 1*, edited by C. Homan, R.K. MacCrone and E. Whalley, pp. 179-182, North-Holland, New York, 1982.

Ducruix, J., V. Courtillot and J. LeMouel, The late 1960s secular variation impulse, the eleven year magnetic variation and the electrical conductivity of the deep mantle, *Geophys. J. R. Astr. Soc., 61*, 73-94, 1984.

Dziewonski, A.M. and D.L. Anderson, Preliminary Reference Earth Model, *Phys. Earth Planet. Inter., 25*, 297-357, 1981.

Edwards, L.R. and W.R. Lynch, The high pressure compressibility and Grüneisen parameter of strontium titanate, *J. Phys. Chem. Solids, 31*, 573-574, 1970.

Fischer, M. and A. Polian, Elastic properties of $BaTiO_3$ at high pressure, *Phase Trans., 9*, 205-213, 1987.

Goodenough, J. B. and J. M. Longo, Crystallographic and magnetic properties of perovskite and perovskite-related compounds, in *Landolt-Bornstein, Zahlenwerte und Funktionen aus Naturwissenschaften und Technik, Neue Serie, Gruppe III*, Band 4, Teil a, pp. 126-314, Springer-Verlag, Berlin, 1970.

Heinz, D.L. and R. Jeanloz, Measurement of the melting curve of $Mg_{0.9}Fe_{0.1}SiO_3$ at lower mantle conditions and its geophysical implications, *J. Geophys. Res. 92*, 11437-14444, 1987.

Hemley, R.J., M.D. Jackson and R.G. Gordon, Theoretical study of the structure, lattice dynamics and equations of state of perovskite-type $MgSiO_3$ and $CaSiO_3$, *Phys. Chem. Minerals, 14*, 2-12, 1987.

Hofmeister, A.M., Q. Williams and R. Jeanloz, Thermodynamic and elastic properties of $MgSiO_3$ perovskite from far-IR spectra at pressure, *EOS, 68*, 1469, 1987.

Horiuchi, H., E. Ito and D.J. Weidner, Perovskite-type $MgSiO_3$: Single crystal x-ray diffraction study, *Amer. Min., 72*, 357-360, 1987.

Ito, E. and Y. Matsui, Synthesis and cystal-chemical characterization of $MgSiO_3$ perovskites, *Earth and Planet. Sci. Lett., 39*, 435-443, 1978.

Ito, E. and H. Yamada, Stability relations of silicate spinels, ilmenites and perovskites, in *High-Pressure Research in Geophysics*, edited by S. Akimoto and M.H. Manghnani, pp. 405-419, Center for Academic Publishing, Toyko, 1982.

Jackson, W.E., E. Knittle, G.E. Brown and R. Jeanloz, Partitioning of Fe in $(Mg,Fe)SiO_3$ perovskite: evidence for unusual geochemistry in the Earth's lower mantle, *Geophys. Res. Lett., 14*, 224-226, 1987.

Jeanloz, R. and D.L. Heinz, Experiments at high temperature and pressure: laser heating through the diamond cell, *J. Physique, 45*, C8-83-C8-92, 1984.

Jeanloz, R. and E. Knittle, Reduction of mantle and core properties to a standard state by adiabatic decompression, in *Chemistry and Physics of Terrestrial Planets*, edited by S.K. Saxena, pp. 275-309, 1986.

Jeanloz, R. and A.B. Thompson, Phase transitions and mantle discontinuities, *Rev. Geophys., 21*, 51-74, 1983.

Kato T. and M. Kumazawa, Garnet phase of $MgSiO_3$ filling the pyroxene-ilmenite gap at very high temperature, *Nature, 316*, 803-805, 1985.

Kay, H.F. and P.C. Bailey, Structure and properties of $CaTiO_3$, *Acta Cryst., 10*, 219-226, 1957.

Knittle, E., R. Jeanloz and G.L. Smith, The thermal expansion of silicate perovskite and stratification of the Earth's mantle, *Nature, 319*, 214-216, 1986.

Knittle, E. and R. Jeanloz, Synthesis and equation of state of $(Mg,Fe)SiO_3$ perovskite to over 100 GPa, *Science, 235*, 669-670, 1987a.

Knittle, E. and R. Jeanloz, The activation energy of the back transformation of silicate perovskite to enstatite, in *High Pressure Research in Mineral Physics*, edited by M.H. Manghnani and Y. Syono, pp. 243-250, American Geophysical Union, Washington, D.C., 1987b.

Kudoh, Y., E. Ito and H. Takeda, Effect of pressure on the crystal structure of perovskite-type $MgSiO_3$, *Phys. Chem. Min., 14*, 350-354, 1987.

Li, X. and R. Jeanloz, Electrical conductivity of $(Mg,Fe)SiO_3$ perovskite and a perovskite-dominated assemblage at lower mantle conditions, *Geophys. Res. Lett., 14*, 1075-1078, 1987.

Liebermann, R.C., L.E.A. Jones and A.E. Ringwood, Elasticity of aluminate, titanate, stannate and germanate compounds with the perovskite structure, *Phys. Earth Planet. Inter., 14*, 165-178, 1977.

Liu, L.-G., Post-oxide phases of olivine and pyroxene and mineralogy of the mantle, *Nature, 258*, 510-512, 1975.

Liu, L.-G., Orthorhombic perovskite phases observed in olivine, pyroxene and garnet at high pressures and temperatures, *Phys. Earth Planet. Inter., 11*, 289-298, 1976.

Liu, L.-G., Phase transformations and the constitution of the deep mantle, in *The Earth: Its Origin, Structure and Evolution*, edited by M.W. McElhinny, pp. 117-202, Academic, New York, 1979.

Lowenstein, W., The distribution of aluminum in the tetrahedra of silicates and aluminates, *Amer. Min., 39*, 92-96, 1954.

Madon, M., P.M. Bell, H.-K. Mao and J.P. Poirier, Transmission electron diffraction and microscopy of synthetic high-pressure $MgSiO_3$ phase with perovskite structure, *Geophys. Res. Lett., 7*, 629-632, 1980.

Madon, M., F. Guyot, J. Peyronneau and J.P. Poirier, Electron microscopy of high-pressure phases synthesized from natural olivine in diamond anvil cell, *Phys. Chem. Min., in press*, 1988.

Ohtani, E. , M. Kumazawa, T. Kato and T. Irifune, Melting of various silicates at elevated pressures, in *High Pressure Research in Geophysics*, edited by M.H. Manghnani and S. Akimoto, pp. 259-270, Center for Academic Publishing, Toyko, 1982.

Okai, B. and J. Yoshimoto, Pressure dependence of the structural phase transition temperature in $SrTiO_3$ and $KMnF_3$, *Proc. 4th Inter. Conf.*

on High Pressure, edited by J. Osugi, pp. 267-270, Phys. Chem. Soc. Japan, Kyoto, 1975.

O'Keeffe, M. and J. -O. Bovin, Solid electrolyte behavior of $NaMgF_3$: geophysical implications, *Science, 206*, 599-600, 1979.

O'Keeffe, M., B.G. Hyde and J.-O. Bovin, Contribution to the crystal chemistry of orthorhombic perovskites: $MgSiO_3$ and $NaMgF_3$, *Phys. Chem. Min., 4*, 299-305, 1979.

Poirier, J.P., J. Peyronneau, J.Y. Gesland and G. Brebec, Viscosity and conductivity of the lower mantle, an experimental study on a $MgSiO_3$ perovskite analog, $KZnF_3$, *Phys. Earth Planet. Int., 32*, 273-287, 1983.

Price, G.D., Perovskites and plate tectonics, *Nature, 319*, 175, 1986.

Rigden, S.M., T.J. Ahrens and E.M. Stolper, Densities of liquid silicates at high pressures, *Science, 226*, 1071-1074, 1984.

Ringwood, A.E., *Composition and Petrology of the Earth's Mantle*, 618 pp., McGraw-Hill, New York, 1975.

Rosenhauer, M., H.-K. Mao and E. Woerman, Compressibility of magnesiowustite $(Fe_{0.4}Mg_{0.6})O$ to 264 kbar, *Carnegie Inst. Wash. Yrbk., 75*, 513-515, 1976.

Samara, G.A., Pressure and temperature dependence of the dielectric properties of the perovskites $BaTiO_3$ and $SrTiO_3$, *Phys. Rev., 151*, 378-386, 1966.

Scarfe, C.M. and E. Takahashi, Melting of garnet peridotite to 13 GPa and the early history of the upper mantle, *Nature, 322*, 354-356, 1986.

Schock, R.N. and A.G. Duba, Point defects and the mechanisms of electrical conduction in olivine, in *Point Defects in Minerals*, edited by R.N. Schock, pp. 88-96, American Geophysical Union, Washington, D.C., 1984.

Shannon, R.D., Revised effective ionic radii and systematic studies of interatomic distances in halides and chalcogenides, *Acta. Cryst. A, 32*, 751-767, 1976.

Shannon, R. D. and C. T. Prewitt, Effective ionic radii in oxides and fluorides, *Act. Cryst. B, 25*, 925-946, 1969.

Suzuki, I., Thermal expansion of periclase and olivine, and their anharmonic properties, *J. Phys. Earth, 23*, 145-159, 1975.

Suzuki, I., S. Okajima and K. Seya, Thermal expansion of single-crystal manganosite, *J. Phys. Earth, 27*, 63-69, 1979.

Takeda, Y., K. Kanno, T. Takada, O. Yamamoto, M. Takano, N. Nakayama and Y. Bando, Phase relation in the oxygen nonstoichiometric system, $SrFeO_x$ $(2.5<x<3.0)$, *J. Sol. State Chem. 63*, 237-249, 1986.

Takeda, Y., S. Naka, M. Takano, T. Shinjo, T. Takada and M. Shimada, Preparation and characterization of stoichiometric $CaFeO_3$, *Mat. Res. Bull. 13*, 61-66, 1978.

Touloukian, Y.S., K.K. Kirby, R.E. Taylor and T.Y.R. Lee, editors, *Thermal Expansion, Nonmetallic Solids (TPRC 13)*, IFI/Plenum Press, New York, 1977.

Wall, A., G.D. Price and S.C. Parker, A computer simulation of the structure and elastic properties of $MgSiO_3$ perovskite, *Min. Mag., 50*, 693-707, 1986.

Wallace, D.C., *Thermodynamics of Crystals*, J. Wiley and Sons, New York, 484 pp., 1972.

Williams, Q. and R. Jeanloz, Spectroscopic evidence for pressure-induced coordination changes in silicate glasses and melts, *Science, 239*, 902-905, 1988.

Williams, Q., R. Jeanloz and P. McMillan, Vibrational spectrum of $MgSiO_3$ perovskite: Zero-pressure Raman and mid-infrared spectra to 27 GPa, *J. Geophys. Res. 92*, 8116-8128, 1987.

Wolf, G.H. and M.S.T. Bukowinski, Theoretical study of the structural properties and equations of state of $MgSiO_3$ and $CaSiO_3$ perovskites: implications for lower mantle composition, in *High Pressure Research in Mineral Physics*, edited by M.H. Manghnani and Y. Syono, pp. 313-331, AGU, Washington, D.C., 1987.

Xiong, D.H., L.C. Ming and M.H. Manghnani, High pressure phase transformations and isothermal compression in $CaTiO_3$ (perovskite), *Phys. Earth Planet. Int., 43*, 244-252, 1986.

Yagi, T., H.-K. Mao and P.M. Bell, Hydrostatic compression of perovskite-type $MgSiO_3$, in *Advances in Physical Geochemistry*, edited by S.K. Saxena, pp. 317-325, Springer-Verlag, New York, 1982.

Yagi, T., H.-K. Mao and P.M. Bell, Structure and crystal chemistry of perovskite-type $MgSiO_3$, *Phys. Chem. Minerals, 3*, 97-110, 1978a.

Yagi, T., H.-K. Mao and P.M. Bell, Lattice parameters and specific volume for the perovskite phase of orthopyroxene composition (Mg,Fe)SiO_3, *Carnegie Inst. Washington Yrbk.*, 612-613, 1978b.

SINGLE-CRYSTAL ELASTIC MODULI OF MAGNESIUM METASILICATE PEROVSKITE

Amir Yeganeh-Haeri and Donald J. Weidner

Department of Earth and Space Sciences, State University of New York
Stony Brook, NY 11794 USA

Eiji Ito

Institute for Study of the Earth's Interior, Okayama University
Misasa, Tottori-Ken 682-02, Japan

Abstract. The single-crystal elastic moduli of magnesium metasilicate $MgSiO_3$ in the perovskite structure, the highest pressure polymorph of $MgSiO_3$ pyroxene so far synthesized in the laboratory, have been determined for the first time under atmospheric pressure and 22°C using Brillouin spectroscopy. The elastic stiffness moduli in (GPa) are: C_{11}=515, C_{22}=525, C_{33}=435, C_{44}=179, C_{55}=202, C_{66}=175, C_{12}=117, C_{13}=117, C_{23}=139. The resulting isotropic aggregate **(VRH)** adiabatic bulk and shear moduli are 246.4 and 184.2 GPa, respectively. Overall, the large magnitude of the single-crystal elastic moduli are consistent with a structural-mechanical model where chains of rigid SiO_6 octahedra form a three dimensional framework.

A comparison of the elastic properties of the spinel and ilmenite phases with those of the perovskite phase suggests that the 670 kilometer seismic discontinuity may be due to either of these phases transforming to the perovskite phase. Furthermore, comparison of our new elasticity data with seismic earth models appropriate to the lower mantle reveals that petrological models having either pyrolite or pyroxene stoichiometries are compatible with the experimental data only if the shear modulus of $MgSiO_3$ perovskite has a very strong negative temperature derivative. Such a scenario would result if the perovskite phase is exhibiting a transverse mode softening (i.e., an acoustic mode instability) such as would be associated with a ferroelastic-paraelastic phase transition.

The resolving power of the seismic data are examined in terms of the trade-offs among the yet undetermined pressure and temperature derivatives of the elastic properties of the perovskite phase. The elasticity data will not be useful to distinguish between models of mantle chemistry without more information concerning the effects of pressure and temperature on the elastic moduli.

Introduction

For well over six decades one of the basic goals of the geophysical community has been to elucidate the structure, composition and state of the earth's mantle. Our primary motivation for determining the dynamical properties of the earth's mantle stems from a simple fact: this region alone, accounts for about 70% of the mass of the planet. An accurate characterization of the dynamic behavior of the earth's mantle is thus of fundamental importance, if we are to place any reasonable constraints on the origin, history, convective state and the mode of evolution of our planet. Toward this end, the variation of seismic wave velocities and density with depth are the most direct probes for examining the structure of earth's interior at depths greater than 200 kilometers.

On the basis of seismic studies the earth's mantle can be separated into three major parts: the upper mantle, the transition zone and the lower mantle. The transition zone (extending from ~350-700 km) is characterized by the presence of two major seismic discontinuities, one at depth of 400 km and the other at 670 km, and anomalously large velocity gradients. On the other hand, in the lower mantle (below ~700 km and extending to the D" layer) the seismic wave velocities vary smoothly and increase continuously with depth. These two sharp velocity discontinuities are either related to gross changes in chemical composition or to the presence of structural phase transitions occurring within a homogenous isochemical mantle. In the absence of laboratory data, neither of these two proposals can be assessed as to how well they account for the observed seismic velocity structure. These two differing view points, however, lead to completely different interpretations regarding the dynamic state of the earth's mantle. For instance, a chemical discontinuity at depth of 670 km would indicate a lack of communication between upper and lower mantle, suggesting distinct geochemical reservoirs and separately convecting cells within the earth's mantle. Specific mantle models are evaluated by comparing calculated acoustic velocities with observed seismic velocities. This test is enabled when laboratory data provide: (1) the equilibrium phase boundaries of transitions for mantle candidate phases, (2) elastic properties of the candidate phases corresponding to mantle pressures and temperatures. A successful model will satisfy both the magnitude and the topology of the seismic discontinuities. The elasticity data, in conjunction with high-pressure phase equilibrium studies, thus allow us to construct a seismic profile. By comparing the calculated profile to the one observed seismically, one can then assess the validity of a proposed petrological model.

Over the last two decades considerable advances have been made in the studies of phase transitions, mainly due to the evolution of ultra-high-pressure large volume multi-anvil systems developed in Japan. These high-pressure devices, capable of reaching pressures of 300 kbar and temperatures in excess of 2000°C, have also been indispensable for the synthesis of single crystals of high pressure phases which are of sufficient size and quality for physical property measurements. The recent synthesis of single-crystals of $MgSiO_3$ perovskite by Ito and Weidner (1986) underscores the importance of such high-pressure systems.

Concurrent with the evolution of high-pressure and high-temperature devices has been the development of Brillouin spectroscopy as a powerful tool for characterizing the mineral sound velocities in microcrystals. This technique is not only capable of defining the complete set of single-crystal elastic moduli for minerals of all symmetry systems but is particularly useful for defining elastic properties for single crystals as small as 50 μm in size. The recent adaptation of this method for measuring the zero pressure acoustic properties of the stishovite, modified spinel (β), spinel (τ), ilmenite and majorite phases—and the integration of these data with high-pressure phase equilibrium studies has provided a much firmer basis for interpretation of the seismic velocity structure of the earth's transition zone.

In this way, Weidner and Ito (1987) and Bina and Wood (1987) have demonstrated that, for example the α to β transition in the olivine component of the earth's upper mantle can adequately describe the 400 km seismic discontinuity depending on the yet unmeasured pressure and temperature derivatives of the elastic properties. Currently, there exists much debate regarding the relative proportions of candidate mantle phases and their pressure and temperature derivatives (Bass and Anderson, 1984; Anderson and Bass, 1986). Nonetheless, such modeling is particularly useful in bringing into focus the experimental studies which are most needed to better constrain the various petrological models which are compatible with the seismic observations.

The perovskite phase of $(Mg,Fe)SiO_3$, aroused a great deal of excitement in the geophysical community when first Liu (1976) reported on the disassociation of the spinel phase of Mg_2SiO_4 (ringwoodite) into perovskite plus periclase. Subsequent work by Ito and Matsui (1978) confirmed Liu's observation. Additional work conducted by Ito and Yamada (1981) and Ito et al. (1984) not only revealed that the equilibrium phase boundaries of the spinel to perovskite plus MgO and the ilmenite to perovskite transitions, occurs in a pressure interval which is comparable to that of the 670-km seismic discontinuity, but that volumetrically, $(Mg,Fe)SiO_3$ perovskite may be by far the most dominant mineral phase under the lower mantle conditions. Knittle and Jeanloz (1987) provided evidence that the stability field of $(Mg,Fe)SiO_3$ perovskite persists throughout the pressure-temperature regime of the entire lower mantle. The elastic properties of $MgSiO_3$ in the perovskite structure therefore, are not only of central importance in evaluating the nature of the 670 km seismic discontinuity but for further geophysical modeling of the earth's mantle.

The purpose of this paper is threefold. First, we wish to present the experimentally determined single-crystal elastic moduli of the perovskite phase of $MgSiO_3$. Second, we will examine our new elasticity data in relation to the elastic properties of other compounds in the perovskite structure from which some additional insights regarding the properties of the $MgSiO_3$ phase might be gained. Third, we wish to assess the resolution that this new data provide in distinguishing between the various petrological models appropriate to the earth's lower mantle, and to further underscore the need for the temperature and pressure derivatives of the elastic properties of the perovskite phase such that a better understanding of the composition and state of the earth's lower mantle can be developed in the future.

Experimental

Sample description. The single-crystals used in this study were grown from a melt at a pressure of 27 GPa and temperature of 1830°C using a large volume high-pressure split-sphere apparatus (USSA-5000) at Okayama University in Japan. The starting material was finely pulverized orthoenstatite crystals. Details of sample synthesis are given elsewhere (Ito and Weidner, 1986). On average the crystals were between 50-200 μm in diameter, optically transparent and colorless, and exhibited {110} and {001} type growth faces. Each specimen was examined for twinning by means of a 4-circle X-ray diffractometer and X-ray precession photographs. Although the majority of the crystals are not twinned some of them are. The twinning is consistent with a unit cell rotation of 90° about the **c** axis, such that the **a** axis in some domains lies parallel to the **b** axis in others. Following Aizu's (1970) and Sapreil's (1975) formulations, these twin domains actually correspond to 2 possible orientation states with equal and opposite resultant strains, and their occurrence is quite common in many distorted perovskite-type structures that are ferroelastics. Indeed, ferroelastic transformation is a well documented phenomenon in the distorted rare-earth orthoferrite and orthoaluminate type perovskites, isostructural with $MgSiO_3$ (Abrahams et al. 1972; 1974). Inasmuch as these domains are energetically equivalent, it is very likely, that twinning can result, during crystal growth, as the crystals minimize their free energy by dividing into a suitable number of twin domains (Forsbergh, 1949). Transparent, twin-free single crystals were selected for Brillouin scattering experiments. The crystals were oriented using an automated, 4-circle X-ray diffractometer, and then transferred to the Brillouin spectrometer with their orientation preserved.

Acoustic measurements. The apparatus for measuring Brillouin spectra has been described by Weidner et al. (1975) and Weidner and Carleton (1977). We employ an argon-ion laser operated in single-mode configuration, and a piezoelectrically-scanned, Fabry-Perot interferometer in triple pass configuration with related photon detecting equipment in 90° scattering geometry. The oriented crystal is mounted on a three-circle single-crystal goniometer and is suspended in a vial filled with a refractive fluid whose index approximates that of the crystal. Initially, the 5145 Å line of the argon ion laser was used to excite the Brillouin signal. However, the perovskite single crystals were found to deteriorate in this light. Laser light levels as low as 5 mW were sufficient to induce local twinning, and the migration of these twins eventually rendered the crystal opaque. X-ray examination of these samples revealed that the once twin-free single crystals become polycrystalline aggregates after a few days of exposure. Moreover, we discovered that the speed of this process correlated directly with the energy of the laser line. The 6471 Å line of a krypton-ion laser was found to cause no damage to the crystals. Acoustic velocities of the undamaged perovskites obtained from laser lines ranging from 4880 Å to 6764 Å did not show any systematic dependence on wavelength.

Compressional and shear acoustic velocities were measured in 26 distinct crystallographic directions, yielding a total of 34 independent mode velocities. The growth faces of the crystal were used to correct for the effect of refractive index mismatch between the crystal and the fluid as outlined by Vaughan and Bass (1983). Structural analysis using both

Table 1. Acoustic velocities in $MgSiO_3$ perovskite.

Wave Normal[1]			V_p(km/s)		V_s(km/s)[2]	
N_a	N_b	N_c	Obs	Cal	Obs	Cal
0.4901	0.1201	-0.8634			7.00	6.73
0.8065	0.4459	-0.3882			6.91	6.87
0.7053	0.7089	-0.0014			6.97	7.00
-0.9922	0.1245	0.0106			6.99	7.01
-0.4549	-0.5475	-0.7023			6.91	6.92
-0.5022	-0.6468	-0.5740			7.00	6.94
0.4875	0.5316	0.6926			6.88	6.92
0.5076	0.5527	0.6609			6.96	6.92
0.5305	0.6264	0.5712			7.14	6.94
-0.5022	0.2684	0.8220			7.00	6.81
0.1926	0.1992	-0.9608			6.83	6.94
-0.7071	0.7071	0.0025	10.89	10.97	6.85	7.00
0.0500	0.5008	-0.9975	10.38	10.30	6.95	7.00
0.9983	0.0139	0.0560			6.49	6.53
0.8892	-0.0209	0.4570			6.40	6.55
0.0352	0.9992	0.0206			6.52	6.61
0.0642	-0.9978	0.0134	11.20	11.30	6.48	6.61
-0.0516	-0.7917	0.6087			6.80	6.72
-0.0680	0.9440	0.3229	11.38	11.24	6.62	6.54
-0.0591	-0.9894	0.1324	11.17	11.29	6.61	6.60
-0.0542	-0.8732	0.4843	11.32	11.16		
-0.0317	-0.4291	0.9027	10.54	10.57	6.40	6.47
0.9960	0.0878	0.0160	11.15	11.19	6.98	7.01
0.5310	0.0232	0.8471			6.54	6.64
0.7761	0.0283	0.6300			6.55	6.60
0.0314	-0.0375	0.9988	10.23	10.29	6.58	6.61

RMS ERROR = .11 km/s; No of Data = 34

1. Wave Normal is given in terms of the direction cosines of the a, b and c crystallographic axes.

2. The polarization of the shear waves are known.

powder and single-crystal specimens have revealed that under ambient conditions magnesium metasilicate perovskite has orthorhombic symmetry (Horiuchi, 1987; Ito and Matsui, 1978). Therefore, a total of nine independent elastic moduli are needed to completely define its elastic properties. The linearized inversion method of Weidner and Carleton (1977) was used to determine the elastic moduli from the observed velocities. For modulus determination, the density of 4.108 g/cm^3 reported by Ito and Matsui (1978) was adopted. The root-mean-square deviation of the final model from the observed velocities is .11 km/s (Table 1). The single-crystal elastic moduli, and the calculated uncertainties along with the predicted aggregate elastic properties are given in Table 2. These data were obtained from 4 crystals.

Discussion

The perovskite structure and ferroelasticity. The perovskite structure is found in an overwhelming number of compounds with ABX_3 stoichiometry. The ideal cubic structure is illustrated in Figure 1. This structure can be imagined as network of regular BX_6 octahedra linked by their corners that extend symmetrically in three dimensions. The A cation occupies the cubical cavity formed by the eight octahedra and each is surrounded by 12

Table 2. Single-crystal elastic moduli of $MgSiO_3$ perovskite.

	C_{ij} (GPa)	S_{ij} (GPa^{-1})
11	515(5)	.00212
22	525(5)	.00214
33	435(5)	.00261
44	179(4)	.00557
55	202(3)	.00494
66	175(4)	.00570
12	117(5)	-.00035
13	117(5)	-.00046
23	139(6)	-.00059

Aggregate elastic properties (GPa) and acoustic velocities (km/s).

	Voigt	Reuss	Hill
K	247.10	245.70	246.4(5.0)
μ	184.90	183.50	184.2(4.0)
V_p	10.96	10.92	10.94
V_s	6.71	6.68	6.69
V_ϕ	7.75	7.73	7.74

Density = 4.108 g/cm^3
Linear Compressilities (GPa^{-1}):
β_a=0.00131; β_b=0.00120; β_c=0.00156
Refractive indices:
n_α=1.722:n_β=1.724:n_τ=1.738.

nearest neighbor oxygens (Glazer, 1972;1975). The cubic structure however, is found only in a few compounds. Due to variations in ionic size and small displacements of atoms, the actual structure found in a majority of perovskites is a pseudosymmetric variant (hettotype) of the ideal (aristotype) arrangement, resulting from small distortions of the unit cell and reduction of the overall symmetry (Megaw, 1973). The above structural description is also applicable to $MgSiO_3$ perovskite, where under ambient conditions, with rotation and tilting of SiO_6 octahedra, it resides in an orthorhombically distorted structure. Such distortions from the ideal structure often have profound effects on the physical properties. It is to such distortions that, cooperative Jahn-Teller, ferroelectric, ferromagnetic, and ferroelastic properties of perovskites have been ascribed (see for example, Rao and Rao, 1977). In fact, a great number of perovskite-type compounds (for example, $SrTiO_3$, $BaTiO_3$, $PrAlO_3$, $SmAlO_3$, $GdAlO_3$, $NaMgF_3$, $KMnF_3$) undergo a variety of structural phase transitions to higher symmetry phases with increasing temperature (Venkataramann, 1979). These transitions are primarily driven by condensation of either zone-boundary or zone-center phonon modes, representing different senses of rotation of the anion octahedra about their respective axes (Figure 1). In addition to softening of optic and acoustic modes, in many cases parallel anomalous behavior in thermal expansion and specific heat also occurs.

As cited previously, the rare-earth orthoferrite and orthoaluminate perovskites, isostructural with $MgSiO_3$, have been found to be ferroelastics. Based upon the observed twin domains and the ease by which untwinned crystals are twinned, we predict that $MgSiO_3$ can be also classified as a ferroelastic phase. Ferroelastic refers to materials which possess at least two stable orientation states, and have a

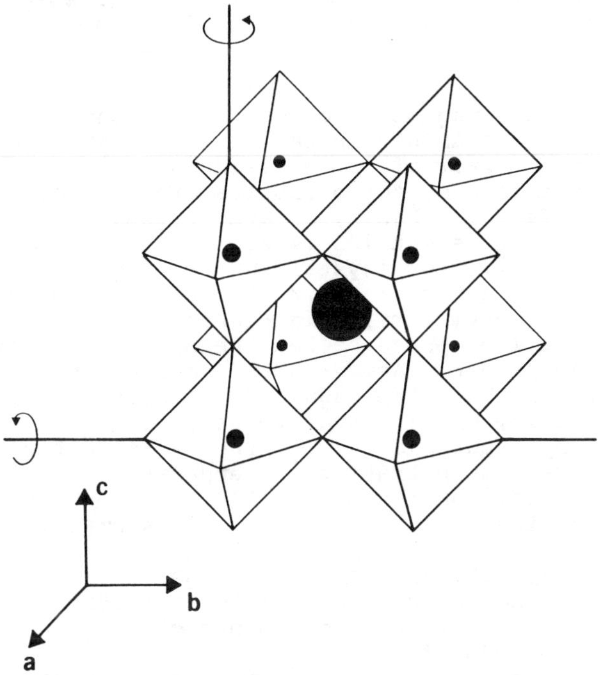

Fig 1. Three dimensional view of the ideal perovskite structure, illustrating chains of BX_6 octahedra. The rotations of these octahedra about their respective axis (as shown by the arrows) results in the distortions of the unit cell and the reduction of the overall symmetry from cubic.

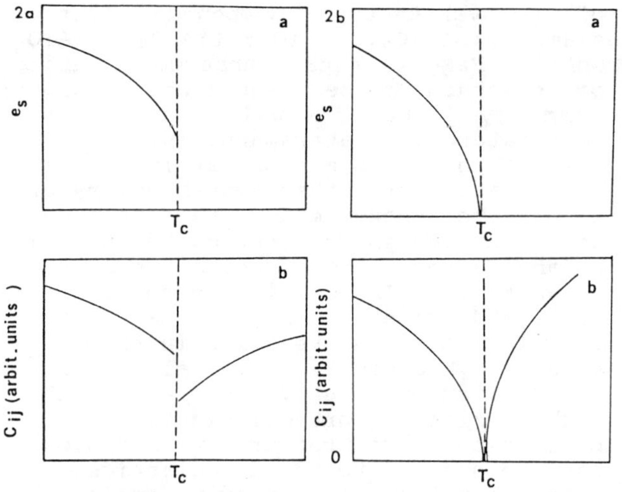

Fig 2. Schematic representation of e_s and the elastic stiffness modulus as a function of temperature for ferroelastic-paraelastic type transition for both first order (Fig. 2a) and second order (Fig. 2b) phase transition. T_c denotes the temperature of transition. Note that in either case the soft elastic modulus becomes stiffer upon further progression from the temperature of transition.

nonzero magnitude of some component of the spontaneous strain tensor in the absence of an external force field (Aizu, 1970). To this end, ferroelastics are a subgroup of ferroic phases, as are ferromagnetics and ferroelectrics (Aizu, 1981). Commonly, for crystals exhibiting ferroelastic transformations, there exists a high-symmetry, high-temperature idealized phase called the paraelastic or the parent phase. In this phase, by definition the magnitude of spontaneous strain, e_s, is equal to zero for temperatures above T_c, the ferroelastic-paraelastic transition temperature (Salje, 1985; Sapriel, 1975). Upon cooling below T_c, some of the symmetry elements of the parent phase are lost (via small distortions of the lattice) and the transition to a ferroelastic phase occurs having a nonzero magnitude of e_s.

The onset of these orientation states (twin domains) is thus related to the distortions from the parent phase, and the magnitude of spontaneous strain is a measure of the degree of this distortion relative to the parent structure. If the crystal is further classified as a proper ferroelastic, then e_s becomes the order parameter inferred in Landau's phenomenological theory of structural phase transitions (Cowely, 1980; Venkataramann, 1979). For such crystals, as T_c is approached from above or below, the lattice becomes increasingly unstable with respect to variations in at least one of the elements of the strain tensor, causing a divergence or "softening" of a particular component of the elastic stiffness tensor whose stiffness modulus is governed by that strain. For true-proper ferroelastics the soft mode is the acoustic phonon whose frequency and acoustic velocity tends towards zero as T_c is approached (Cowely, 1976; Taledano and Taledano, 1980). Figures 2a and 2b, illustrate the expected behavior of e_s and the elastic moduli as function temperature for first and second order transitions.

In order to define correctly the form of the strain tensor in the ferroelastic phase it is essential to know the space group and/or the point group symmetry of the paraelastic phase. The most common way of determining the symmetry is through high-temperature X-ray, structural analysis. In absence of such data, or if the crystal melts or disassociates before the critical temperature of transition is reached, the study of domain walls can be instrumental in deducing the symmetry of the paraelastic phase (Dvorak, 1978). Furthermore, theoretical and experimental study has shown that the resultant lattice strain (formation of domain walls) in the low-symmetry ferroelastic phase is the manifestation of the freezing in of the soft modes just below T_c. This indicates that the crystallographic directions in which the domain walls are found can be used to obtain the displacement and propagation directions of the soft acoustic modes which have condensed in these walls (David, 1983 and the references therein).

In the following, we wish to examine the consequences of ferroelasticity on the elastic properties of several oxide perovskites. Our objective is to put forth a hypothesis that might explain the trends observed in the elastic properties of these crystals, from which additional insights regarding the behavior of $MgSiO_3$ might be gained. We have selected the orthoaluminates for this comparison for several reasons: (1) to our knowledge they are the only crystals that have the same space group as $MgSiO_3$ for which single-crystal elastic moduli exist, (2) they are structurally very similar to $MgSiO_3$, and

Table 3. Single-crystal elastic properties of several oxide perovskites.

	$SmAlO_3$[1]		$GdAlO_3$[1]		$ScAlO_3$[1]		$MgSiO_3$	
	(1)	(2)	(1)	(2)	(1)	(2)	(1)	(2)
11	403	334	391	336	424	438	515	493
22	328	334	302	336	420	438	525	493
33	334	334	343	343	395	395	435	435
44	152	148	155	148	179	154	179	190
55	144	148	142	148	129	154	202	190
66	94	125	101	112	163	146	175	201
12	115	146	124	134	129	112	117	143
13	147	125	154	140	174	186	117	128
23	103	125	126	140	198	186	139	128
Distortion Parameters.[2]								
Φ°	2.49		8.02		19.35		14.35	
ϕ°	0.0		3.97		14.75		11.70	
(e_s)[3]	$7.0x10^{-4}$		$7.4x10^{-3}$		$4.1x10^{-2}$		$2.2x10^{-2}$	

1. Data from Bass (1984).
2. The tilt angles were calculated using the relation: $\Phi=\cos^{-1}a/b$; and $\phi=\cos^{-1}\sqrt{2}a/c$.
3. The magnitude of e_s at ambient conditions was calculated using the convention of Aizu (1970) for the ferroelastic species m3mFmmm.

(3) they offer an extended range of lattice distortions from cubic symmetry which makes them useful candidates for elucidating the effects of structural strains on elasticity.

In Table (3) we give the single crystal elastic moduli of $SmAlO_3$, $GdAlO_3$, $MgSiO_3$ and $ScAlO_3$ perovskites. Note that for each crystal two sets of elasticity data are presented. The single-crystal elastic moduli listed under column (1) correspond to the orthorhombic coordinate system, while those listed under column (2) have been rotated by 45° to coincide with the pseudocubic coordinate system. Our observation are summarized as follows: (1) For Sm and Gd orthoaluminate perovskites we note that the difference between the pure compressional moduli $C_{11} - C_{22}$ is large in the orthorhombic coordinate system. By contrast, the moduli C_{11} and C_{22} are very similar in magnitude for $ScAlO_3$ and $MgSiO_3$. In the rotated coordinate system (column 2), even though C_{11} and C_{22} are forced to be equal, C_{33} is very similar to them for all materials and the elastic moduli and have a pseudocubic form.

(2) As seen in Table 3 (column 1) modulus C_{66} decrease significantly upon going from $ScAlO_3$ to $SmAlO_3$.

(3) Overall, the relative magnitudes of the single-crystal elastic moduli of the two least distorted perovskites, $SmAlO_3$ and $GdAlO_3$, display a pattern which deviates far from cubic symmetry (that is, $C_{11}=C_{22}=C_{33}$, and so on), while the pattern observed in the elastic moduli of the two most distorted perovskite, $MgSiO_3$ and $ScAlO_3$, have the appearance of pseudocubic symmetry. Thus, the departure from cubic symmetry of the elastic moduli seems to increase substantially with decreasing degree of structural distortion from cubic symmetry.

A physical interpretation of the behavior described above, can be afforded on the basis of Landau's theory of phase transitions, and as schematically illustrated in Figures 2a and 2b. For this analysis, however several assumptions are required. First, we have assumed that the space group of our hypothetical parent phase is Pm3m. The highest conceivable space group for all perovskite-structured compounds. In fact, Coutures and Coutures (1984) have observed that as a function of temperature, $SmAlO_3$ ultimately transforms to a cubic phase. The crystallization of the orthoaluminates and $MgSiO_3$ in the orthorhombically distorted perovskite structure, space group Pbnm, can be imagined as a reduction in the symmetry operations of our parent phase. These perovskites thus belong to the ferroelastic species m3mFmmm, and the form of the strain tensor is given by Aizu (1970). The calculated magnitude of spontaneous strain at ambient conditions for each of the 4 perovskites considered is also listed in Table 3. Another key assumption in our analysis is that the progression from $ScAlO_3$ to $SmAlO_3$ is equivalent to raising the temperature towards T_c since the degree of distortion from cubic symmetry decreases systematically. This assumption is required since at this time no data exists on the temperature dependencies of the single-crystal properties of these crystals.

As can be seen, the magnitude of e_s is very small for $SmAlO_3$ and $GdAlO_3$ (Table 3). In the case of $MgSiO_3$ and $ScAlO_3$, however, e_s is quite large. On a comparative basis, therefore, it is very likely that the small magnitude of e_s has lead to a certain degree of pretransitional "softening" of the elastic properties of Sm and Gd orthoaluminate perovskites, thus giving rise the observed divergence of moduli C_{11} and C_{22} and the reduction in C_{66}. No such "softening" however, is observed in $ScAlO_3$ and $MgSiO_3$, consistent with the large magnitude of e_s for these material. Therefore it can be argued that the crystals of $SmAlO_3$ and $GdAlO_3$, have become increasingly more unstable with respect to strain of the type e_{xy} and/or $e_{xx}-e_{yy}$ (in the cubic coordinate system), than those of $MgSiO_3$ and $ScAlO_3$.

The crystallographic directions in which the twin domains are found in $SmAlO_3$, $GdAlO_3$ and $MgSiO_3$ is consistent with the displacement and propagation directions of acoustic velocities which correspond to these elastic moduli. Although we do not find anomalous behavior in the zero-pressure and room-temperature elastic properties of $MgSiO_3$, we speculate that it is quite conceivable that $MgSiO_3$ perovskite may exhibit critical shear-mode softening at elevated temperatures. In fact, recent theoretical calculations of Wolf and Bukowinski (1987) support this proposal. A surprising observation that we make is that $MgSiO_3$ has a large value of the shear modulus under ambient conditions. This high value may have significant geophysical implications. As will be discussed below an extremely strong temperature dependence of the shear modulus is required if we are to extrapolate the elasticity data from ambient conditions to mantle temperatures and pressures.

Comparison With Previous Results. While these are the first reported velocity measurements for the perovskite phase of $MgSiO_3$, there have been both theoretical and experimental studies in determining the elastic properties of this phase to which our data can be compared.

Pressure-volume data for perovskite have previously been reported by Yagi et al. (1982), Knittle and Jeanloz (1987) and Kudoh et al. (1987). The isothermal bulk modulus K_T and the assumed value of its pressure derivative K'_T, along with our new bulk modulus, K_S are summarized in Table 4. No correction was made in converting our adiabatic bulk modulus to isothermal conditions. To a first approximation, at room temperature the term $(1+T\alpha\gamma)$ needed for the conversion of K_S to K_T, can be taken as unity.

Table 4. Bulk modulus (GPa) and pressure derivative of bulk modulus. DAC=Diamond anvil cell. Max.press= Maximum pressure attained.

K	K'	Max.Press	Method	Ref
258±20	3-5	8.2(GPa)	DAC	(1)
266±6	3.9	127(GPa)	DAC	(2)
247±14	4	9.6(GPa)	DAC	(3)
246.4±5	---	0.1(MPa)		(4)

1. Yagi et al. (1982).
2. Knittle and Jeanloz (1987).
3. Kudoh et al. (1987).
4. This study.

Overall, the results of these experiments compare reasonably well with our value. When comparing these data to each other, however, several points must be taken into consideration. First, the pressure-volume relations reported by Yagi et al. (1982) and Knittle and Jeanloz (1987) were determined using polycrystalline specimens of composition $MgSiO_3$ and $(Mg_{88},Fe_{12})SiO_3$, respectively. In principle, the bulk modulus can be determined by differentiation of the pressure-volume curve obtained from powder X-ray diffraction data. However, it is important to emphasize that the resolution of the data must be extremely high to allow the determination of the bulk modulus and its pressure derivative with any degree of certainty, for materials that have a large bulk modulus such as $MgSiO_3$ perovskite. Second, it is common practice to assume a value for dK/dP and then fit the pressure-volume data via an equation of state to obtain the bulk modulus. Thus, different assumptions of dK/dP give rise to different values for the bulk modulus. For a review of the sources of error that may affect the determination of K and K' from the differentiation of pressure-volume data see the work of Bass et al. (1981) and Bell et al. (1987).

The bulk modulus reported by Kudoh et al. (1987), however, was determined using single-crystals of composition $MgSiO_3$. The specimens used by the latter investigators were from the same synthesis run as the crystals used in our study. In fact, the bulk modulus, the trends, and the magnitudes of the linear compressibilities reported by these authors are in close agreement to our results (Table 2).

Several previous investigators have attempted to estimate the elastic properties of magnesium silicate perovskite via empirical elasticity systematics. In Table 5, we summarize and compare the elastic properties of variety of compounds in the perovskite structure to those obtained from this study. It is clear from this table that the shear modulus of the $MgSiO_3$ perovskite is by far much larger than that observed in other oxide perovskites. On the basis of single-crystal structural refinements, Horiuchi et al. (1987) suggest that the structure of $MgSiO_3$ can be represented as being only composed of chains of rigid and relatively regular SiO_6 octahedra that extend in three dimensions. To a first approximation, such an arrangement should inherently result in a lack of substantial degree of anisotropy in the elastic properties. Furthermore, following structural-mechanical formulations (Weidner et al. 1982) it is expected that such an arrangement would lead to an elastically stiff structure, inasmuch as it would be very hard to deform chains of SiO_6 octahedra that form the three dimensional framework. Inspection of the single-crystal elastic moduli of Table 2 reveals that this is indeed the case. The extremely large magnitudes of the single-crystal elastic moduli attest to the fact that silicon octahedron is indeed quite stiff under compression and shear. Consequently, this property leads to an extremely large shear modulus for $MgSiO_3$ perovskite as compared with analogue compounds. Moreover, the high shear rigidity of the SiO_6 octahedra is best reflected in the ratio of bulk to shear modulus, K/μ. By comparison, this ratio is about ~12 to ~60% smaller for $MgSiO_3$ perovskite than is for other oxide perovskites (Table 5). This behavior is consistent with previous observations on the elastic properties of stishovite, SiO_2, (Weidner et al. 1982) and ilmenite, $MgSiO_3$, (Weidner and Ito, 1985) where silicon is in 6-fold coordination and which also exhibit a large shear modulus as compared with their analogue counterparts. On the basis of bulk modulus systematics, however, Liebermann et al. (1977), were able to accurately estimate the bulk modulus of $MgSiO_3$ perovskite.

Table 5. Aggregate elastic (GPa) and acoustic velocities (km/s) for several compounds with perovskite type crystal structure.

	K	μ	V_p	V_s	K/μ	Ref
$MgSiO_3$	246	184	10.93	6.70	1.34	(1)
$SmAlO_3$	198	123	7.7	4.14	1.60	(2)
$GdAlO_3$	203	120	7.44	4.02	1.69	(2)
$ScAlO_3$	249	140	10.08	5.72	1.77	(2)
$CaTiO_3$	177	104	8.84	5.07	1.70	(3)
$CdTiO_3$	212	98	8.36	3.99	2.16	(3)
$CaGeO_3$	198	109	8.15	4.59	1.81	(3)
$SrTiO_3$	174	117	8.03	4.77	1.49	(4)

1. This study.
2. Bass (1984).
3. Liebermann et al. (1977).
4. Bell and Rupprecht (1963).

Recently, Cohen (1987) predicted the single-crystal elastic moduli of $MgSiO_3$ perovskite using the potential induced breathing model. The high value of the shear modulus and the aggregate elastic properties predicated by the PIB model are in agreement with our experimental data. According to Cohen (1987) and Hemley et al. (1988), the PIB model can yield pressure derivatives of the elastic moduli that are rather accurate. With the inclusion of our new data, it may now be possible to improve the PIB model such that more accurate pressure derivatives of the elastic moduli of $MgSiO_3$ can be determined in the future.

The 670-km Seismic discontinuity. The 670-km seismic discontinuity which separates the upper from the lower mantle has often been associated with the break-down of the spinel phase of $(Mg,Fe)_2SiO_4$ to perovskite plus magnesiowustite, and the transformation of the ilmenite phase to perovskite. Table 6 compares the aggregate elastic properties (VRH) of the spinel phase of Mg_2SiO_4 and the ilmenite phase of $MgSiO_3$ to those of MgO and the perovskite phase of $MgSiO_3$ under ambient conditions. As can be seen from this table, the transformation of ilmenite to perovskite is associated with an increase of ~8% in the compressional and ~13% in the shear wave velocities. By the same token, the disproportionation of the spinel phase to perovskite plus MgO increases the longitudinal and the shear wave velocities, by ~9% and ~14%, respectively. Both of these phase transformations can contribute to the 670 km seismic discontinuity. Furthermore,

Table 6. Isotropic aggregate elastic properties (VRH), of spinel, ilmenite, periclase and perovskite. V_p= longitudinal velocity; V_s= shear velocity; K = bulk modulus; μ= shear Modulus.

	spinel	ilmenite	MgO	perov.
V_p (km/s)	9.79	10.11	9.70	10.94
V_s (km/s)	5.77	5.90	6.05	6.69
K (GPa)	184	212	163	246.4
μ (GPa)	119	132	131	184.2

the transformation of complex garnet (majorite) to perovskite, which occurs over a broader pressure interval would significantly add to the complexity of this discontinuity (Ito and Takashi, 1987).

Compositional models of the earth's lower mantle range from pyrolite (silica poor) to pyroxene (silica rich) stoichiometries. Under lower mantle conditions, this variation corresponds to petrological assemblages of 80% (by volume) perovskite with 20% magnesiowustite to pure perovskite. Tests of compositional models can be afforded by comparing the elastic properties implied by the compositions with those inferred from seismology. While a large data base is available for the temperature and pressure variation of the elasticity of magnesiowustite, we have insufficient data to extrapolate the perovskite data from ambient conditions to mantle pressures and temperatures. If, in fact, our proposal regarding ferroelastic phase transitions in the perovskite phase is true, then there is no justification in accurately predicting the temperature and pressure derivatives of the elastic moduli from empirical systematics (because no detailed experimental elasticity studies have been performed through these transitions) at this time.

Instead, we have chosen to calculate the combinations of $(d\mu/dT)_P$, $(d\mu/dP)_S$, $(dK/dP)_S$ and $(dK/dT)_T$ for the perovskite phase which are required to match the elastic properties of the PREM model (Dziewonski and Anderson, 1981) at the depth of 1071 km, for both the pyrolite and the pyroxene compositional models. The results are illustrated in Figure 3, assuming a ratio of Mg/(Mg+Fe) of 0.9 and an adiabatic mantle temperature gradient which is 1400°C at zero pressure. It is clear form this figure that differences between the pyrolite and pyroxene lower mantle models are very difficult to resolve on the basis of the existing elastic properties. In fact, very accurate pressure and temperature

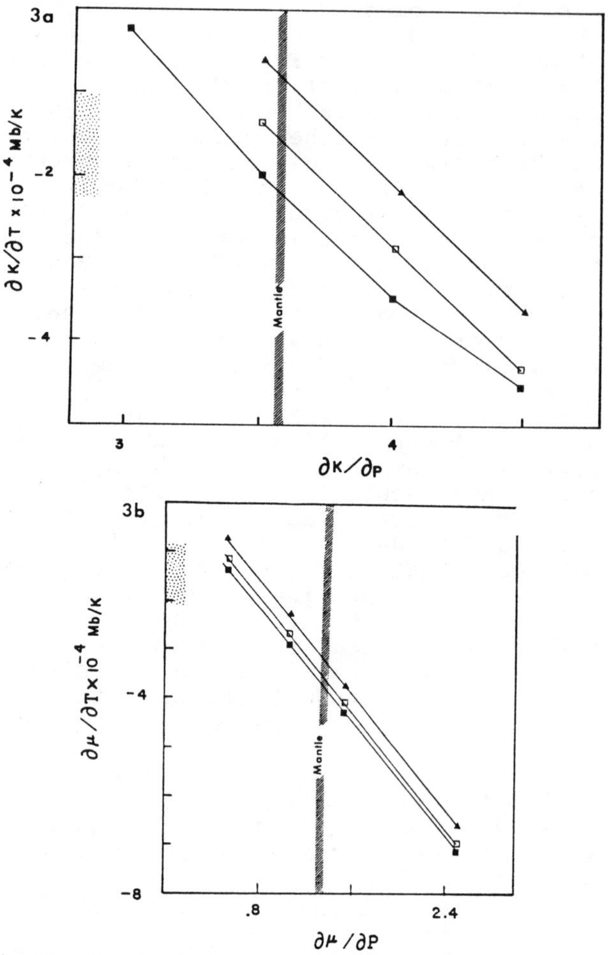

Fig 3. Trade-off curves for various compositional models of the earth's lower mantle. Filled triangle= pyrolite; open square= chondrite; filled square= pyroxene. Figures 3a and 3b illustrate these curves for $(dK/dT)_P$ vs $(dK/dP)_S$ and $(d\mu/dT)_P$ vs $(d\mu/dP)_S$. The stippled area on the vertical axis denotes the range of temperature derivatives for various silicate phases. The region hatched as "Mantle" illustrates the pressure and temperature dependence of the bulk and shear moduli for these petrological models which are required to match the elastic properties of the PREM model, at the depth of 1071 km.

derivatives of the elastic properties of the perovskite must be measured in order to choose one composition in favor of the other.

The variation of the elastic modulus with depth can be expressed as an adiabatic pressure derivative if indeed the temperature profile at that depth is adiabatic. Those pressure derivatives deduced from the PREM model are illustrated as hatched areas in Figure 3. The stippled region on the y axis, denotes the range of temperature derivatives for various silicate minerals. Additionally, in Table 7 we compare the zero pressure elastic properties of the perovskite phase to the elastic properties deduced from the PREM model at depth of 1071 kilometers. We have selected this depth for our comparison so as to avoid the complexity of velocity gradients in the vicinity of the 670-km discontinuity (Walck, 1984). We feel that, at this depth the mineralogy of the earth's lower mantle should be primarily dominated by an assemblage of pure perovskite to perovskite plus magnesiowustite. In both Table 7 and Figure 3 we note that, the pressure and temperature dependence of the bulk modulus which is required to match the seismic data are very similar to what would be expected on the basis of behavior of other silicates. On the other hand, the required magnitude of the temperature derivative of the shear modulus of perovskite is several times greater than is normally observed. We speculate that such a large negative value results only if the mantle temperature is sufficiently close to that required for a paraelastic-ferroelastic phase transformation as to induce significant transverse mode softening.

Table 7. Comparison of the zero pressure elastic properties of $MgSiO_3$ perovskite to the elasticity data deduced from the PREM model at depth of 1071 km.

	PREM	perovskite
Depth (km)	1071.00	0.000
V_p (km/s)	11.5522	10.94
V_s (km/s)	6.4074	6.69
K (GPa)	363.70	246.4
μ (GPa)	189.70	184.2
dK/dP	3.60	
dμ/dP	1.36	

This conclusion is independent of the assumed bulk composition of the mantle. Such behavior would also explain the large variation of shear velocity relative to longitudinal velocity (i.e., $d\ln V_s/d\ln V_p$) as observed by tomography, since the shear modulus would be much more sensitive to lateral variations in temperature than would the bulk modulus. If in fact, the perovskite phase in the earth's lower mantle is suffering from such strain induced phase transitions the geological implications could indeed be far-reaching as we could expect anomalous behavior in thermal expansion, heat capacity, and diffusion controlled properties such as electrical conductivity and rheology.

Summary

The nine adiabatic single-crystal elastic moduli of $MgSiO_3$ in the perovskite structure, now widely accepted as the most abundant mineral phase in the earth's mantle, have been determined under ambient conditions for the first time. The large magnitude of the elastic moduli indicate that silicate perovskite is significantly stiff under compression and shear. The high value of the shear modulus of perovskite is geophysically interesting, since it signals that an extremely large negative temperature derivative of the shear modulus is needed to match the mantle values. Furthermore, our data also imply that the 670-km seismic discontinuity can be associated with either the ilmenite or the spinel to perovskite phase transitions, and that the compositional models of the earth's lower mantle can encompass either pyrolite or pyroxene stoichiometries. We conclude that, it is very likely that the perovskite phase $(Mg,Fe)SiO_3$ under the lower mantle conditions may be influenced by ferroelastic phase transitions to higher symmetry phases (e.g., tetragonal and/or cubic) which are driven by condensation of unstable transverse acoustic modes.

Acknowledgments. The authors thank an anonymous reviewer for helpful suggestions on the manuscript.

References

Abrahams, S. G., J. L. Bernstein, and J. P. Remeika, Ferroelastic transformation in Samarium orthoaluminate, Mat. Res. Bull, 9, 1513-1616, 1974.

Abrahams, S. C., R. L. Barns, and J. L. Bernstein, Ferroelastic effect in $LaFeO_3$, Solid State Comm., 10, 379-381, 1972.

Anderson, D. L., and J. D. Bass, Transition region of the Earth's upper mantle, Nature, 320, 321-328, 1986.

Aizu, K, Determination of the state parameters and formulation of spontaneous strain for ferroelastics, J. Phy. Soc. Japan, 28, 706-716, 1970.

Aizu, K, Theoretical investigation of the variety of first-order transitions between prototypic and ferroic phases, Phys. Rev. B, , 23, 1293-1299, 1981.

Bass, J. D., R. C. Liebermann, D. J. Weidner and S. J. Finch, Elastic properties from acoustic and volume compression experimental data, Phys. Earth Planet. Inter., 25, 140-158, 1981.

Bass, J. D., and D. L. Anderson, Composition of the upper mantle: Geophysical tests of two petrological models, Geophys Res. Lett., 11, 237-240, 1984.

Bass, J. D., Elasticity of single-crystal $SmAlO_3$, $GdAlO_3$ and $ScAlO_3$ perovskites, Phys. Earth Planet. Inter., 36, 145-156, 1984.

Bell, P. M., H-K., Mao, and J. A. Xu, Error analysis of parameter-fitting in equation of state for mantle minerals, in High Pressure Research in Mineral Physics, eds. M. H. Manghanin and Y. Syono, Terra. Sci. Publ. Tokyo: American Geophysical Union, Washington, D. C. pp.447-454, 1987.

Bell, R. O. and G. Rapprecht, Elastic constants of strontium titanate, Phys. Rev., 129, 90-94, 1963.

Bina, C. R. and B. J. Wood, Olivine-spinel transition: Experimental and thermodynamic constraints and implications for the nature of the 400 km seismic discontinuity, J. Geophys. Res., 92, 4853-4866, 1987.

Cohen, R. E., Elasticity and equation of state of $MgSiO_3$ perovskite, Geophys. Res. Lett., 14, 1053-1056,

Cowley, R. A., Acoustic phonon instabilities and structural phase transitions, Rev. B., 13, 4877-4885, 1976.

Cowely, R. A., Structural phase transitions: I. Landau Theory, Adv. Phys., 29, 1-110, 1980.

Coutures, J., and J. P. Coutures, Etude par rayons x a haute temperature des transformations polymorphiques des perovskite $LnAlO_3$ (Ln = element lonthanidique) J. Sol. State Chem., 52, 95-100, 1984.

David, W. I. F., Structural relationships between spontaneous strain and acoustic properties in ferroelastics, J. Phys. C., 16, 2455-2470, 1983.

Dvorak, V., On the structural phase transitions from a hypothetical phase, Czech. J. Phys., B28, 989-994, 1978.

Dziewonski, A. M. and D. L. Anderson, Preliminary reference earth model, Phys. Earth. Planet. Inter., 25, 297-356, 1981.

Forsbergh, Jr., P. W., Domain structuris and phase transitions in barium titanate, Phys. Rev., 76,1187-1201,1949.

Knittle, E., and R. Jeanloz, Synthesis and equation of state of $(Mg,Fe)SiO_3$ perovskite to over 100 gigapascals, Science, 235, 666-670, 1987.

Kudoh, Y., E. Ito, and H. Takeda, Effect of pressure on the crystal structure of perovskite-type $MgSiO_3$, Phys. Chem.

Glazer, A. M., The classification of tilted octahedra in perovskites, Acta. Cryst., B28, 3384-3394, 1975.

Glazer, A. M., Simple ways of determining perovskite structures, Acta. Cryst., A31, 756-762, 1972.

Hemley, R.J., R.E. Cohen, A. Yeganeh-Haeri, H.K. Mao, D.J. Weidner, and E.Ito, Raman spectroscopy and lattice dynamics of $MgSiO_3$ perovskite at high-pressure, this volume, 1988.

Horiuchi, H., E. Ito, and D. J. Weidner, Perovskite-type $MgSiO_3$: Single-crystal x-ray diffraction study, A. Min., 172, 357-360, 1987.

Ito, E. Y. and Matsui, Synthesis and crystal chemical characterization of $MgSiO_3$ perovskite, Earth Planet. Sci. Lett., 38, 443-450, 1978.

Ito, E., E. Takahashi, and Y. Matsui, The mineralogy and chemistry of the lower mantle: An implication of the ultra high-pressure phase relation in the system MgO-FeO-SiO_2, Earth Planet. Sci. Lett., 67, 238-248, 1984.

Ito, E., and D. J. Weidner, Crystal growth of $MgSiO_3$ perovskite, Geophys. Res. Lett., 13, 464-466, 1986.

Ito, E. and E. Takahashi, The phase equilibria and the constitution in the deep mantle, in High Pressure Research in Mineral Physics, eds. M. H. Manghnani and Y. Syono, Terra. Sci. Publ. Tokyo: American Geophysical Union,Washington,D.C.,pp. 221- 229,1987.

Ito, E., and H. Yamada, Stability relations of silicate spinels, ilmenites and perovskite, in High Pressure Research in Geophysics, eds.

S. Akimoto and M. H. Manghnani, Center for Publications, Tokyo, pp., 405-419, 1981.
Minerals, 14, 350-354, 1987.

Liebermann, R. C., L. Jones, and A. E. Ringwood, Elasticity of aluminate, titanate, stannate and germanate compounds with the perovskite structure, Phys. Earth and Planet. Inter., 14, 165-178, 1977.

Liu, L. G., The post-spinel phase of forsterite, Nature, 262, 770-772, 1976.

Megaw, H. D., Origin of ferroelasticity in barium titanate and other perovskite-type crystals, Acta. Cryst., 5, 739-749, 1952.

Megaw, H. D., Crystal Structures: A Working Approach, W. B. Sanders Company, Philadelphia, 1973.

Rao, R., and L. K. Rao, Phase Transitions in Solids, McGraw-Hill, New York,1977.

Salje, E., B. Kuscholke, and B. Wruck, Domain wall formation in minerals: I. Theory of twin boundary shapes in Na. feldspar, Phys. Chem. Minerals, 12, 132-140, 1985.

Sapriel, J., Domain-wall orientation in ferroelastics, Phys. Rev. B., 12, 5128-5140, 1975.

Taledano, J. C. and P. Taledano, Order paromameter symmetrics and free energy expansions of purely ferroelastic transitions, Phys. Rev. B, 21, 1139-1172, 1980.

Vaughan, M. T., and J. D. Bass, Single-crystal elastic properties of protoenstatite: A comparison with orthoenstatite, J. Phys. Chem. Minerals, 10, 62-68, 1983.

Venkataraman, G., Soft modes and structural phase transitions, Bull. Matt. Sci., 1, 129-163, 1979.

Walck, M. C., The P wave upper mantle structure beneath an active spreading center: The Gulf of California, Geophys. J. R. astr. Soc., 76, 697-723, 1984.

Weidner, D. J., and E. Ito, Mineral physics constraints on a uniform mantle composition, in High Pressure Research in Mineral Physics, eds. M. H. Manghnani and Y. Syono, Terra Sic Publ, Tokyo; American Geophysical Union, Washington, D. C. ,pp., 439-446, 1987.

Weidner, D. J., and E. Ito, Elasticity of $MgSiO_3$ in the ilmenite phase, Geophys. Res. Lett., 12, 417-420, 1985.

Weidner, D. J., K. Swyler, and H. R. Carleton, Elasticity of microcrystals, Geophys. Res. Lett., 2, 189-192, 1975.

Weidner, D. J. and H. R. Carleton, Elasticity of coesite, J. Geophys. Res., 82, 1334-1345, 1977.

Weidner, D. J., J. D. Bass, A. E. Ringwood, and W. Sinclair, The single-crystal elastic moduli of stishovite, J. Geophys. Res., 87, 4740-4746, 1982.

Weidner, D. J., H. Sawamoto, H. Sasaki, and M. Kumazawa, Singel-crystal elastic properties of the spinel phase of Mg_2SiO_4, J. Geophys. Res., 89, 7582-7860, 1984.

Weidner, D. J., J. D. Bass, and M. T. Vaughan, The effect of crystal structure and composition on elastic properties of silicates, in High Pressure Research: Applications in Geophysics, eds. S. Akimoto and M. H. Manghnani, pp. 275-283, 1982.

Wolf, G. and M. S. T. Bukowinski, Theoretical study of the structural properties and equation of state of $MgSiO_3$ and $CaSiO_3$ perovskite: implivation for lower mantle composition, in High Pressure Research in Mineral Physics, eds. M. H. Manghnani and Y. Syono, Terra. Sci. Publ., Tokyo; American Geophysical Union, Washington, D. C. pp.,315-331, 1987.

Yagi, T., H-K. Mao, and M. Bell, Hydrostatic compression of perovskite-type $MgSiO_3$, in Advances in Physical Geochemistry, 2 ed. S.Saxena, pp. 317-325, Springer-Verlag, New York, 1982.

E. Ito, Institute for Study of the Earth's Interior, Okayama University, Misasa, Tottori-Ken 682-02, Japan.

D.J. Weidner and A. Yeganeh-Haeri, Department of Earth and Space Sciences, State University of New York at Stony Brook, Stony Brook, NY 11794 USA.

STABILITY RELATIONS OF SILICATE PEROVSKITE UNDER SUBSOLIDUS CONDITIONS

Eiji Ito

Institute for Study of the Earth's Interior, Okayama University,
Misasa, Tottori-ken 682-02, Japan

Abstract. Stability relations of silicate perovskite under subsolidus conditions are summarized for the systems $MgO-SiO_2$ and $MgO-FeO-SiO_2$, and $CaSiO_3-MgSiO_3-Al_2O_3$. The phase boundary curves of the ilmenite-perovskite transformation in $MgSiO_3$ and the dissociation of spinel into perovskite and periclase in Mg_2SiO_4 are P(GPa) = 26.8 - 0.0025T(°C) and P(GPa) = 27.6 - 0.0028T(°C), respectively, in the temperature range 1000 °C to 1600 °C. In the system $MgO-FeO-SiO_2$, perovskite and perovskite-bearing assemblages are stabilized at 1600 °C and pressures higher than 23 GPa. The maximum solubility of $FeSiO_3$ component in the perovskite phase is 8 mol% at 1100 °C and 11 mol% at 1600 °C. The iron content of magnesiowüstite in coexistence with perovskite and stishovite also increases from 46 mol% at 1100 °C to 58 mol% at 1600 °C. The $(Mg,Fe)_2SiO_4$ spinel phase with iron content of less than 22 mol% dissociates into perovskite and magnesiowüstite within a quite narrow pressure interval (less than 0.15 GPa at 1600 °C). In the system $CaSiO_3-MgSiO_3-Al_2O_3$, the stability field of majorite was found to expand rapidly towards the $CaSiO_3-MgSiO_3$ join in the pressure range 10-18 GPa, and to contract towards the grossular-pyrope join at pressures higher than 23 GPa, dissociating $MgSiO_3$-rich perovskite and an unquenchable phase (Ca-P) which is diopsidic in composition. The complete dissociation of majorite, however, requires a fairly large pressure interval (ca. 3 GPa), and produces a small amount of stishovite and an Al_2O_3-rich phase in addition to the above two phases. Dissociation of both spinel and majorite could be responsible for the 670 km discontinuity. The spinel reaction makes the discontinuity very sharp in its initial stage, while the majorite reaction broadens it in the deeper region. The lower mantle mineralogy is inferred to be $MgSiO_3$-rich perovskite (70% in volume), magnesiowüstite (18%), Ca-P (8%), and trace amounts of stishovite and Al-P, for a peridotitic or pyrolitic composition.

Introduction

Synthesis of the orthorhombic perovskite modification of $MgSiO_3$ [Liu, 1976a; Ito, 1977] and the dissociation of Mg_2SiO_4 spinel into $MgSiO_3$ perovskite and MgO periclase [Liu, 1976b; Ito, 1977] were important findings of high-pressure geophysics in the 1970s. As $MgSiO_3$ perovskite is stable at pressures higher than 20 GPa and the perovskite structure is the densest known modification of ABO_3 type compounds, silicate perovskite has been accepted as a major phase of the lower mantle. Therefore, detailed knowledge of the stability of perovskite phases for systems of mantle composition is indispensable for better understanding of the 670 km seismic discontinuity and the structure of the deep mantle.

In this article, a summary is given of recent results of high-pressure experiments on chemical systems relevant to the mantle, focusing on the stability relations of silicate perovskites. Based on these results, comments are made on the character of the 670 km discontinuity and the mineralogy of the lower mantle.

Experimental Procedure

High-pressure and high-temperature experiments were conducted using a uniaxial split-sphere apparatus (USSA-5000) at the Institute for Study of the Earth's Interior, in which a cubic tungsten carbide assembly is compressed with the aid of a 5000 ton hydraulic press. Details of the experimental procedure are described elsewhere [Ito and Yamada, 1982; Ito et al., 1984; Ito and Takahashi, 1988].

Pressure was monitored using the oil pressure of the press and calibrated up to 23 GPa (metallic transition of GaP) at room temperature. High temperature corrections to the calibration curve were made at temperatures of 1000 °C or higher [e.g., Ito and Takahashi, 1988], based on the transition pressures of several silicates, all of which are consistent with thermochemical data. The accuracy of pressure determinations at temperatures higher than 1000 °C is probably within ±0.5 GPa. More accurate measurement of pressure is quite difficult because of the lack of a reliable pressure scale under these conditions.

Powdered starting materials were put directly into a hollow cylindrical tantalum heater embedded in the center of an octahedron made of semi-sintered magnesia. Experimental temperature was monitored with a Pt/Pt-13%Rh thermocouple with cold junctions at ambient conditions. No correction was made for the effect of pressure on the thermocouple emf. The sample was held at the desired pressure and temperature for a certain duration, and then quenched by shutting off the electric power supply. The pressure was then released, and the run product was recovered.

Identification of phases present in the run product was made by the X-ray diffraction. Small chips next to the thermocouple junction were taken out from some run products and examined by optical microscope, SEM, and EPMA.

Stability relations of silicate perovskites

Mg-silicates

Ito and Yamada [1982] examined the equilibria between ilmenite and perovskite in $MgSiO_3$ and between spinel and an assemblage of $MgSiO_3$ perovskite and MgO periclase in Mg_2SiO_4. They demonstrated that the phase boundary curves for both reactions have negative slopes (dP/dT= -0.002 GPa/deg). They also suggested that the two phase boundaries are located close to each other. In their experiments, however, the pressure determinations contain rather large uncertainties due to the use of baked pyrophyllite as a pressure medium [Suito, 1986]. Also Ito and Yamada carried out phase identification by ordinary powder X-ray diffractometry. For proper identification, several milligrams of sample were needed, and the coexistence of high- and low-temperature phases was inevitably observed for products quenched at conditions near the phase boundary because of the steep temperature gradient in a small heater. Therefore, the results of their study are not sufficient to define the precise locations of the phase boundaries.

Recently Ito and Takahashi [1988] thoroughly reexamined both reactions based on the new calibrations, using semi-sintered magnesia as a pressure medium. In their work, the phases just adjacent to the thermocouple junction were specifically identified by micro-focused X-ray diffractometry using a 100 μm diameter beam of Cr K_α radiation. Therefore the ambiguity in phase identification due to the temperature gradient was almost completely eliminated. The phase boundary curves thus determined in the temperature interval between 1000 °C and 1600 °C are represented by the following equations: P(GPa)=26.8-0.0025T(°C) for the ilmenite-perovskite transformation in $MgSiO_3$ and P(GPa)=27.6-0.0028T(°C) for the dissociation of Mg_2SiO_4 spinel.

The system $MgO-FeO-SiO_2$

The effect of the iron component on the phase equilibria is an important factor in constraining the possible mineralogy of the deep mantle. Liu [1976c] first studied the systems $MgSiO_3$-$FeSiO_3$ and Mg_2SiO_4-Fe_2SiO_4 at pressures up to 30 GPa and temperatures of 1400-1800 °C. He suggested that the perovskite phase is stable for compositions ranging from $MgSiO_3$ to $(Mg_{0.3}Fe_{0.7})SiO_3$. Yagi et al. [1978] proposed a set of phase relations for the system $MgO-FeO-SiO_2$ at 1000 °C for pressures up to 50 GPa. According to their results, the maximum solubility of the $FeSiO_3$ component in $MgSiO_3$ perovskite is limited to $(Mg_{0.8}Fe_{0.2})SiO_3$, with the appearance of $(Mg_{0.24}Fe_{0.76})O$ magnesiowüstite and stishovite for more iron-rich compositions. Their work, which used a diamond anvil cell coupled with a laser heating system, was somewhat ambiguous, because the temperature measurements had large uncertainties, and because the chemical compositions of the phases present in the run products were not well-defined.

We have performed more systematic investigations on the systems $MgSiO_3$-$FeSiO_3$ [Ito and Yamada, 1982], Mg_2SiO_4-Fe_2SiO_4 [Ito and Takahashi, 1988], and $MgO-FeO-SiO_2$ [Ito et al., 1984]. In these studies, we have tried to locate the phase boundaries accurately by carrying out runs on several compositions bracketing the phase boundaries closely.

The psuedobinary diagram of the system $MgSiO_3$-$FeSiO_3$ at 1100 °C made by Ito and Yamada [1982] is reproduced in Fig. 1. The pressures were

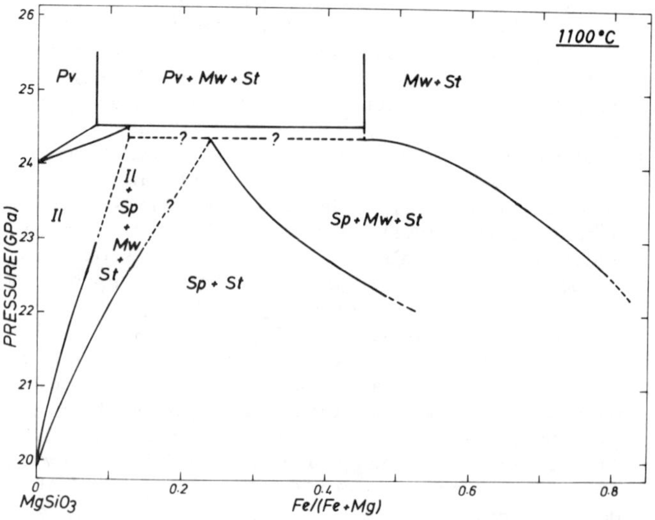

Fig. 1 Pseudobinary diagram of the system $MgSiO_3$-$FeSiO_3$ after Ito and Yamada(1982). Pv: perovskite; I: ilmenite; Mw: magnesiowüstite; Sp: spinel; St: stishovite.

reduced by about 1 GPa, consistent with our current calibration [Ito and Takahashi, 1988], and the three-phase field of perovskite + magnesiowüstite + stishovite has been shifted slightly towards magnesium-rich compositions, taking more recent results into account. In the magnesium rich compositions, ilmenite and perovskite coexist across a narrow transitional loop, which widens slightly with increasing iron content. The maximum iron content of perovskite is 8 mol% at 1100 °C. Neither ilmenite nor perovskite is present for bulk compositions more iron-enriched than $(Mg_{0.55}Fe_{0.45})SiO_3$; instead, the coexistence of magnesiowüstite and stishovite is stable. In the intermediate compositional range, the spinel- and ilmenite-bearing assemblages transform into $(Mg_{0.92}Fe_{0.08})SiO_3$ perovskite, $(Mg_{0.55}Fe_{0.45})O$ magnesiowüstite, and stishovite, passing through the univariant boundary at 24.5 GPa.

Recently Ito and Takahashi [1988] have investigated the system Mg_2SiO_4-Fe_2SiO_4. The phase relations determined at 1100 °C and 1600 °C are shown as the psuedobinary diagrams in Fig. 2. At 1100 °C, a spinel phase with compositions from Mg_2SiO_4 to $(Mg_{0.78}Fe_{0.22})_2SiO_4$ dissociates into perovskite and magnesiowüstite, with a narrow transitional loop. The spinel phase decomposes into magnesiowüstite and stishovite for compositions more ion-rich than $(Mg_{0.54}Fe_{0.46})_2SiO_4$. The univariant boundary is located at about 24.5 GPa and is observed for a compositional range from Fe/(Mg+Fe)=0.22 to 0.46. The compositional relation in the three-phase field of perovskite + magnesiowüstite + stishovite is identical to that in the $MgSiO_3$-$FeSiO_3$ system.

The overall topology of the phase boundaries at 1600 °C is similar to that at 1100 °C. However, all phase boundaries of perovskite-forming reactions shift towards both lower pressures and higher Fe/(Mg+Fe) ratios with increasing temperature.

Fig. 2 shows that the pressure value of the univariant reaction is very close to that for the dissociation of Mg_2SiO_4 spinel and that the three phase loop of spinel + perovskite + magnesiowüstite is quite narrow. Ito and Takahashi [1988] showed that the dissociation of $(Mg_{0.8}Fe_{0.2})_2SiO_4$ spinel is completed within a pressure interval smaller than 0.15 GPa at 1600 °C.

The stability relations in the system MgO-FeO-SiO_2 at 25 GPa and 1600 °C are shown in Fig. 3 [cf. Ito et al., 1984]; those at 1100 °C are also shown for comparison. The maximum iron content of perovskite increases from 8 mol% at 1100 °C to 11 mol% at 1600 °C. The composition of the magnesiowustite coexisting with perovskite and stishovite also shifts towards FeO with increasing temperature (46 mol% FeO at 1100 °C and 58 mol% at 1600 °C). The compositions of coexisting perovskite and magnesiowüstite at 1600 °C are connected by tie lines with the iron-magnesium partition coefficient $K' = (Fe/Mg)^{Pv}/(Fe/Mg)^{Mw}$ showing a strong compositional dependence.

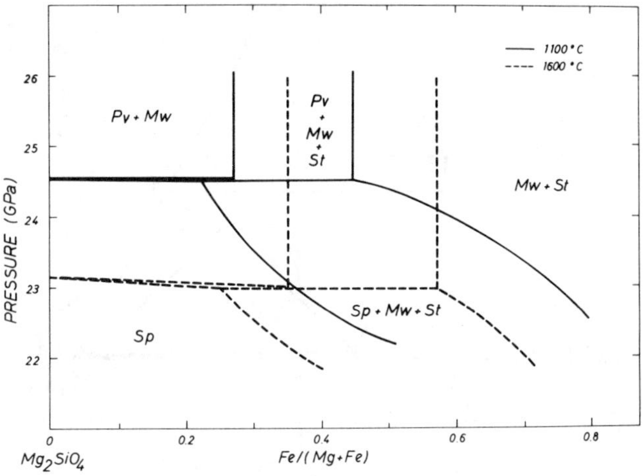

Fig. 2 Pseudobinary diagrams of the system Mg_2SiO_4-Fe_2SiO_4 after Ito and Takahashi [1988]. Pv: perovskite; Mw: magnesiowüstite; Sp: spinel; St: stishovite.

The system $CaSiO_3$-$MgSiO_3$-Al_2O_3

Ringwood and Major [1966] and Ringwood [1967] first discovered that the pyroxene components $M_4Si_4O_{12}$ (M=Mg, Fe, Ca) dissolve extensively into the garnet structure at pressures higher than 10 GPa. More detailed studies were carried out on the systems $Mg_4Si_4O_{12}$-$Mg_3Al_2Si_3O_{12}$ [Akaogi and Akimoto, 1977; Kanzaki, 1987] and $Ca_2Mg_2Si_4O_{12}$-$Ca_{1.5}Mg_{1.5}Al_2Si_3O_{12}$ [Akaogi and Akimoto, 1979]. All of these investigations confirmed the extensive dissolution of pyroxenes into garnet at

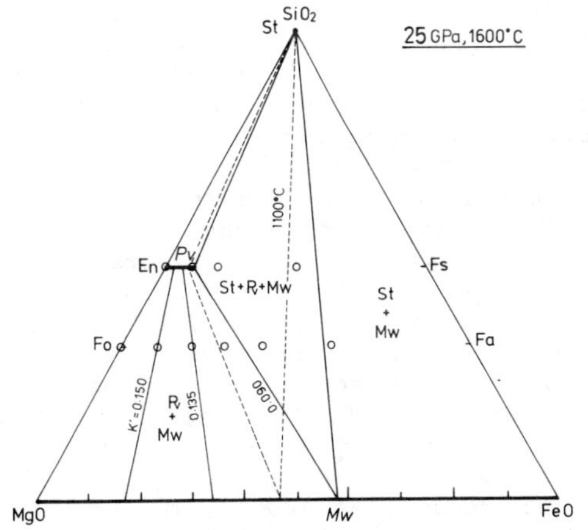

Fig. 3 Stability relations in the system MgO-FeO-SiO_2 at 25 GPa [cf. Ito et al., 1984]. Open-circles denote the compositions of starting materials used. Coexisting perovskite and magnesiowüstite are connected by tie lines with K' values.

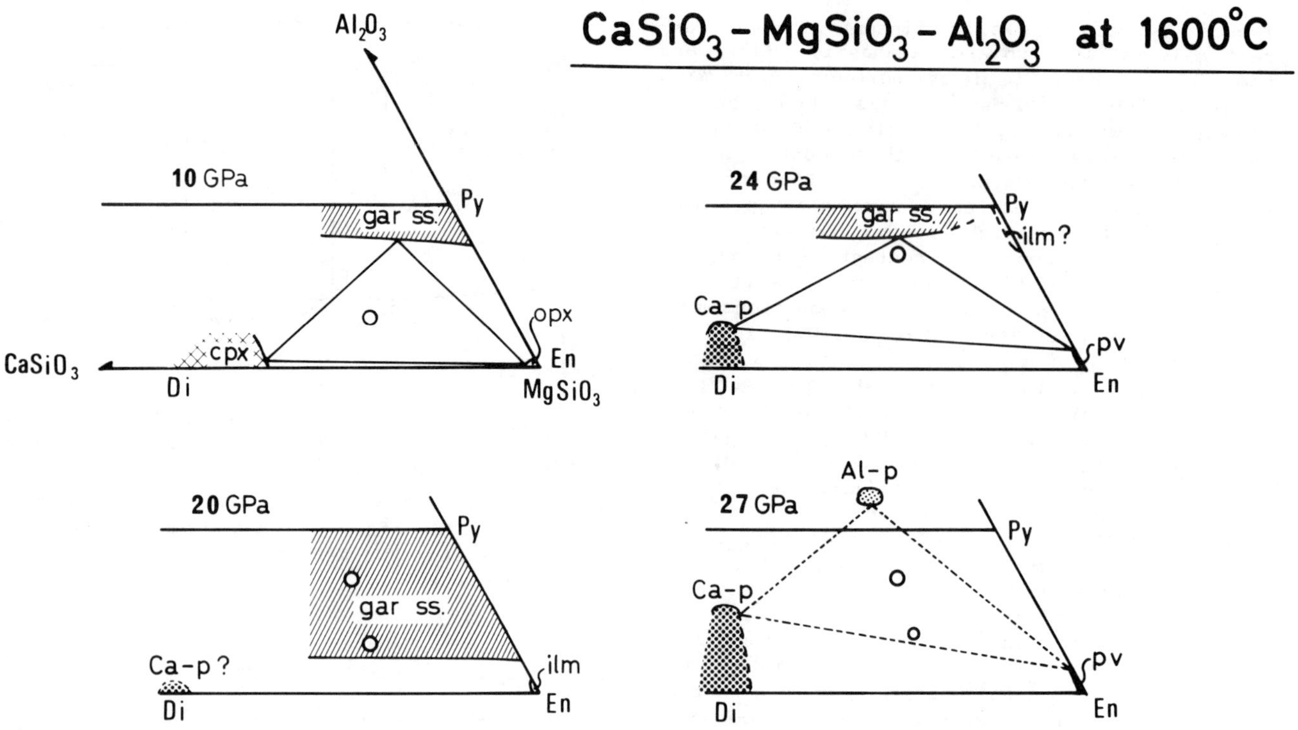

Fig. 4 Phase diagrams in the system $CaSiO_3$-$MgSiO_3$-Al_2O_3 at 1600 °C and pressures up to 27 GPa [Ito et al, 1987]. Open circles denote the compositions of starting materials. En, Di, Py, and Go represent enstatite, diopside, pyrope, and grossular compositions, respectively. gar ss: garnet solid solution; Ca-P and Al-P: see text; Ilm: ilmenite; Pv: perovskite.

pressures higher than 12 GPa, suggesting that the resultant garnet solid solution, majorite, would be a dominant constituent of the transition zone.

It was also revealed that the dissolution of Ca-rich pyroxene in garnet takes place at pressures 2-3 GPa higher than that for the case of Ca-poor pyroxene [Akaogi and Akimoto, 1979]. Therefore, investigation on the stability of majorite in the ternary system $CaSiO_3$-$MgSiO_3$-Al_2O_3 may provide more suitable information about the mantle mineralogy. In the system $Mg_4Si_4O_{12}$-$Mg_3Al_2Si_4O_{12}$, majorite transforms into ilmenite [Liu, 1977, Kanzaki, 1987] and subsequently into perovskite [Liu, 1977], under increasing pressure. However, the effect of CaO on the post-majorite transformation was not previously known. In this context, Ito and Takahashi [1987] studied the stability relations of the system $CaSiO_3$-$MgSiO_3$-Al_2O_3 up to 27 GPa, and their results at 1600 °C are summarized in Fig. 4, in the order of increasing pressure. The open circles denote the compositions of glasses used as the starting materials; EWC ($0.725MgSiO_3$ + $0.2CaSiO_3$ + $0.075Al_2O_3$ in molar ratio) and EWC10 ($0.64MgSiO_3$ + $0.18CaSiO_3$ + $0.18Al_2O_3$). EWC is very close to the olivine-subtracted peridotitic mantle compositions, if iron is considered together with magnesium [Yamada and Takahashi, 1984].

At 10 GPa, an assemblage of orthopyroxene, clinopyroxene, and garnet solid solution is stable. The stability of the garnet solid solution expands towards the $CaSiO_3$-$MgSiO_3$ join with increasing pressure, and the EWC composition contains majorite at pressures higher than 20 GPa. At pressures under which $MgSiO_3$ perovskite is stable, however, the stability field of majorite regresses towards the grossular-pyrope join, dissociating into $MgSiO_3$-rich perovskite (Mg-Pv) and an unquenchable phase with a diopsidic composition (Ca-P), along with a trace amount of stishovite (see the diagram at 24 GPa in Fig. 4). With further increase in pressure, the stability field of majorite continues to diminish and the dissociation products become more dominant. The Ca-P might be related to a $CaMgSi_2O_6$ perovskite phase recently reported by Liu [1978]. Powder x-ray diffraction on the run products of EWC10, however, revealed that a new crystalline phase appears and becomes pronounced as majorite diminishes. The crystal structure of the new phase is not yet known. Chemical analysis by

EPMA and TEM suggested that the composition of the phase is richer in Al_2O_3 than the pyrope and grossular join, and therefore the phase is designated as Al-P in Fig. 4. The measured Si/(Mg+Ca) ratio of Al-P is less than unity, which is consistent with the presence of stishovite.

It is noteworthy that the dissociation of majorite proceeds over a fairly large pressure interval of about 3 GPa. The dissociation product is an assemblage of Mg-Pv, Ca-P, stishovite, and Al-P. For the EWC composition, the former two phases predominate over the latter two by an order of magnitude.

The results are in general agreement with those of Irifune and Ringwood [1987] on the pyrolite minus olivine composition. However, they reported that $CaSiO_3$-rich perovskite [cf. Liu and Ringwood, 1975] exsolves from the majorite at pressures higher than 20 GPa and that, at still higher pressures, this phase is a dominant dissociation product of majorite, rather than the Ca-P of Ito and Takahashi. This disagreement might be partly due to the effect of TiO_2. Irifune and Ringwood's Ca-rich perovskite contained 5 mol% $CaTiO_3$ and the cubic $CaSiO_3$ perovskite might be stabilized by dissolution of $CaTiO_3$ component [Ringwood and Major, 1971]. More recently, however, Tamai and Yagi [1988] observed a dissociation of diopside into $CaSiO_3$ perovskite and high-pressure forms of $MgSiO_3$ at pressures higher than 17 Gpa, in contrast to the conversion to single phase of $CaMgSi_2O_6$ perovskite [Liu, 1978]. Therefore, further investigation is clearly needed to reveal the stability relation between $CaSiO_3$ perovskite and the Ca-P.

The 670 km discontinuity and the lower mantle

It is generally accepted that MgO, FeO, and SiO_2 are the major components of the mantle, the sum of these three components being greater than 90 % of the total, and that the Si/(Mg+Fe) and Fe/(Mg+Fe) ratios are in the ranges of 0.5-1.0 and 0.1-0.15, respectively [e.g., Ringwood, 1975; Zindler and Hart, 1986]. According to the stability relations in the system MgO-FeO-SiO_2 shown in the last section, a material of this composition exists as the assemblage of spinel and stishovite at 22 GPa, which undergoes successive transformations under increasing pressure, and is replaced by an assemblage of perovskite and magnesiowüstite at pressures higher than 24.5 GPa. Therefore it is indicated that the stabilization of a perovskite-bearing assemblage might characterize the 670 km discontinuity regardless of the variation of composition within reasonable limits. Ito et al. [1984] proposed the assemblage of $(Mg_{0.96}Fe_{0.04})SiO_3$ perovskite and $(Mg_{0.76}Fe_{0.24})O$ magnesiowüstite as the lower mantle mineralogy based on the density restriction from geophysics, and the Fe-Mg partioning between perovskite and magnesiowüstite.

However, the remaining components should be taken into consideration in order to make a more realistic mantle model. Among them, a small amount of Al_2O_3 (about 4 %) causes a remarkable change in the mantle mineralogy, by facilitating the stabilization of garnet solid solution. Almost all pyroxene components are incorporated into garnet to form majorite at pressures higher than 17 GPa. The stabilities of modified spinel and spinel, on the other hand, are little affected by the existence of remaining components [Akaogi and Akimoto, 1979; Takahashi and Ito, 1987]. In the lower half of the transition zone, therefore, spinel and majorite are by far the dominant constituents.

As shown in the last section, both spinel and majorite dissociate into perovskite-bearing assemblages at pressures higher than 23 GPa, accompanied by density increases of some 10 percent. Although elastic data are not available for all the phases concerned, an increase in seismic velocity greater than 5 percent is expected for both dissociations, judging from the recent results of velocity measurements by means of Brillouin spectroscopy on Mg_2SiO_4 spinel [Weidner et al., 1984], $MgSiO_3$ perovskite [Yeganeh-Haeri et al., 1987], and majorite [Weidner et al., 1987]. Therefore it is proposed that the 670 km discontinuity may be caused by the dissociations of spinel and majorite. It is noteworthy, however, that the dissociation of spinel is complete within a very small depth range (less than 4 km) [Ito and Takahashi, 1988], while that of majorite proceeds over a fairly large depth range of about 80 km. These features are consistent with the characteristic of the discontinuity shown in recent seismic models in which the sharp velocity increase is followed by a zone of high velocity gradient zone extending down to about 750 km depth [Dziewonski and Anderson, 1981; Grand and Helmberger, 1984].

Assuming a peridotitic mantle composition such as pyrolite [Ringwood, 1975], the resultant lower mantle mineralogy is inferred to be 70 % Mg-Pv, 16 % magnesiowüstite, 8 % Ca-P with a small amount of stishovite and Al-P (volume proportions). However, $CaSiO_3$-rich perovskite [Irifune and Ringwood, 1987] in stead of Ca-P cannot be ruled out.

Acknowledgments. I am grateful to E. Takahashi for his help and discussion. I thank Y. Matsui and S. Akimoto for their support and encouragement. Critical reading of manuscript by K. Leinenweber is much appreciated. I am grateful to A. Navrotsky and C. Hertzberg for their careful review.

References

Akaogi, M. and S. Akimoto, Pyroxene-garnet solid solution equilibria in the $Mg_4Si_4O_{12}$ -$Mg_3Al_2Si_3O_{12}$ and $Fe_4Si_4O_{12}$-$Fe_3Al_2Si_3O_{12}$ at high pressures and temperatures, Phys. Earth Planet. Inter., 15, 90-106, 1977.

Akaogi, M. and S. Akimoto, High-pressure phase equilibria in a garnet lherzolite, with special reference to Mg^{2+}-Fe^{2+} partitioning among constituent minerals, Phys. Earth Planet. Inter., 19, 31-51, 1979.

Dziewonski, A. M. and D. L. Anderson, Preliminary reference earth model, Phys. Earth Planet. Inter., 25, 297-356, 1981.

Grand, S. and D. V.Helmberger, Upper-mantle shear structure of North America, Geophys. J. R. Astron. Soc., 76, 399-438, 1984.

Irifune, T. and A. E. Ringwood, Phase transformations in primitive MORB and pyrolite compositions to 25 GPa and some geophysical implications, in High-Pressure Research in Mineral Physics, edited by M. H. Manghnani and Y. Syono, pp. 231-242, TERRA PUB/Am. Geophys. Union, Tokyo/Washington, D. C., 1987.

Ito, E., The absence of oxide mixture in high-pressure phases of Mg-silicates, Geophys. Res. Lett., 4, 72-74, 1977.

Ito, E. and E. Takahashi, Post-spinel transformations in the system Mg_2SiO_4-Fe_2SiO_4 and their geophysical implications, to be submitted to J. Geophys. Res., 1988.

Ito, E. and E. Takahashi, Ultrahigh-pressure phase transformations and the constitution of the deep mantle, in High-Pressure Research in Mineral Physics, edited by M. H. Manghnani and Y. Syono, pp. 221-229, TERRA PUB/Am. Geophys. Union, Tokyo/Washington, D. C., 1987.

Ito, E., E. Takahashi, and Y. Matsui, The mineralogy and chemistry of the lower mantle:an implication of the ultrahigh-pressure phase relations in the system MgO-FeO-SiO_3, Earth Planet. Sci. Lett., 67, 238-248, 1984.

Ito, E. and H. Yamada, Stability relations of silicate spinels, ilmenites, and perovskites, in High-Pressure Research in Geophysics, edited by S. Akimoto and M. H. Manghnani, pp. 405-419, Center Acad. Pub. Japan/Reidel, Tokyo/Dordrecht, 1982.

Kanzaki, M., Ultrahigh-pressure phase relations in the system $Mg_4Si_4O_{12}$-$Mg_3Al_2Si_3O_{12}$, Phys. Earth Planet. Inter., 49, 168-175, 1987.

Liu, L., The high-pressure phases of $MgSiO_3$, Earth Planet. Sci. Lett., 31, 200-208, 1976a.

Liu, L., The post-spinel phases of forsterite, Nature, 262, 770-772, 1976b.

Liu, L., Orthorhombic perovskite phases observed in olivine, pyroxene and garnet at high pressures and temperatures, Phys. Earth Planet. Inter., 11, 289-298, 1976c.

Liu, L., The system enstatite-pyrope at high pressures and temperatures and the mineralogy of the earth's mantle, Earth Planet. Sci. Lett., 36, 237-245, 1977.

Liu, L., New silicate perovskites, Geophys. Res. Lett., 14, 1079-1082, 1978.

Liu, L. and A. E. Ringwood, Synthesis of a perovskite-type polymorph of $CaSiO_3$, Earth Planet. Sci. Lett., 28, 209-211, 1975.

Ringwood, A. E., The pyroxene-garnet transformation in the earth's mantle, Earth Planet. Sci. Lett., 2, 255-263, 1967.

Ringwood, A. E., Composition and Petrology of the Earth's Mantle, 618pp., McGrow-Hill, New York, 1975.

Ringwood, A. E. and A. Major, Synthesis of majorite and another high pressure garnets and perovskites, Earth Plant. Sci. Lett., 12, 411-418, 1971.

Suito, K., Disproportionation of pyrophyllite into new phases at high pressure and temperature, Physica, 139 & 140B, 246-250, 1986.

Takahashi, E. and E. Ito, Mineralogy of mantle peridotite along a model geotherm up to 700 km depth, in High-Pressure Research in Mineral Physics, edited by M. H. Manghnani and Y. Syono, pp. 427-437, TERRA PUB/Am. Geophys. Union, Tokyo/Washington, D. C., 1987.

Tamai, H. and T. Yagi, High-pressure and high-temperature phase relations in $CaSiO_3$ and $CaMgSi_2O_6$ and elasticity of perovskite-type $CaSiO_3$, Phys. Earth Planet. Inter., in press, 1988.

Weidner, D. J., H. Sawamoto, S. Sasaki, and M. Kumazawa, Single-crystal elastic properties of spinel phase of Mg_2SiO_4, J. Geophys. Res., 89, 7852-7860, 1984.

Weidner, D. J., A. Yeganeh-Haeri, and E. Ito, majorite, EOS, 68, 410, 1987.

Yamada, H. and E. Takahashi, Subsolidus phase relations between coexisting garnet and two pyroxenes at 50 to 100 kbar in the system Ca-MgO-Al_2O_3-SiO_2, in Kimberlites, II: The Mantle and Crust-Mantle Relationship, edited by J. Kornprobst, pp. 247-255, Elsevier, Amsterdam, 1984.

Yeganeh-Haeri, A., D. J. Weidner, and E. Ito, Single-crystal elestic properties of perovskite: $MgSiO_3$, EOS, 68, 1469, 1987.

Zindler, A. and Hart, S. R. A., Chemical geodynamics, in Ann Rev. Earth Planet. Sci., 14, 493-571, 1986.

HIGH-PRESSURE STRUCTURAL STUDY ON PEROVSKITE-TYPE $MgSiO_3$ - A SUMMARY

Yasuhiro Kudoh, Eiji Ito*, Hiroshi Takeda

Mineralogical Institute, Faculty of Science, University of Tokyo, Tokyo 113, Japan. *Institute for Study of the Earth's Interior, Okayama University, Misasa, Tottori 682-02, Japan

Introduction

The orthorhombic modification of $MgSiO_3$ perovskite was first found by Liu (1974) and was synthesized at about 300 kbar and 1000°C in a diamond-anvil press coupled with laser heating (Liu,1975). The crystal structure of perovskite-type $MgSiO_3$ at 1 bar was determined by Yagi et al. (1978) and Ito and Matsui (1978) by powder X-ray diffraction analysis. The structure is a distorted rare-earth, orthoferrite-type (orthorhombic space group Pbnm), isostructural with $ScAlO_3$ perovskite (Reid and Ringwood, 1975). Details of the crystal structure were refined by a single crystal X-ray diffraction method (Horiuchi et al. ,1987).

This paper summarizes the results of our high-pressure structural study to 96 kbar with a single-crystal X-ray diffraction method using a diamond anvil, high-pressure cell (Kudoh et al., 1987) on perovskite-type $MgSiO_3$ synthesized by Ito and Weidner (1986).

Bulk modulus of $MgSiO_3$ perovskite

Based on the measured unit cell volumes with the single crystal specimens at high pressures up to 96 kbar using diamond anvil cell, the parameters K_0 and K_0' in the Murnaghan-Birch equation of state were calculated (Kudoh et al., 1987). The bulk modulus thus obtained is given in Table 1. The value of bulk modulus observed in our work agrees well with that obtained from Brillouin spectroscopy (Yeganeh-Haeri et al., 1988). The bulk sound velocity calculated from the values of bulk modulus of present work and the density reported by Ito and Matsui (1978) is 7.8 km/second, which is in general agreement with the value 7.9 km/second predicted by Liebermann et al. (1977).

Linear compressibilities of unit cell axes

Knittle and Jeanloz (1987) reported that the axial ratios b/a and c/a of $Mg_{0.88}Fe_{0.12}SiO_3$ perovskite were constant up to 1119 kbar. However, the linear compressibilities of unit cell axes observed in our work indicate that the c axis is more compressible than the a axis, and the a axis is more compressible than the b axis. Thus the axial ratio b/a increases with increasing pressure and the c/a decreases with increasing pressure. The trend of variation of axial ratios observed in the present work agrees well with those calculated from the predicted values of unit cell axes compressibilities by Matsui et al. (1987). In the crystal structure of orthorhombic $MgSiO_3$ perovskite, there are four shorter Mg-O bonds. Out of these, three shorter Mg-O bonds are almost nearly parallel to the b axis, one almost nearly parallel to the a axis and none parallel to the c axis. This anisotropic distribution of shorter Mg-O bonds can qualitatively interpret the observed anisotropy of linear compressibility of unit cell axes.

SiO_6 and MgO_8 polyhedron

Based on the observed variation of bond lengths with pressure, linear compressibilities of mean bond lengths of the MgO_8 and the SiO_6 polyhedra to 96 kbar were calculated to be 0.18 $Mbar^{-1}$ for Mg-O and 0.09 $Mbar^{-1}$ for Si-O. from the consideration of the sizes and compressibilities of cations, O'Keeffe et al. (1979) suggested that the effect of increased pressure would be to decrease the coordination of Mg in $MgSiO_3$. In comparing the bond lengths around Mg, a distinct difference was observed between the linear compressibilities of shorter 8 bonds and longer 4 bonds. The average linear compressibilities of the shorter 8 bonds were around 0.2 $Mbar^{-1}$ and those of the longer 4 bonds around 0.0 $Mbar^{-1}$. From this fact ,it can be seen that pressure

Table 1. Observed bulk modulus (Mbar)

	K_0	K_0'
Yagi et al.(1982)[a]	2.60(20)	3-5
Knittle and Jeanloz (1987)[b]	2.66(6)	3.9
Kudoh et al.(1987)[c]	2.47(14)	4
Yeganeh-Haeri et al.(1988)[d]	2.45(5)	-

a:Powder X-ray with DAC up to 82 kbar.
b:Powder X-ray with DAC up to 1119 kbar.
c:Single crystal X-ray with DAC up to 96 kbar.
d:Brillouin spectroscopy.

moves coordination of Mg in the direction of 8-fold coordination rather than 12-fold coordination, and that the unit cell compression is controlled mainly by the tilting of relatively rigid SiO_6 octahedra. This situation can easily be interpreted by the fact that the Si-O-Si angles decrease with increasing pressure.

References

Horiuchi, H., E. Ito and D.J. Weidner, $MgSiO_3$(perovskite-type): single crystal X-ray diffraction study. Am. Mineral. 72, 357-360, 1987.

Ito, E. and Y. Matsui, (1978) Synthesis and crystal-chemical characyerization of $MgSiO_3$ perovskite. Earth Planet. Sci. Lett., 38, 443-450, 1978.

Ito, E. and D.J. Weidner, Crystal growth of $MgSiO_3$ perovskite. Geophys. Res. Lett., 13, 464-466, 1986.

Knittle, E. and R. Jeanloz, Synthesis and equation of state of $(Mg,Fe)SiO_3$ perovskite to over 100 gigapascals, Science, 235, 668-670,1987.

Kudoh, Y., E. Ito and H. Takeda, Effect of pressure on the crystal structure of perovskite-type $MgSiO_3$, Phys. Chem. Minerals, 14, 350-354, 1987.

Liebermann, R.C., L.E.A. Jones and A.E. Ringwood, Elasticity of aluminate, titanate, stannate and germanate compounds with the perovskite structure. Phys. Earth Planet. Inter., 14, 165-178, 1977.

Liu, L.G., Silicate perovskite from phase transformations of pyrope-garnet at high pressure and temperature. Geophys. Res. Lett., 1, 277-280, 1974.

Liu, L.G., Post-oxide phases of forsterite and enstatite. Geophys. Res. Lett., 2, 417-419, 1975.

Matsui,M., M. Akaogi and T. Matsumoto, Computational model of the structural and elastic properties of the ilmenite and perovskite phases of $MgSiO_3$. Phys. Chem. Minerals, 14, 101-106, 1987.

O'Keeffe, M.,B.G. Hyde and J.O. Bovin, Contribution to the crystal chemistry of orthorhombic perovskites: $MgSiO_3$ and $NaMgF_3$. Phys. Chem. Minerals, 4, 299-305, 1979.

Reid, A.F. and A.E. Ringwood, High-pressure modification of $ScAlO_3$ and some geophysical implications. J. Geophys. Res., 80, 3363-3370, 1975.

Yagi, T., H.K. Mao and M. Bell, Structure and crystal chemistry of perovskite-type $MgSiO_3$. Phys. Chem. Minerals, 3, 97-110, 1978.

Yagi, T., H.K. Mao and M. Bell, Hydrostatic compression of perovskite-type $MgSiO_3$. In S.Saxena, Ed., Advances in Physical Geochemistry, Vol.2, pp.317-325, Springer-Verlag, New York,1982.

Yeganeh-Haeri, Y., D.J. Weidner and E. Ito, Elasticity of $MgSiO_3$ in the perovskite structure. Nature, (in press).

RAMAN SPECTROSCOPY AND LATTICE DYNAMICS OF $MgSiO_3$-PEROVSKITE AT HIGH PRESSURE

R. J. Hemley[1], R. E. Cohen[2], A. Yeganeh-Haeri[3], H. K. Mao[1], D. J. Weidner[3], and E. Ito[4]

Abstract. Vibrational Raman spectra have been obtained for 50 to 100 μm single crystals of $MgSiO_3$ perovskite in situ at high pressure. Seven bands were tracked as a function of pressure to 26 GPa using a diamond-anvil high-pressure cell with rare-gas pressure-transmitting media. The frequency shifts with pressure are positive, and no soft modes were observed, in agreement with the present and previous lattice dynamics calculations. Zero-pressure frequency shifts $(d\nu_i/dP)_0$ vary between 1.7 and 4.2 cm^{-1}/GPa, which contrasts with the uniform shift of 2.6 cm^{-1}/GPa for modes measured in high-pressure mid-infrared spectra. The mode-Grüneisen parameters γ_i determined from the present data span the range 1.6 - 1.9, and are generally higher than those reported in the infrared study. The Raman data are interpreted using the lattice dynamics calculated from the potential-induced breathing (PIB) model, a Gordon-Kim approach that includes the effects of charge relaxation on the dynamics. Good agreement with the experimentally determined frequencies is obtained, particularly in the lower frequency range, in comparison with previous rigid-ion results. The high thermal expansivity for $MgSiO_3$-perovskite is shown to be due to the comparatively high values for γ_i associated with the lower frequency modes. Thermal weighting of the individual γ_i is required for an accurate calculation of the thermal Grüneisen parameter γ_{TH} and thermal expansivity.

Introduction

A consideration of the lattice dynamics of perovskites is essential for understanding their thermal and high-pressure properties, including soft-mode behavior, structural phase transitions, entropy, and thermal expansivity. In particular, the lattice dynamics of magnesium silicate perovskite at high pressure is of interest because of the likely abundance of this material (and structurally related phases) in the Earth's lower mantle [Liu, 1976; Yagi et al., 1978, 1982; Knittle and Jeanloz, 1987]. To obtain useful information on thermal properties of high-pressure phases it is essential that measurements be carried out at pressure conditions appropriate for the stability field of the phase, which for $MgSiO_3$-perovskite correspond to pressures in the 20-GPa range and above. Vibrational measurements of samples under these conditions are possible using high-pressure spectroscopic methods, techniques which provide useful structural and thermodynamic constraints on high-pressure silicate phases [Hemley, 1987].

In the last few years theoretical models based on non-empirical approximate methods have become useful for understanding static and dynamical properties of high-pressure oxide and silicate phases containing closed-shell ions. These methods are based on the Gordon-Kim model for the short-range forces [Gordon and Kim, 1972], implemented at various levels of approximation for crystals [Cohen and Gordon, 1976; Hemley et al., 1985; Mehl et al., 1986]. Several calculations of the properties of orthorhombic $MgSiO_3$-perovskite have been performed [Wolf and Bukowinski, 1985; 1987; Hemley et al., 1987a; Cohen, 1987]. The more recent calculations show that this non-empirical approach is capable of quantitative predictions of changes in structural properties as a function of pressure and temperature, phase transformations, and elasticity. The results may also be compared with lattice-dynamical calculations using empirical potentials [Wall et al., 1986; Matsui et al., 1987].

In the present study, we have measured the Raman spectrum as a function of pressure of recently grown single crystals of $MgSiO_3$-perovskite [Ito and Weidner, 1986]. These data complement previously measured zero-pressure Raman and high-pressure mid-infrared spectra [Williams et al., 1987], recent high-pressure far-infrared spectra [Hofmeister et al., 1987], and zero-pressure single-

---

[1] *Geophysical Laboratory, Carnegie Institution of Washington*
[2] *Condensed Matter Physics Branch, Naval Research Laboratory*
[3] *Department of Earth and Space Sciences, SUNY Stony Brook*
[4] *Institute for Study of the Earth's Interior, Okayama University*

crystal Brillouin scattering spectra [Yeganeh-Haeri et al., 1988] of this material. The results are analyzed in terms of lattice dynamics calculated using the potential-induced breathing model (PIB). The theoretical results are also used to interpret the recent high-pressure infrared data.

Experimental Method

Single crystals of $MgSiO_3$-perovskite with maximum dimensions of 50-100 μm were synthesized in the uniaxial split-sphere apparatus following the technique described by Ito and Weidner [1986]. For each high-pressure spectroscopic run, a crystal was mounted in a 200 μm diameter pressure chamber of a Mao-Bell-type diamond-anvil cell [Mao and Bell, 1978] along with argon gas to serve as a pressure-transmitting medium [Jephcoat et al., 1987]. The argon, which solidifies at 1.2 GPa, provides a quasi-hydrostatic pressure environment for the crystal in the diamond-anvil cell over the pressure range of the present experiments.

The single-crystals were probed with a micro-optical spectrometer system described elsewhere [Hemley, 1987]. Both Ar^+ and Kr^+ gas lasers were used to excite the Raman spectrum. In most of the runs the pressure was determined by the ruby fluorescence method [Mao et al., 1986]. For the measurements performed with the Kr^+ laser we used only a small amount of ruby in order to minimize interference in the weak Raman spectrum from the intense ruby fluorescence. For these runs we determined the pressure to 15 GPa by measuring the shift in the strong 464 cm^{-1} mode in the Raman spectrum of α-quartz placed in the diamond-anvil cell with the perovskite sample. The pressure shift of this Raman band has been calibrated against the ruby scale [Hemley, 1987]. A back-scattering geometry was also used for the measurements carried out at zero-pressure (0.1 MPa) in order to obtain polarization information [Mao et al., 1987].

Experimental Results

It is useful first to consider the group theoretical predictions for the number and symmetries of optical vibrations for this crystal [Fateley et al., 1972]. For orthorhombic $MgSiO_3$-perovskite (space group Pbnm; Z = 4 [Yagi et al., 1978; Horiuchi et al., 1987], the irreducible representation of the optical vibrations is

$$\Gamma_{opt} = 7A_g + 7B_{1g} + 5B_{2g} + 5B_{3g}$$
$$+ 8A_u + 7B_{1u} + 9B_{2u} + 9B_{3u}.$$

The centrosymmetric unit cell forces a separation of the Raman- and infrared-active vibrations in the perfect crystal. The gerade vibrations, A_g, B_{1g}, B_{2g}, and B_{3g} are all active in the first-order Raman spectrum. Of the ungerade modes, the B_{1u}, B_{2u}, and B_{3u} vibrations are active in infrared absorption, whereas the A_u modes are optically inactive.

The initial measurements were performed at 0.1 MPa using an Ar^+ laser (457.9 - 514.5 nm). With these wavelengths considerable absorption of laser light by the sample was observed. This absorption resulted in strong fluorescence and ultimately decomposition of the sample. Similar problems were observed in Brillouin scattering studies carried out using Ar^+ laser excitation on crystals from the same synthesis [Yeganeh-Haeri et al., 1988]. We found that the use of a Kr^+ laser operating longer wavelengths (647.1 nm) significantly reduced these effects in the lower pressure studies. We found that at high pressure the spurious fluorescence obtained with Ar^+ excitation is diminished. In addition, the signal-to-noise ratio of the spectra improves because the index of refraction of the solid argon begins to approach that of the perovskite. Finally, as the stability field of the perovskite is approached at higher pressure (>20 GPa), higher laser intensity can be used without decomposition of the sample under the beam.

Representative spectra measured from 150 to 650 cm^{-1} are shown in Fig. 1. Six well-resolved peaks were

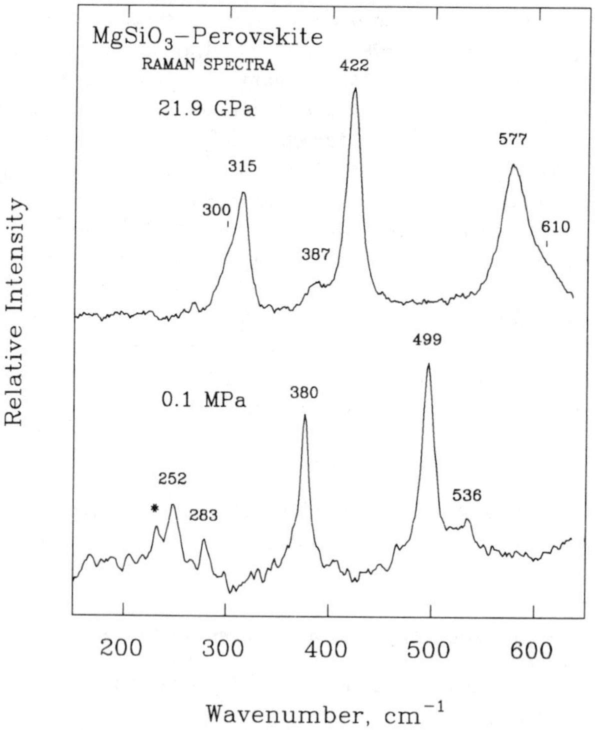

Fig. 1. Raman spectra (low frequency region) of $MgSiO_3$-perovskite at 0.1 MPa (zero pressure) and 21.9 GPa. The spectra were measured at 298 (± 2) K. Plasma lines from the ion laser are indicated by an asterisk.

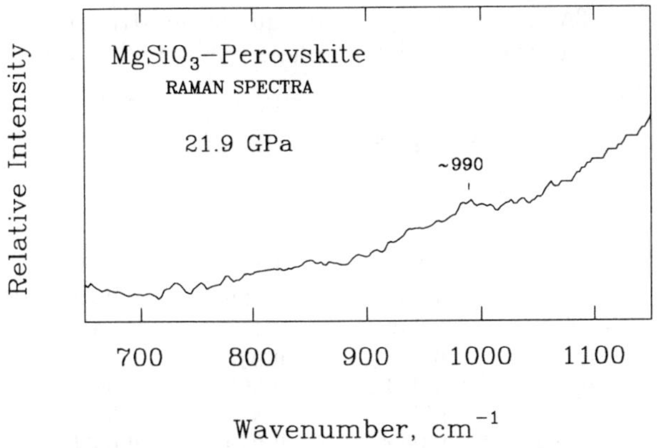

Fig. 2. Raman spectrum (high frequency region) of $MgSiO_3$-perovskite at 21.9 GPa (298 K).

observed under ambient pressure. A spectrum measured at high pressure (21.9 GPa) is also shown in the figure. With the improved spectra at higher pressure two additional peaks become evident (estimated zero-pressure frequencies of ~370 and ~900 cm^{-1}). All peaks have a positive frequency shift with pressure. However, the magnitude of the shift varies among the bands. In addition, there is evidence that the relative intensities of the principal bands may change with pressure. The likelihood of weaker first-order Raman peaks in the low frequency range is discussed below.

A high-pressure Raman spectrum measured in the higher frequency region is shown in Fig. 2. The weak 900 cm^{-1} peak shifts to near 1000 cm^{-1} at high pressure. Broader features in the higher frequency region were also observed in spectra measured under intense laser irradiation (e.g., >50 mW power). These bands may arise from second-order scattering from the perovskite due to heating under the laser beam. Most of this intensity, however, appears to be associated with a glassy material formed by thermal decomposition of the perovskite (at lower pressures). This reaction appears to proceed via thermally-induced micro-twinning of the single-crystals [Yeganeh-Haeri et al., 1988], but the specific mechanism has not yet been explored in detail. We have reproduced the broad bands present in the spectrum of Williams et al. [1987] by measuring zero-pressure Raman spectra of $MgSiO_3$ glass samples that had been laser-heated under pressure. The broad features can be identified as Raman bands of $MgSiO_3$ glass recovered from high-pressure and temperature [Hemley et al., 1987b].

The numerical results are summarized in Table 1, and the pressure shifts are shown in Fig. 3. The four peaks reported in the earlier zero-pressure Raman study of Williams et al. [1987] are confirmed; differences in peak positions are within the estimated error of both measurements. Preliminary polarization measurements

TABLE 1. $MgSiO_3$-perovskite Raman spectrum: zero-pressure vibrational frequencies, pressure derivative, and mode-Grüneisen parameters

ν_i, cm^{-1}	$d\nu_i/dP$, cm^{-1}/GPa	$\gamma_i = -d\ln\nu_i/d\ln V$
252 [251][a]	2.9(2)	3.0(2)[b]
283 [282]	1.7(1)	1.6(1)
370[c]	2.4(5)	1.7(3)
380 [378]	2.4(1)	1.9(1)
499 [499]	4.2(2)	3.5(2)
536	3.2(1)	1.6(2)
~690		
~900[c]	2.5(8)	0.7(2)

a. Williams et al. [1987].
b. Calculated using K_0 = 260 GPa [Yagi et al., 1982].
c. Frequency at zero pressure determined by extrapolation from high pressure.

are consistent with the assignment of these bands to A_g fundamentals. Fig. 3 includes the pressure dependence of the mid-infrared modes measured in the previous study. The large variation in the pressure shifts obtained from the high-pressure Raman spectra contrasts with that

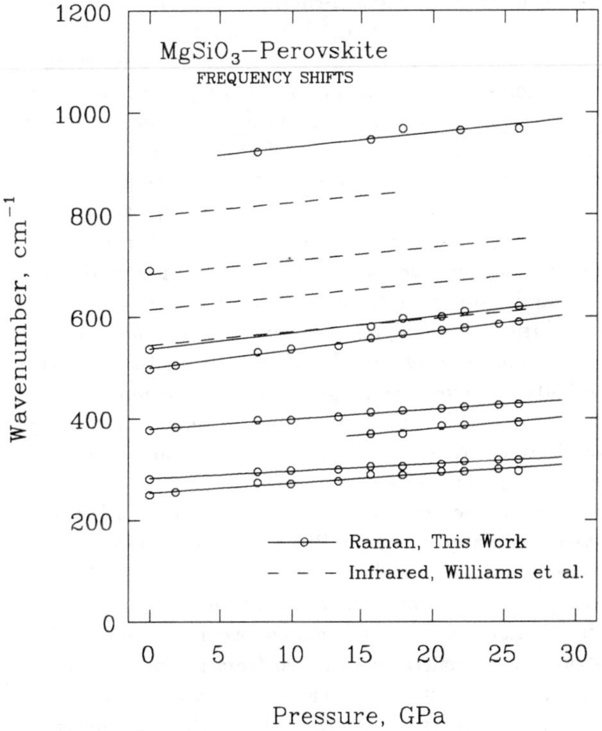

Fig. 3. Pressure shifts of the observed Raman and infrared bands of $MgSiO_3$-perovskite.

found in the IR study, in which the same shift for each band was reported.

Theoretical Model

We now analyze the experimental data by calculating the full vibrational spectrum as a function of pressure using the potential-induced breathing (PIB) model. In this Gordon-Kim type model, the electron density of the crystal is assumed to be composed of a superposition of electron density of the component ions, each of which is calculated using the local-density approximation with self-interaction corrections [e.g., Mehl et al., 1986]. A Watson sphere is included in each atomic calculation to simulate the electrostatic field of the crystal. The radius of the Watson sphere is chosen to give an electrostatic potential within the sphere equal to the Madelung potential at the site in the crystal. This prescription has been shown to give excellent agreement at low pressures with highly precise charge densities calculated using the Linearized Augmented Plane Wave (LAPW) method for MgO and CaO [Mehl et al., 1988]. A comparison of the PIB charge density with the LAPW charge density for cubic $MgSiO_3$ is given in another paper in this volume [Cohen et al., 1988].

The total energy for the PIB charge density is the sum of three parts: the long-range Madelung energy between the ions, the self-energy of each ion, and the short-range overlap energy between ions. The self-energy is fit to a fifth-order polynomial as a function of Madelung potential. The short-range overlap energy between ions i and j are calculated as a function of interionic distance r_{ij} for a series of charge densities for each ion calculated with different Watson-sphere potentials P_i and P_j. The calculated interionic potentials are then fit as a function of r_{ij}, P_i, and P_j for use in the crystal calculation. We use the Thomas-Fermi and Hedin-Lundquist functionals for the kinetic and exchange-correlation energies, respectively. The dependence of the short-range energy and the self-energy on the Madelung potentials introduces many-body contributions to the total energy. The lattice-dynamical contributions were determined by the calculation of the dynamical matrix for this model [Cohen et al., 1987].

The single-crystal elastic moduli and equation of state calculated from the PIB model have been published previously [Cohen, 1987; see also Cohen et al., 1988]. Before proceeding with the calculation of the optic (Raman and infrared) modes, we first examine the structural properties calculated from the model. In this calculation the structure was determined by minimization of the free energy at 298 K, with the thermal contribution calculated from the quasiharmonic approximation. The results are listed in Table 2, along with the experimental data and previous calculations. Although the model tends to underestimate the degree of distortion of $MgSiO_3$ from the cubic aristotype, the agreement with experiment is reasonable. The coupling of internal strain and ion breathing, which was neglected in our previous calculations, is included here. The effect on the structural properties is found to be small.

TABLE 2. Zero-pressure equilibrium structure of $MgSiO_3$-perovskite: orthorhombic, Pbnm (Z=4)

	SSMEG[a]	PIB[b]	Exp.[c]
V, Å^3	166.60	165.05	162.75
a, Å	4.849	4.877	4.780(1)
b, Å	4.937	4.896	4.933(1)
c, Å	6.959	6.912	6.902(1)
Mg x	0.995	0.999	0.974(7)
Mg y	0.020	0.005	0.063(5)
Mg z	1/4	1/4	1/4
O(1) x	0.080	0.049	0.096(10)
O(1) y	0.480	0.493	0.477(11)
O(1) z	1/4	1/4	1/4
O(2) x	0.711	0.726	0.696(7)
O(2) y	0.288	0.273	0.291(7)
O(2) z	0.042	0.025	0.056(4)

a. Hemley et al. [1987]; T=298 K.
b. Equilibrium structure determined by Cohen [1987]; T=298 K.
c. Yagi et al. [1978]; T=295 K. See also the recent single-crystal x-ray diffraction study of Horiuchi et al. [1987].

Theoretical Results

The frequencies of the optical ($\underline{k}$ = 0) vibrations calculated from the PIB model at zero-pressure and 298 K are listed in Table 3. The present Raman frequencies and the recent infrared data of Williams et al. [1987] and Hofmeister et al. [1987] are also listed. There is a close correspondence in frequency for a number of the modes, although the symmetries of the weaker bands in the spectra have not been determined. A low-frequency A_g mode at 95 cm^{-1} is predicted theoretically, along with B_{1g} and B_{3g} modes at 101 and 138 cm^{-1}, respectively. However, the lowest frequency Raman bands that are well-resolved appear at 252 and 283 cm^{-1}. The latter are close in frequency to calculated A_g modes at 203 and 278 cm^{-1}, and preliminary polarization measurements are consistent with this assignment. We therefore propose that an additional set of lower frequency Raman bands is present in $MgSiO_3$ but is not observed experimentally as a result

TABLE 3. Calculated and measured vibrational frequencies ($\underline{k} = 0$) at zero pressure

RAMAN			INFRARED			INACTIVE	
Γ_i	Theory	Exp[a]	Γ_i	Theory	Exp	Γ_i	Theory
A_g	95		B_{1u}	232	225[b]	A_u	242
B_{1g}	101		B_{3u}	238	250	A_u	294
B_{3g}	138		B_{2u}	250	279	A_u	399
A_g	203	252	B_{2u}	292		A_u	502
A_g	278	283	B_{1u}	307	316[b]	A_u	648
B_{1g}	303		B_{3u}	327	354	A_u	719
B_{2g}	338		B_{2u}	360	385	A_u	889
B_{1g}	342		B_{3u}	382		A_u	908
A_g	373	380	B_{2u}	400			
B_{3g}	428		B_{3u}	418			
B_{2g}	458		B_{2u}	443			
A_g	461	499	B_{3u}	470	544[c]		
B_{1g}	482		B_{2u}	641	614		
B_{3g}	515	536	B_{3u}	704	683		
B_{2g}	678		B_{1u}	710			
B_{3g}	679		B_{2u}	760			
A_g	683	~690	B_{1u}	771	797[c]		
B_{2g}	770		B_{3u}	773			
A_g	771		B_{1u}	900			
B_{1g}	771		B_{2u}	909			
B_{2g}	773		B_{1u}	917			
B_{1g}	1070	900	B_{3u}	920			
B_{3g}	1113		B_{2u}	1010			
B_{2g}	1114		B_{3u}	1019			
			B_{2u}	1363			

a. This work.
b. Hofmeister et al. [1987].
c. Williams et al. [1987].

of low scattering cross-section. Very weak features were observed in some of our spectra in this region (e.g., zero-pressure frequencies of approximately 130 and 190 cm^{-1}); however, the intensities of these bands are approximately equal to the noise level so that positive identification and assignments cannot be made at this time.

The calculation of specific atomic displacements associated with each vibrational mode details the degree to which internal and external vibrations of the component polyhedra may be mixed in the lower orthorhombic symmetry structure (relative to that in the cubic aristotype). In most previous investigations the arguments used to assign vibrational bands within a given symmetry species to specific atomic displacements have been only qualitative. The displacements for the seven A_g modes calculated from the PIB lattice dynamics are shown in Fig. 4. For these (totally symmetric) modes there can be no motion involving cations located at special positions in the unit cell (e.g., the Si ions). In general, a large degree of coupling of octahedral libration and Mg-ion translation is observed, particularly at intermediate frequencies (200-400 cm^{-1}). We note, however, that the 95 cm^{-1} mode is associated with nearly pure SiO_6 octahedral librational (rocking) motion, with little coupling with cation displacements and internal octahedral motions. The 278 and 373 cm^{-1} modes are derived from the same Mg-ion translations, but with orthogonal displacements. The 461 cm^{-1} mode (499 cm^{-1} experimental) is largely an octahedral deformation mode, with contributions from both cation displacements and octahedral libration. The 771 and 683 cm^{-1} modes are symmetric combinations of nearly pure octahedral deformations.

The calculated pressure derivatives and mode-Grüneisen parameters are listed in Table 4. We note that particularly large γ_i are found for the lower frequency modes (e.g., $\gamma_i = 6.42$ for the 95 cm^{-1} mode). The pressure dependencies of the calculated mode frequencies at $\underline{k} = 0$ are shown to 150 GPa in Fig. 5 along with the available Raman and infrared data (at lower pressures). Close inspection of the results for for the pressure shifts of the Raman modes indicates that the agreement between theory and experiment is only fair (Fig. 5a). On

TABLE 4. Calculated vibrational frequencies and mode-Grüneisen parameters ($\underline{k} = 0$) at zero pressure

RAMAN			INFRARED			INACTIVE		
Γ_i	ν_i	γ_i	Γ_i	ν_i	γ_i	Γ_i	ν_i	γ_i
A_g	95	6.42	B_{1u}	232	1.03	A_u	242	1.02
B_{1g}	101	5.37	B_{3u}	238	2.06	A_u	294	1.97
B_{3g}	138	6.78	B_{2u}	250	2.03	A_u	399	2.00
A_g	203	4.11	B_{2u}	292	2.29	A_u	502	0.30
A_g	278	3.10	B_{1u}	307	3.09	A_u	648	0.48
B_{1g}	303	2.11	B_{3u}	327	2.97	A_u	719	0.59
B_{2g}	338	2.10	B_{2u}	360	1.18	A_u	889	1.31
B_{1g}	342	2.75	B_{3u}	382	1.28	A_u	908	1.20
A_g	373	2.80	B_{2u}	400	1.16			
B_{3g}	428	0.45	B_{3u}	418	1.60			
B_{2g}	458	1.03	B_{2u}	443	2.25			
A_g	461	0.56	B_{3u}	470	1.87			
B_{1g}	482	1.08	B_{2u}	641	0.40			
B_{3g}	515	1.32	B_{3u}	705	0.45			
B_{2g}	678	1.37	B_{1u}	710	0.50			
B_{3g}	679	1.35	B_{2u}	760	0.33			
A_g	683	1.35	B_{1u}	771	0.61			
B_{2g}	770	0.61	B_{3u}	773	0.46			
A_g	771	0.69	B_{1u}	900	1.25			
B_{1g}	771	0.73	B_{2u}	909	1.28			
B_{1g}	773	1.00	B_{1u}	917	1.26			
B_{1g}	1070	0.84	B_{3u}	920	1.19			
B_{3g}	1113	0.88	B_{2u}	1010	0.47			
B_{2g}	1114	0.97	B_{3u}	1019	0.49			
			B_{2u}	1363	0.61			

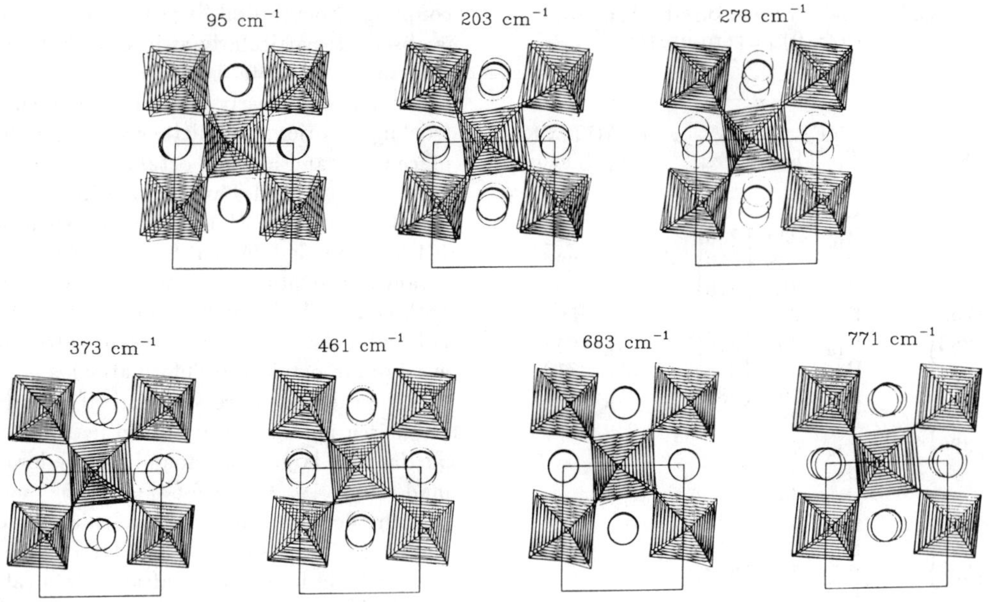

Fig. 4. Mode displacements for A_g vibrations calculated from the theoretical model. The darker lines give the equilibrium structure; the lighter lines show the atomic displacements (×5) corresponding to the eigenvector of the normal mode. The orthorhombic unit cell is indicated.

the other hand, there appears to be very good agreement with the recent high-pressure far-infrared measurements of Hofmeister et al. [1987] (Fig. 5b). We also note that avoided crossings of certain sets of curves as a function of pressure are apparent (e.g., the optically inactive A_u vibrations at 500-600 cm^{-1} at ~80 GPa in Fig. 5c). For the optically active modes, this mixing will in general cause a change in the relative intensities measured spectroscopically, which is consistent with the intensity variations in the Raman spectra noted above.

Thermodynamic Properties

The zero-pressure volume V_0, Grüneisen parameter γ_{TH} and bulk modulus K_0 calculated from the PIB model are in good agreement with experiment. However, Cohen [1987] found a low value for the thermal expansivity compared with experiment [Knittle et al., 1986]. In addition, the calculated thermal expansivity was lower than that expected on the basis of previous calculations [Hemley et al., 1987a] in which an explicit tradeoff between the the bulk modulus and α was observed; i.e., there is a small volume dependence of αK_T ($= \gamma_{TH} C_v/V$). We since have determined that the low thermal expansivity found previously was due to a number of unphysical imaginary frequencies arising in the current PIB model due to the Thomas-Fermi approximation for the kinetic energy. The imaginary frequencies arise from a problematic term in PIB lattice dynamics in which effective charges appear [Cohen et al., 1987] (see Eqs. A7-A12, A41), which are the first derivatives of the total energy with respect to the Madelung potentials at each site. Two calculations were performed to remove the effect of the unstable modes for the purpose of estimating the thermodynamic properties for $MgSiO_3$-perovskite from the model.

In the first approach, the thermodynamic Grüneisen parameter was calculated from the available spectral data and the lattice-dynamical model by a direct averaging over the individual modes, which formally gives the high-temperature limit of γ. The calculation was also performed by explicitly including a thermal contribution of each mode using the expression $\langle\gamma_i\rangle = \Sigma C_{vi}\gamma_i/\Sigma C_{vi}$, where C_{vi} is the contribution of each mode i to the heat capacity at constant volume. Because of instabilities occurring for some of the $\underline{k} \neq 0$ modes in this model, we sum over only a limited set of modes. For the enlarged unit cell of orthorhombic perovskite (Z = 4), we found that a sufficiently converged calculation was obtained by summing over the $\underline{k} = 0$ vibrations (60 modes). We obtain $\langle\gamma_i\rangle = 2.0$ at 298 K. This value decreases to $\langle\gamma_i\rangle = 1.63$ at 1000 K, with $\langle\gamma_i\rangle = 1.59$ as the high temperature limit. From the identity, $\alpha = \gamma_{TH} C_v/K_T V$, the thermal expansivity was estimated using the thermally weighted $\langle\gamma_i\rangle$, and the heat capacity was calculated from the $\underline{k} = 0$ modes, to-

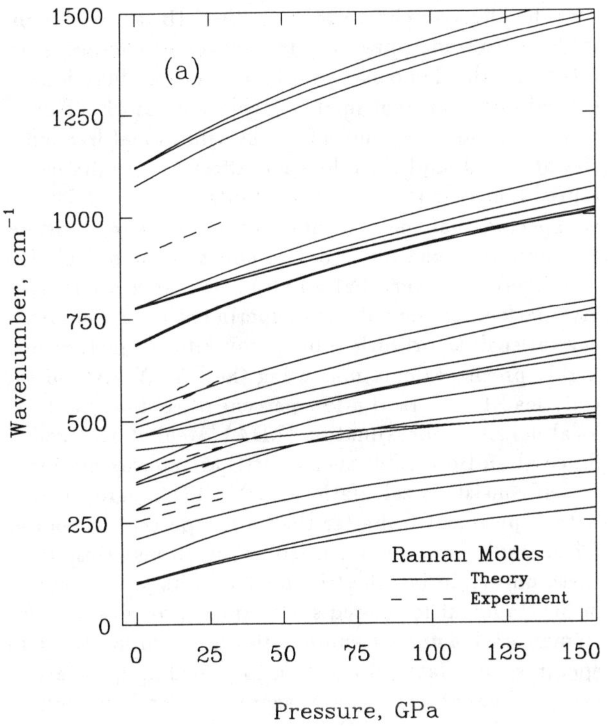

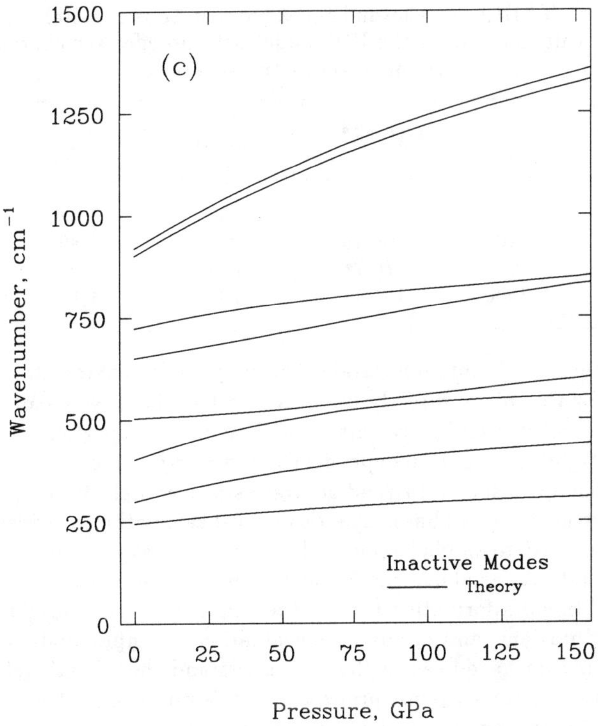

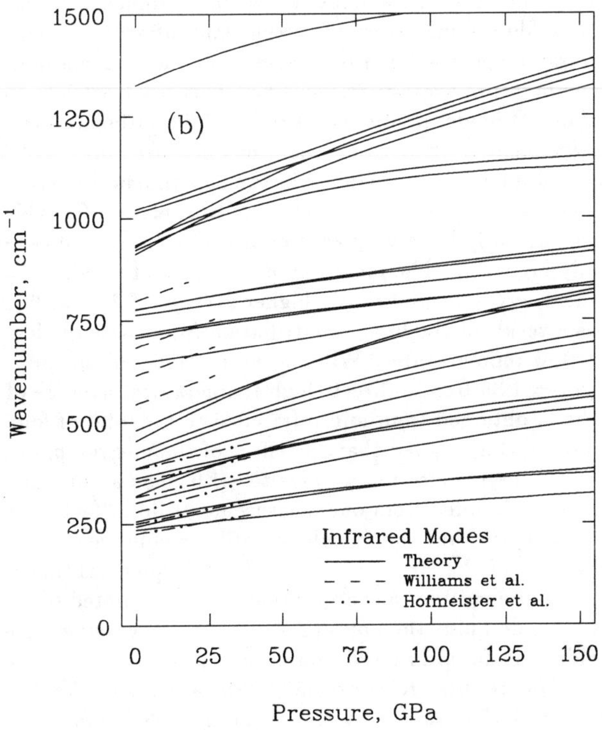

Fig. 5. Calculated and measured pressure dependence of the $\underline{k} = 0$ modes of $MgSiO_3$-perovskite (298 K). (a) Raman modes. (b) Infrared modes. (c) Inactive modes.

gether with the calculated molar volumes and bulk moduli reported previously [Cohen, 1987]. From this calculation we obtain a value for α of 23 $\times$ 10^{-6} K^{-1} at 298 K, increasing to 31$\times$ 10^{-6} K^{-1} at 1000 K. The higher temperature result represents a lower bound on α, because the decrease in K_T with temperature has been neglected.

In the second approach for examining corrections to the thermal expansivity, the thermal equation of state was calculated after setting the PIB effective charges to zero, which makes all the PIB frequencies real at positive pressures. We do, however, find a mode instability at negative pressures (high temperatures). We have fit the free energies for positive static pressures and present the equation of state parameters in Table 5. We find a mean thermal expansivity from 298 to 1000 K of 44 $\times$ 10^{-6} K^{-1}. The room-temperature volume differs by only 0.3% from the value obtained by Cohen [1987]; this agreement indicates that the zero-point contribution was estimated accurately in spite of the imaginary frequencies.

Discussion

In the present study, the frequencies of eight Raman bands have been determined for $MgSiO_3$-perovskite at zero pressure, and pressure shifts for seven bands have been measured to 26 GPa. The much larger number of modes (24) predicted to be Raman-active for this crystal complicates both definitive assignments of the modes and

TABLE 5. Thermal equation of state parameters calculated from the PIB model with no effective charge terms in the lattice dynamics

T	V_0, Å^3	K_0, GPa	K_0'
298	165.56	253	3.88
400	166.19	248	3.89
1000	170.78	218	3.98
2000	181.11	167	4.15

the calculation of accurate thermodynamic constraints for this phase directly from the spectra. Thus we have found it useful to calculate the dynamical properties for $MgSiO_3$ from an independent parameter-free model that has been shown to yield accurate static and elastic properties for this phase. The theoretical calculation provides insights to the behavior of this compound at pressures that are far above current measurements, provides additional information such as the character of the normal vibrations, and permits the evaluation of simple models that have been applied to understand the vibrational and thermodynamic properties of this important phase. Finally, comparison between theory and experiment shed light on certain fundamental approximations used in the model.

We have tentatively assigned the stronger bands in the spectrum (252, 283, 380, and 499 cm^{-1} at zero pressure) to the A_g symmetry species. The PIB calculation with dynamical charge relaxation effects are in good agreement with experiment. This interpretation is also consistent with analyses of Raman spectra of isostructural orthorhombic perovskites, which indicate that the spectra are dominated by (totally symmetric) A_g modes [e.g., Scott and Remeika, 1970; Alain and Piriou, 1975]. The weak band at ~900 cm^{-1} is likely to be a B_{ng} mode, correlating with the 965 cm^{-1} octahedral deformation mode observed in the Raman spectrum of SiO_2 stishovite [Hemley et al., 1986]. Positive frequency shifts with pressure are obtained theoretically for all modes; i.e., no soft modes were observed over the range of pressures corresponding to the lower mantle. The PIB calculation indicates that there is strong coupling of octahedral librational and Mg translations, particularly in the 200-500 cm^{-1} region. On the basis of our previous experience, we expect that the calculated modes may not be found in the same order as the measured modes. For BeO, for example [Jephcoat et al., 1988], we found better agreement for the pressure shifts than for the mode frequencies. Thus in the more complicated case of orthorhombic $MgSiO_3$-perovskite, our mode assignments are only tentative, but we expect the averages over the modes, such as the frequency moments and average γ's, to be reliable.

The effective charge term on the PIB lattice dynamics leads to some imaginary frequencies in complex crystals when the Thomas-Fermi kinetic energy functional is used in the current model [Mehl et al., 1986]. This term, however, does not affect the vibrational frequencies at $\underline{k} = 0$, and thus does not affect the results presented above or the elastic constants [Cohen, 1987]. It is expected that more accurate results for $\underline{k} \neq 0$ will be available with continued development of the model. It is unlikely, however, that an ab initio model would ever give perfect agreement with experiment for a complex crystal such as orthorhombic perovskite. In fact, even highly precise calculations using the LAPW method typically lead to errors of a few percent in volume due to the local density approximation (LDA) [Mehl et al., 1988]. Since ab initio models such as PIB are approximations to self-consistent calculations within LDA, agreement with experiment to better than a few percent in volume, for example, must be regarded as fortuitous. Despite these uncertainties, the PIB model, in its present state of development for closed-shell oxides, provides a useful framework for understanding vibrational properties and spectroscopic data, in addition to providing quite accurate P-V equations of state, aggregate elastic moduli, and pressure derivatives of elastic moduli.

The measured mode-Grüneisen parameters obtained from the present Raman data, when combined with those determined from the recent mid-infrared and far-infrared spectra, span the range 0.8-3.5 and give a high-temperature average $\langle \gamma_i \rangle \approx 1.65$. This result is close to the PIB model prediction of $\langle \gamma_i \rangle = 1.59$, although the agreement is likely to be fortuitous because only 17 of 60 $\underline{k} = 0$ modes were included in the experimental analysis. The acoustic and low-frequency optical modes, for which we currently have no pressure measurements, have generally higher γ_i. The inclusion of the thermal weighting of the γ_i gives a significantly higher average of 2.0 at 298 K as a result of the larger contribution from these modes at low temperatures. Williams et al. [1988] obtained $\langle \gamma_i \rangle = 1.36$ from a direct (high-temperature) average of the γ_i obtained from mid-infrared spectra (total of four modes), but argued that the effect of the missing modes should be to raise the average to ~1.9. The additional modes obtained from our measurements do in fact raise the average, but the calculation still remains incomplete because of the large number of mode frequencies that are still missing. We thus argue that the γ_i predicted from the model, plus the appropriate thermal averaging, provide a useful alternative constraint on γ_{TH}.

Our results are in excellent agreement with the $\gamma_{TH} = 1.77$ obtained by Knittle et al. [1986] from analyses of thermal expansion data from 300 to 860 K; the theoretical average over this range is 1.83. The thermal expansivity calculated from the PIB model either using the two approaches described above is in accord with the value obtained by these investigators. The mean thermal expansivity from 298 to 1000 K of 44×10^{-6} K^{-1} calculated from the modified treatment of the lattice dynamics is close to the experimental value of 40×10^{-6} K^{-1} [Knittle et al., 1986]. We note that the thermal expansivity has a significant temperature dependence, with α calculated to be 23×10^{-6} K^{-1} at 298 K. This value is in excellent agreement with the results of recent low-temperature single-crystal x-ray diffraction measurements of Hazen and Ross [1988].

Acknowledgments. We are grateful to L. W. Finger for assistance with the calculations and to A. M. Hofmeister for permission to quote her unpublished far-infrared data. We also thank R. M. Hazen, N. L. Ross, A. Navrotsky, and K. D. Leinenweber for comments on the manuscript. This work was supported by NSF grant EAR-8608946.

References

Alain, P. and B. Piriou, High temperature Raman scattering and phase transitions in $EuAlO_3$, *Solid State Comm. 17*, 35-39, 1975.

Cohen, A. J., and R. G. Gordon, Modified electron-gas study of the stability, elastic properties, and high-pressure behavior of MgO and CaO crystal, *Phys. Rev. B, 14*, 4593-4605, 1976.

Cohen, R. E., Elasticity and equation of state of $MgSiO_3$ perovskite, *Geophys. Res. Lett. 14*, 1053-1056, 1987.

Cohen, R. E., L. L. Boyer, and M. J. Mehl, Lattice dynamics of the potential-induced breathing model: phonon dispersion in the alkaline-earth oxides, *Phys. Rev. B. 35*, 5749-5760, 1987.

Cohen, R. E., L. L. Boyer, M. J. Mehl, and W. E. Pickett, Electronic structure and total energy calculations for oxide perovskites and superconductors, *this volume*, 1988.

Fateley, W. G., F. R. Dollish, N. T. McDevitt, and F. F. Bentley, *Infrared and Raman Selection Rules for Molecular and Lattice Vibrations: The Correlation Method*, Wiley-Interscience, New York, 1972.

Gordon, R. G., and Y. S. Kim, Theory for the forces between closed-shell atoms and molecules, *J. Chem. Phys. 56*, 3122-3133, 1972

Hazen, R. M., and N. L. Ross, Thermal expansion study of $MgSiO_3$ perovskite *EOS Trans. Am. Geophys. Union, 69*, 473, 1988.

Hemley, R. J., Pressure dependence of Raman spectra of SiO_2 polymorphs: α-quartz, coesite, and stishovite, in *High-Pressure Research in Mineral Physics*, edited by M. H. Manghnani and Y. Syono, (Terra Scientific-AGU, Washington, D.C.), pp. 347-359, 1987.

Hemley, R. J., M. D. Jackson, and R. G. Gordon, First-principles theory for the equations of state of minerals to high pressures and temperatures: application to MgO, *Geophys. Res. Lett. 12*, 247-250, 1985.

Hemley, R. J., H. K. Mao, and E. C. T. Chao, Raman spectrum of natural and synthetic stishovite, *Phys. Chem. Minerals, 13,* 285-290, 1986.

Hemley, R. J., M. D. Jackson, and R. G. Gordon, Theoretical study of the structure, lattice dynamics, and equations of state of perovskite-type $MgSiO_3$ and $CaSiO_3$, *Phys. Chem. Minerals, 14*, 2-12, 1987a.

Hemley, R. J., J. D. Kubicki, and H. K. Mao, In situ high-pressure Raman spectroscopy of $MgSiO_3$, $CaSiO_3$, and $CaMgSi_2O_6$ Glasses, *EOS Trans. Am. Geophys. Union, 68*, 1456, 1987b.

Hofmeister, A. M., Q. Williams, and R. Jeanloz, Thermodynamic and elastic properties of $MgSiO_3$ perovskite from far-IR spectra at pressure, *EOS Trans. Am. Geophys. Union, 68*, 1469, 1987.

Horiuchi, H., E. Ito, and D. J. Weidner, Perovskite-type $MgSiO_3$: single-crystal x-ray diffraction study, *Am. Mineral. 72*, 357-360, 1987.

Ito, E., and D. J. Weidner, Crystal growth of $MgSiO_3$ perovskite, *Geophys. Res. Lett. 13*, 464-466, 1986.

Jephcoat, A. P., H. K. Mao, and P. M. Bell, Operation of the megabar diamond-anvil cell, in *Hydrothermal Experimental Techniques*, edited by G. C. Ulmer and H. L. Barnes, (Wiley-Interscience, New York) pp. 469-506, 1987.

Jephcoat, A. P., R. J. Hemley, H. K. Mao, R. E. Cohen, and M. J. Mehl, Raman spectroscopy and theoretical modelling of BeO at high pressure, *Phys. Rev. B, 37*, 4727-4734, 1988.

Knittle, E., and R. Jeanloz, Synthesis and equation of state of $(Mg,Fe)SiO_3$ perovskite to over 100 gigapascals, *Science, 235*, 668-670, 1987.

Knittle, E., R. Jeanloz, and G. L. Smith, Thermal expansion of silicate perovskite and stratification of the Earth's mantle, *Nature, 319*, 214-216, 1986.

Liu, L.-G., Orthorhombic perovskite phases observed in olivine, pyroxene, and garnet at high pressures and temperatures, *Phys. Earth Planet. Inter. 11*, 289-298, 1976.

Mao, H. K., J. Xu, and P. M. Bell, Calibration of the ruby pressure gauge to 800 kbar under quasihydrostatic conditions, *J. Geophys. Res. 91*, 4673-4676, 1986.

Mao, H. K., R. J. Hemley, and E. C. T. Chao, The application of micro-Raman spectroscopy to analysis and identification of minerals in thin section, *Scanning Microsc., 1*, 495-501, 1987.

Matsui, M., M. Akaogi, and T. Matsumoto, Computational model of the structural and elastic properties of the ilmenite and perovskite phases of $MgSiO_3$, *Phys. Chem. Minerals, 14*, 101-106, 1987.

Mehl, M. J., R. J. Hemley, and L. L. Boyer, Potential-induced breathing model for the elastic moduli and high-pressure behavior of the cubic alkaline-earth oxides. *Phys. Rev. B, 33*, 8685-8696, 1986.

Mehl, M. J., R. E. Cohen, and H. Krakauer, LAPW electronic structure calculations for MgO and CaO, *J. Geophys. Res., 93* 8009-8022, 1988.

Scott, J. F., and J. P. Remeika, High-temperature Raman study of samarium aluminate, *Phys. Rev. B, 1*, 4182-4185, 1970.

Wall, A., G. D. Price, and S. C. Parker, A computer simulation of the structure and elastic properties of $MgSiO_3$ perovskite, *Mineral. Mag. 50*, 693-707, 1986.

Williams, Q., R. Jeanloz, and P. McMillan, Vibrational spectrum of $MgSiO_3$ perovskite: zero-pressure Raman and mid-infrared spectra to 27 GPa, *J. Geophys. Res. 92*, 8116-8128 (1987).

Wolf, G. and M. Bukowinski, Ab initio structural and thermoelastic properties of orthorhombic $MgSiO_3$ perovskite, *Geophys. Res. Lett. 12*, 809-812, 1985.

Wolf, G. and M. Bukowinski, Theoretical study of the structural properties and equations of state of $MgSiO_3$ and $CaSiO_3$ perovskites: implications for lower mantle composition, in *High-Pressure Research in Mineral Physics*, edited by M. H. Manghnani and Y. Syono, (Terra Scientific-AGU, Washington, D.C.), pp. 313-331, 1987.

Yeganeh-Haeri, A., D. J. Weidner, and E. Ito, Single-crystal elastic moduli of magnesium metasilicate perovskite, *this volume*, 1988.

Yagi, T., H. K. Mao, and P. M. Bell, Structure and crystal chemistry of perovskite-type $MgSiO_3$, *Phys. Chem. Minerals, 3*, 97-110, 1978.

Yagi, T., H. K. Mao, and P. M. Bell, Hydrostatic compression of perovskite-type $MgSiO_3$, in *Advances in Physical Geochemistry,* Vol. 2, ed. S. K. Saxena (Springer-Verlag), pp. 317-325, 1982.

R. E. Cohen, Condensed Matter Physics Branch, Code 4684, Naval Research Laboratory, Washington, DC 20375 USA.

R. J. Hemley and H. K. Mao, Geophysical Laboratory, Carnegie Institution of Washington, Washington, DC 20008 USA.

E. Ito, Institute for Study of the Earth's Interior, Okayama University, Misasa, Tottori-ken 682-02 Japan.

D. J. Weidner and A. Yeganeh-Haeri, Department of Earth and Space Sciences, State University of New York at Stony Brook, Stony Brook, NY 11794 USA.

DEFECTS AND DIFFUSION IN $MgSiO_3$ PEROVSKITE: A COMPUTER SIMULATION

Alison Wall[1] and Geoffrey D. Price

Department of Geological Sciences, University College London,
Gower Street, London WC1E 6BT, England

Abstract. We use an atomistic computer model to calculate the energies of point defects in $MgSiO_3$ perovskite. A magnesium and oxygen Schottky pair defect is predicted to be the lowest energy defect in this structure, while Si Frenkel pair defects are the most difficult to form. The diffusion of oxygen in the perovskite lattice is predicted to occur most readily along <100> of the orthorhombic cell, and to have an activation energy of 452 $kJmol^{-1}$ for intrinsic diffusion and 81 $kJmol^{-1}$ for extrinsic diffusion. The predicted activation volume for this process is 1.6 cm^3mol^{-1}. All of these values are in excellent agreement with inferred or measured behaviour in other perovskite and related phases. This study lays the foundations for further work aimed at determining the rheological behaviour of silicate perovskites and hence of the Earth's lower mantle.

1. Introduction

Progress in understanding mantle dynamics and the thermal evolution of the Earth is limited by our lack of information on the high pressure and temperature properties of most mantle-forming phases. Thus, for example, although it is believed that the bulk of the lower mantle is composed of magnesium silicate perovskite, virtually nothing is known about the likely rheological behaviour of this major Earth-forming phase. Furthermore, it is unlikely that it will be possible to carry out any experiments to determine the critical parameters controlling the deformation behaviour of magnesium silicate perovskite in the near future, because of the high pressures and temperatures needed to stabilize this phase.

As an alternative to direct study, therefore, we have attempted to use computer simulations to suggest the nature of defects in magnesium silicate perovskite. The study of structural defects is vital if we are to establish the rheological properties of a material, as it is these defects that provide the mechanism by which solid state transport occurs. In this preliminary study, only intrinsic defects in $MgSiO_3$ perovskite have been considered; the effects of the substitution of, for example Fe^{2+}, Ca^{2+}, Al^{3+}, Fe^{3+}, into the perovskite lattice are not treated and, therefore, the results presented in this paper are likely only to be applicable to a real system at high temperature, where the concentration of intrinsic defects would be expected to be much higher than the concentration of extrinsic defects. It has also been assumed that the defect concentration in $MgSiO_3$ perovskite will be small (less than 0.1 atomic percent), and hence that the defects can be considered to be isolated and non-interacting [Lasaga, 1981]. In the following sections, we will outline the details of the technique used to simulate the defects in $MgSiO_3$, and will discuss the results of these calculations in terms of the few experimental data that do exist.

2. Computer simulation techniques

Ideally the computer simulation of a solid would involve an explicit solution of the Schrödinger equation to a high degree of sophistication. This is, however, impractical for the study of complex silicates and, therefore, a simpler, more approximate approach must be used. The approach adopted in this work is to define an algorithm, or interatomic potential, to describe the total energy of the system in terms of the atomic positions. Using such potential models, minimization of the energy of the crystal system with respect to the atomic coordinates enables the prediction of the minimum energy structure and of properties, such as the elastic constants, dielectric constants and phonon frequencies, which depend on the second derivative of the energy. Such atomistic simulation techniques have been successfully used to predict crystal structures, elastic constants, lattice dynamics, defect energetics, phase transitions, transport properties and equations of state for a range of ionic and semi-ionic materials [Catlow et al., 1984; Lewis and Catlow, 1986; Matsui and Busing, 1984; Matsui et al., 1987; Parker,

---

[1] Current Address; The Atlas Centre, Rutherford Appleton Laboratory, Chilton, Didcot, Oxfordshire OX11 0QX, England

TABLE 1. Short range potential parameters for potentials THB1, TPV1 and TPV2.

Potential	THB1	TPV1	TPV2
A_{Mg-O}	1428.5	1233.8	1018.9
A_{Si-O}	1283.9	1383.7	1283.9
A_{O-O}	22764.3	22764.3	20443.2
B_{Mg-O}	0.2945	0.2925	0.2950
B_{Si-O}	0.3205	0.3205	0.3305
B_{O-O}	0.1490	0.1490	0.1490
C_{Si-O}	10.66	10.66	23.74
C_{O-O}	27.88	27.88	31.83
q_{Mg}	+2.000	+2.000	+1.756
q_{Si}	+4.000	+4.000	+3.605
q_{O-core}	+0.848	+0.848	-1.787
$q_{O-shell}$	-2.848	-2.848	-
k_s	74.920	74.920	-
k_{O-Si-O}	2.097	2.097	-

Short range potential:

$$U_{ij} = A_{ij} \exp(-r / B_{ij}) - C_{ij} r^{-6} + k_s r^2_{core-shell} + k_{O-Si-O}(\theta - 1.57)^2$$

Units: eV, Å, radians.

1983a,b; Price and Parker, 1984; Rahman, 1979; Walker et al., 1981, Wall and Price, 1988a].

The interatomic potential functions are defined to model the net forces acting between atoms in a solid, which include contributions from ionic, covalent and van der Waals bonding. One of the simplest and most widely used expression for an interatomic potential has a rigid ion, central force, pair-wise additive form. The static cohesive energy (U_T) is calculated by a summation of all the interactions between pairs of ions in the unit cell. In an ionic solid the dominant contribution to the static cohesive energy is made by the long range Coulombic term (U_C), but it is also necessary to include a term, U_{ij}, to model the short range repulsive and attractive forces known to act between atoms [see, for example, Cochran, 1973]:

$$U_T = U_C + U_{ij} \quad (1)$$

A rigid ion potential may be extended to include a shell model, that simulates ionic polarizability, and bond-bending and other three body terms, that simulate, for example, the directional component of covalent bonding. The terms used in atomistic models are discussed more fully in several of our recent papers [e.g. Wall et al. 1986; Price et al., 1987a, b]. Three interatomic potentials, THB1, TPV1 and TPV2, defined in Table 1, were used in these defect calculations because they predict the experimentally determined structure and elastic moduli reasonably well (Table 2), and are dynamically stable.

In addition to perfect lattice properties, atomistic simulations can be used to predict the energy of defects in a crystal lattice. In calculating the energy of a defect it is vital that the atoms surrounding the defect are allowed to relax; in semi-ionic and ionic solids defects are generally charged and hence have long range Coulombic forces

TABLE 2. Observed and simulated structural and elastic properties of $MgSiO_3$ perovskite.

	Observed	THB1	TPV1	TPV2
a	4.775	4.824	4.819	4.847
b	4.929	4.847	4.893	4.909
c	6.897	6.844	6.893	6.921
I*		0.063	0.033	0.045
V	162.3	160.0	162.5	164.7
Si-O(2)	1.783	1.737	1.779	1.775
Si-O(2)	1.796	1.739	1.783	1.779
Si-O(2)	1.801	1.739	1.790	1.787
I*		0.055	0.010	0.014
Mg-O(1)	2.014	2.123	1.997	2.046
Mg-O(2)	2.052	2.134	2.015	2.061
Mg-O(1)	2.096	2.318	2.206	2.229
Mg-O(2)	2.278	2.401	2.385	2.397
Mg-O(2)	2.427	2.414	2.449	2.458
Mg-O(1)	2.846	2.568	2.780	2.758
Mg-O(1)	2.961	2.708	2.858	2.832
Mg-O(2)	3.120	2.768	2.891	2.945
I* [8]		0.115	0.070	0.078
I* [12]		0.203	0.115	0.109
K	247	347	333	255
G	150	213	199	151
r.i.	1.839	1.669	1.641	-

*root mean square error in Å. Unit cell parameters and bond lengths in Å or Å^3 as appropriate. Observed structure from Horiuchi et al. (1987). Bulk modulus and shear modulus in GPa. Observed values from Kudoh et al. (1987) and Liebermann et al. (1977).

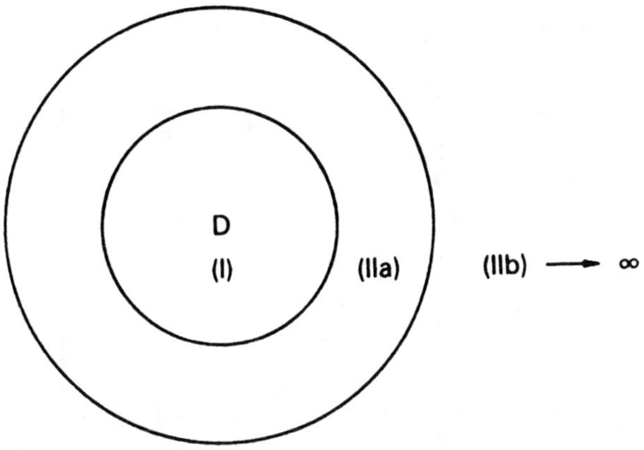

Fig. 1. The defect model; showing regions 1, 2A and 2 around the defect, D.

associated with them. The CASCADE code calculates defect energies using the approach originally developed by Mott and Littleton [1938] in which the crystal is split into two spherical regions, shown in Figure 1 [Catlow, 1986; Parker, 1983b; Catlow and Mackrodt, 1982].

Region 1, immediately surrounding the defect, is defined to contain 100 or more ions. An atomistic simulation is explicitly carried out to adjust the ionic coordinates until they are at positions of zero net force. More approximate methods are used to treat the outer regions in which the forces arising from the defect are relatively weak.

In the Mott-Littleton approach, a defect of charge q is considered to cause a polarization **P**, at a distance **r** from the defect, of:

$$\mathbf{P} = qr^{-3}(1 - \varepsilon_0^{-1})\mathbf{r} \tag{2}$$

where ε_0 is the static dielectric constant. (This relation is only strictly applicable to cubic crystals, but see Catlow and Mackrodt [1982] for more details). The defect energy E_D is expressed as:

$$E_D = E_1(\mathbf{r}_1) + E_{1,2}(\mathbf{r}_1,\delta\mathbf{r}_2) + E_2(\delta\mathbf{r}_2) \tag{3}$$

where E_1 is the energy from region 1 with a coordinate vector $\mathbf{r}_1$, E_2 is the energy of region 2 with vector $\delta\mathbf{r}_2$ of coordinate displacements and $E_{1,2}$ is the interaction energy between regions 1 and 2.

If the magnitude of the vector $\delta\mathbf{r}_2$ is sufficiently small, then E_2 may be expressed, using the harmonic approximation, as:

$$E_2 = 1/2.\delta\mathbf{r}_2.\mathbf{A}.\delta\mathbf{r}_2 \tag{4}$$

where **A** is the force constant matrix, and at equilibrium:

$$\left.\frac{dE_{1,2}(\mathbf{r}_1,\delta\mathbf{r}_2)}{d\delta\mathbf{r}_2}\right|_{\delta r_2=\delta r_2^*} = -\mathbf{A}.\delta\mathbf{r}_2 \tag{5}$$

where $\delta\mathbf{r}_2^*$ is the equilibrium value for $\delta\mathbf{r}_2$. Hence, substituting equation (5) into equation (4) and then in to equation (3), the defect energy becomes:

$$E_D = E_1(\mathbf{r}_1) + E_{1,2}(\mathbf{r}_1,\delta\mathbf{r}_2) - 1/2.\delta\mathbf{r}_2.\left.\frac{dE_{1,2}(\mathbf{r}_1,\delta\mathbf{r}_2)}{d\delta\mathbf{r}_2}\right|_{\delta r_2=\delta r_2^*} \tag{6}$$

This expresses E_D in terms of E_1 and $E_{1,2}$ only and so E_2 does not need to be explicitly evaluated. E_1 is calculated by an explicit atomistic summation of the ions in region 1. To evaluate $E_{1,2}(\mathbf{r}_1,\delta\mathbf{r}_2)$ and its derivatives, it has been found necessary to define a subdivision of region 2, region 2A, in which the interaction of the ions with region 1 is considered in more detail than for the rest of region 2. The displacements in region 2A are calculated using the Mott-Littleton procedure as the sum of displacements due to all the component defects in region 1. The energy $E_{1,2}$ and its derivatives are calculated by direct summation of the interactions between the two regions. For the rest of region 2, the interaction is assumed to arise purely from the net effective charge of the defect in region 1.

3. Results

The dominant type of point defect present at a particular temperature and pressure will be the one that is energetically most favorable. Therefore, the free energies of formation of some possible point defects in $MgSiO_3$ perovskite were evaluated at zero pressure and temperature and also at 2625 K and 100 GPa (lower mantle conditions). If bulk diffusion in the intrinsic regime occurs by vacancy or interstitial jumps, the activation energy for diffusion is dependent on the free energy required to form defects (ΔG_f) and the free energy required to enable the defect to migrate to a new site (ΔG_m). Hence, the free energies of migration of oxygen and magnesium at zero pressure and temperature and the free energy of migration of oxygen at 2625 K and 100 GPa were also calculated.

3.1 Defect formation energies at 0 K - 0 GPa

The free energies of formation of some isolated defects in $MgSiO_3$ perovskite were calculated using the computer code CASCADE. The defects considered were intrinsic Schottky and Frenkel defects. The free energy of formation of a defect (ΔG_f) is given by [e.g. Poirier, 1985]:

$$\Delta G_f = \Delta H_f - T\Delta S_f \tag{7}$$

where ΔS_f is the entropy of formation and ΔH_f is the enthalpy of formation, defined in terms of the formation energy (ΔE_f) and the formation volume (ΔV_f) as:

$$\Delta H_f = \Delta E_f + P\Delta V_f \tag{8}$$

The ΔV_f and ΔS_f terms are assumed to be negligible at low

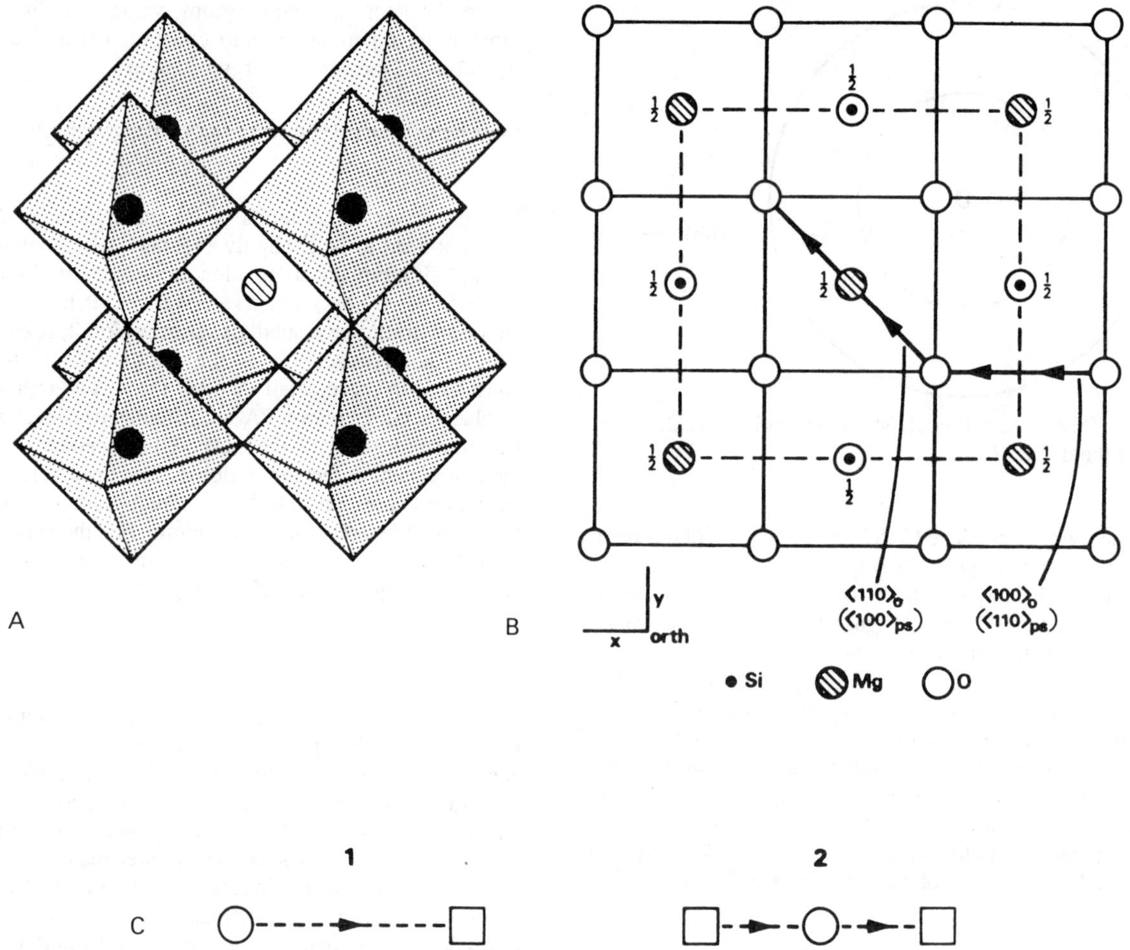

Fig. 2. (a) The ideal cubic perovskite structure; perspective view of the framework of SiO_6 octahedra. Hatched circles, Mg; black circles, Si. (b) Projection of the perovskite structure along $[001]_O$ showing two possible paths of oxygen diffusion. Broken lines outline the orthorhombic unit cell and the O atoms around the edges of the SiO_6 octahedra are joined by solid lines. (c) Configuration of atoms used to calculate the migration energies of O and Mg. A square represents a vacant site and a circle an atom.

pressure and temperature. Hence, the component energy terms required to evaluate these defect free energies are the individual vacancy and interstitial defect energies (ΔE_f), derived from the CASCADE code, and the appropriate lattice energies. The quartz polymorph was used to evaluate the lattice energy of SiO_2 as this is the stable phase at these pressure and temperature conditions.

Three interatomic potentials, THB1, TPV1 and TPV2, were used in these defect calculations. However, because of the requirement to maintain charge neutrality, it was not possible to calculate the MgO and SiO_2 Schottky formation energies using the partially ionic potential TPV2. The perfect atomic coordinates and lattice vectors were derived by relaxing the structure at constant pressure (effectively at zero pressure and zero temperature). The effect of the orthorhombic distortion was ignored for this preliminary investigation of the defect energetics of the $MgSiO_3$ perovskite structure, and only one interstitial site was considered. This was the largest interstitial site in the $MgSiO_3$ perovskite structure, and is centered in an Mg_2O_4 octahedral cluster (see Figure 2a). During each interstitial energy calculation, an ion was initially placed at the centre of this site, but it was allowed to move during the relaxation procedure to find a minimum energy position.

$MgSiO_3$ perovskite has a very high density and hence the region 1 radius was restricted by the amount of computer memory available. During each CASCADE calculation, the region 1 and 2A radii were increased to check the effect of the region sizes on the defect energy. The maximum radius of region 1 used was 0.76 nm which

TABLE 3. Vacancy, interstitial and lattice energies at 0 K-0 GPa.

Potential	THB1	TPV1	TPV2
V_{Si}	96.34	94.74	78.25
V_{Mg}	25.14	26.17	19.64
V_O	23.71	23.86	19.41
Si_i	-72.33	-72.84	-56.98
Mg_i	-12.00	-13.64	-9.93
O_i	-12.64	-12.50	-11.88
MgO	-41.30	-42.35	--
Quartz	-128.65	-125.96	--
$MgSiO_3$	-171.30	-169.65	-135.38

Energies in eV.

contained 183 ions, and the total number of ions considered explicitly (in regions 1 and 2A) was about 2800. The degree of convergence obtained from increasing the region 1 radii was better than 0.1 eV for the vacancy energies, but up to 0.5 eV for some interstitial energies.

The energies of isolated vacancies and interstitials in the $MgSiO_3$ perovskite lattice, together with the required lattice energies, are given in Table 3 and the resulting energies of defect formation are presented in Table 4. There is little difference in the defect formation energies predicted by potentials THB1 and TPV1, but the energies simulated by potential TPV2 are lower because of the partial ionic charges in this potential. However, all the relative total defect energies are similar.

By comparing the formation energies per defect, the MgO Schottky pair is predicted to be the lowest energy intrinsic defect in $MgSiO_3$ perovskite, and hence this is the defect type likely to predominate at high temperature. The highest energy defect (and, therefore, the defect least likely to occur) is the Si Frenkel pair.

The site fraction (X) of MgO Schottky defects is related to the formation energy and the temperature (assuming that the formation energy is independent of temperature) by the relationship:

$$X = \exp(-\Delta G_f / 2RT) \qquad (9)$$

where ΔG_f is the formation energy per mole [see, for example, Lasaga, 1981] and R is the gas constant. The number of MgO Schottky pairs in $MgSiO_3$ perovskite was, therefore, calculated at various temperatures; the results are presented in Table 5. Compared to the calculated concentration of Schottky defects in MgO and NaCl [see Lasaga, 1981], the concentration of defects in $MgSiO_3$ perovskite is very low. For example, MgO has a melting point very similar to the estimated melting point of $MgSiO_3$ perovskite [Heinz and Jeanloz, 1987], but at 2273 K the concentration of Schottky defects is 8.6 x 10^{23} m^{-3} and NaCl, which melts at 1074 K, has a defect concentration at 800 K of 1.5 x 10^{23} m^{-3}. A higher intrinsic defect concentration may be predicted, however, if the vibrational entropy is also considered; for example, Harding [1986] calculated that the energy, ΔE_f, of a Schottky defect in KCl decreases with increasing temperature while the defect entropy, ΔS_f, increases.

3.2 Defect migration energies at 0 K - 0 GPa

Since it is predicted that the MgO Schottky defect will be the predominant intrinsic defect in $MgSiO_3$ perovskite, it is probable that Mg and O diffusion will occur by a vacancy migration mechanism. In this section, the migration energies and pathways for O diffusion are investigated.

Assuming that O migration occurs by a vacancy hopping mechanism, there are two possible paths for O migration through the perovskite lattice; along the orthorhombic $<100>_O$ direction (around the edges of the SiO_6 octahedra) and along the $<110>_O$ direction (through the Mg_2O_4 octahedral cluster). These two routes are illustrated in Figure 2b.

The free energies of migration (ΔG_m) were calculated for each of these paths using potential TPV1 to assess which is the more favorable (lower energy) route. The free energy of defect migration is defined in a similar way to the free energy of defect formation, namely:

TABLE 4. Defect formation energies in $MgSiO_3$ perovskite at 0 K - 0 GPa.

Potential	THB1	TPV1	TPV2
Schottky			
MgO	7.6	7.7	--
SiO_2	15.1	16.0	--
$MgSiO_3$	21.3	22.8	20.7
Frenkel pair			
Mg	13.1	12.5	9.7
Si	24.0	21.9	21.3
O	11.1	11.3	7.5

Energies in eV.

TABLE 5. Number of MgO Schottky pairs in $MgSiO_3$ perovskite at 0 GPa.

T (K)	X	n
300	2.1×10^{-65}	5.2×10^{-37}
1000	4.0×10^{-20}	1.0×10^{9}
2000	2.0×10^{-10}	5.0×10^{18}
2750	8.8×10^{-8}	1.7×10^{21}

X = site fraction; n = number per m^3

$$\Delta G_m = \Delta H_m - T\Delta S_m \tag{10}$$

$$\Delta H_m = \Delta E_m + P\Delta V_m \tag{11}$$

Calculations of the energy levels along each of these routes were made by placing two vacancies on adjacent O sites to represent the site vacated by the hopping ion and the destination site, and an interstitial oxygen ion was placed at chosen fixed points between these two vacancies. This configuration is sketched in Figure 2c. The migration pathway may not follow the shortest route between the two vacancies, but will depend on the relative forces exerted on the diffusing atom by the surrounding atoms and, therefore, possible deviations of the migration path were also investigated. The energy of migration (ΔE_m) is derived by subtracting the energy of the oxygen ion on its starting site from the point of highest energy along the migration path (the saddle point energy).

The calculated migration pathways are shown in Figures 3 and 4. For migration along the $<110>_O$ direction, the diffusing oxygen ion is predicted to follow almost a straight path, possibly deviating very slightly towards one of the two surrounding Mg ions. The ΔE_m is 3.21 eV (310 $kJmol^{-1}$). During diffusion along the $<100>_O$ direction, the oxygen ion is predicted not to follow a straight path, but to move away from the Si atom in the centre of its SiO_6 octahedra. The ΔE_m is predicted to be 0.84 eV (81 $kJmol^{-1}$). Hence, oxygen migration along the $<100>_O$ direction is predicted to be energetically much more favorable. This result is confirmed in a molecular dynamics simulation reported elsewhere [Wall and Price, 1988b].

The activation energy for oxygen diffusion in the intrinsic regime is given by a combination of the energy required to form the oxygen vacancy (assumed to be half the MgO Schottky pair energy) and the oxygen migration energy (assuming that the oxygen atom will jump along the $<100>_O$ direction):

$$\Delta G_D = \Delta G_f / 2 + \Delta G_m \tag{12}$$

[see, for example, Lasaga, 1981; Poirier, 1985]. The activation energy for oxygen diffusion is, therefore, calculated to be 452 $kJmol^{-1}$. In the extrinsic regime,

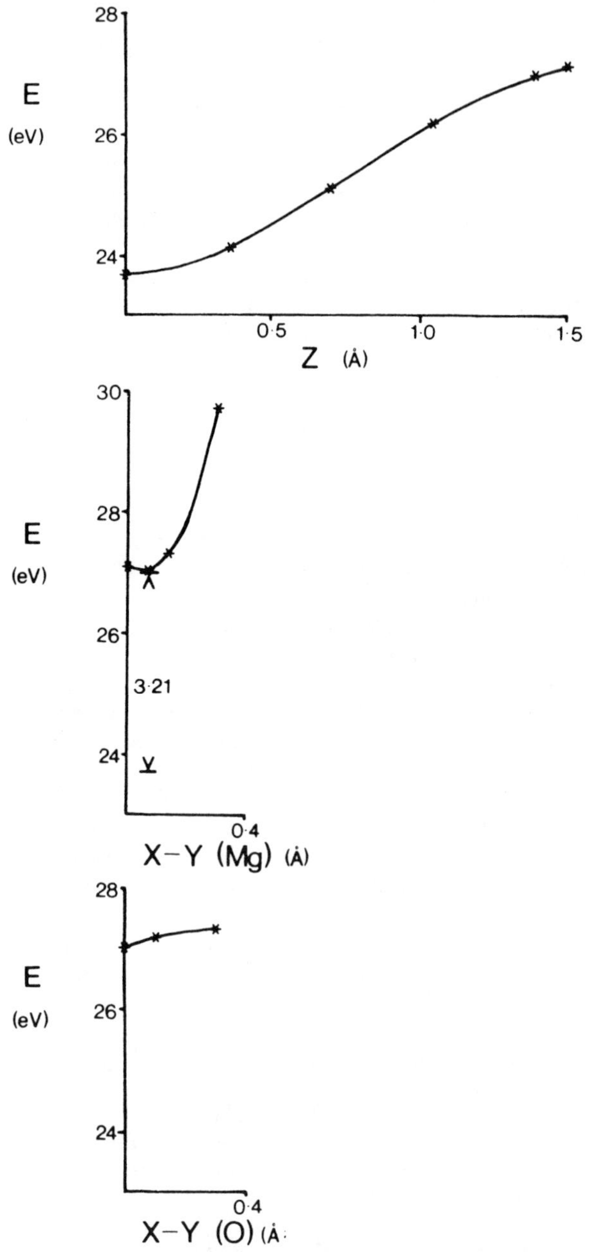

Fig. 3. Oxygen defect energies sampled at various points along a path in approximately the $<110>_O$ direction between two oxygen lattice sites. a) Along the Z direction to the mid-point; b) at the mid-point along Z and in the X-Y direction towards the adjacent Mg atom; c) at the mid-point along Z and in the X-Y direction towards the adjacent O atom.

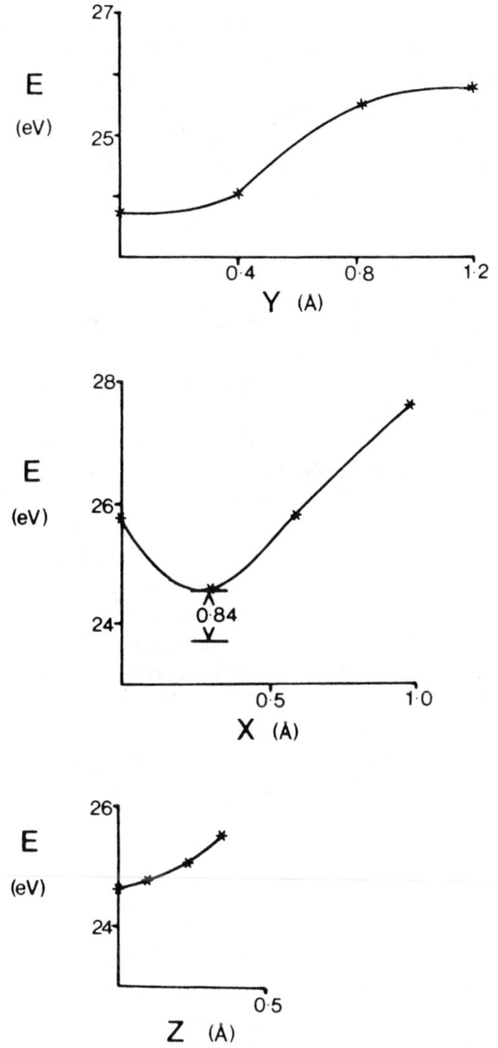

Fig. 4. Oxygen defect energies sampled at various points along a path in approximately the <100>$_O$ direction between two oxygen lattice sites. a) Along the Y direction to the mid-point; b) at the mid-point along Y and away from the adjacent Si atom; c) at the mid-point along Y and towards the adjacent Mg and O atoms.

however, the energy for oxygen diffusion is equivalent to the energy for migration alone: 81 $kJmol^{-1}$.

To calculate the activation energy for magnesium diffusion, the migrating magnesium atom has been assumed to move along a straight path through the centre of the Mg_2O_4 octahedron. The ΔE_m was calculated to be 4.56 eV (440 $kJmol^{-1}$), giving an activation energy for magnesium diffusion of 812 $kJmol^{-1}$, considerably higher than that required for oxygen diffusion. The activation energy for magnesium migration in the extrinsic regime is calculated to be 440 $kJmol^{-1}$, again, considerably higher than that predicted for oxygen diffusion.

3.3 Defect energies at 2625 K - 100GPa

The defect formation energy calculations above were repeated with potential TPV1 and a structural data set minimised at 2625 K and 100 GPa; conditions which are thought to occur in the lower mantle.

The vacancy, interstitial and lattice energies, listed in Table 6, were much more difficult to evaluate with this data set. The structure is even denser than at 0 K - 0 GPa and, therefore, the region 1 size in the CASCADE calculations was reduced to a radius of 0.62 nm and contained 155 ions. The total number of ions considered explicitly was about 2900.

During CASCADE runs to evaluate the interstitial energies, the interstitial ion and other neighboring ions moved large distances away from their lattice sites preventing successful energy minimisation. For example, the insertion of a Si ion at the centre of an Mg_2O_4 octahedral cluster, caused an adjacent Mg ion to move 0.045 nm away from its lattice site and the Si interstitial to move 0.028 nm away from its starting position, thus creating a split interstitial configuration. These configurations are, therefore, predicted to be more stable than an isolated interstitial defect and are used in the calculation of defect formation energies.

The defect formation energies were evaluated and are listed in Table 7. Again the MgO Schottky pair energy is predicted to be the most stable intrinsic defect. At this increased density, it is energetically less favorable to have Frenkel pairs in the structure while, conversely, the Schottky defects have marginally lower energy than under ambient conditions. This result for Schottky defects is

TABLE 6. Vacancy, interstitial and lattice energies at 100GPa-2625K.

Potential	TPV1
V_{Si}	97.23
V_{Mg}	25.27
V_O	22.90
Si_i	-65.80
Mg_i	-7.74
O_i	-9.45
MgO	-40.93
Stishovite	-126.96
$MgSiO_3$	-167.07

Energies in eV.

TABLE 7. Defect formation energies in $MgSiO_3$ perovskite at 100GPa - 2625K.

Potential	TPV1 eV	TPV1 eV per defect
Schottky		
MgO	7.3	3.6
SiO	16.1	5.4
$MgSiO_3$	24.1	4.8
Frenkel pair		
Mg	17.5	8.8
Si	31.4	15.7
O	13.5	6.7

surprising as increasing pressure usually has the effect of lowering the vacancy concentration in ionic crystals. However, the difference between the Schottky energies at 0 K - 0 GPa and at 2625 K and 100 GPa is only small and may be due, in part, to errors in the defect energies caused by the restrictions on the region 1 size.

The migration energy for O diffusion in the orthorhombic $<100>_O$ direction was calculated in a similar way to that described above. However, the assumption that the saddle point energy would occur in about the same relative position was made, and hence less points in the perovskite structure were sampled. Oxygen diffusion along the $<100>_O$ direction is predicted to have a migration energy of 2.14 eV (206.5 $kJmol^{-1}$), much higher than at 0 K and 0 GPa. The corresponding activation energy for diffusion, assuming a vacancy hopping mechanism in the intrinsic regime, was calculated to be 556 $kJmol^{-1}$.

We have also performed calculations that enable us to estimate the activation volume for O diffusion in perovskite. The activation volume describes the variation of the free energy with hydrostatic pressure. It can be defined for the free energy of defect formation or the activation energy for diffusion as:

$$V^* = d\Delta G/dP \qquad (13)$$

where ΔG is given by:

$$\Delta G = \Delta E + P\Delta V - T\Delta S \qquad (14)$$

The activation volumes for MgO Schottky defect formation and oxygen diffusion are calculated by assuming that the ΔE does not vary with pressure and the $T\Delta S$ term is zero at 0 K and thus:

$$d\Delta G = dP.d\Delta V \qquad (15)$$

where $d\Delta V$ is equivalent to V^*. The defect energies used were those predicted at 0 K and 0 and 100 GPa.

At 0 K and 100 GPa, the energy of formation of an MgO Schottky defect is predicted to be 7.08 eV (683.0 $kJmol^{-1}$) and hence the activation volume for defect formation is -0.6 cm^3mol^{-1}. The free energy of oxygen diffusion along the $<100>_O$ direction at 0 K and 100 GPa was calculated to be 608 $kJmol^{-1}$, giving an activation volume of 1.6 cm^3mol^{-1}.

4. Discussion and conclusion

In our calculations, the MgO Schottky pair is predicted to be the lowest energy intrinsic defect in $MgSiO_3$ perovskite both at 0 K - 0 GPa and at 2625 K and 100 GPa. This result is in accordance with other calculations of defect formation energies in the perovskites $BaTiO_3$ [Lewis and Catlow, 1986] and $KMnF_3$.

At 0 K - 0 GPa, the lowest energy path for oxygen migration is in the orthorhombic $<100>_O$ direction with an activation energy for intrinsic diffusion of 452 $kJmol^{-1}$. Magnesium, assumed to hop along a straight path, has a higher activation energy for diffusion of 812 $kJmol^{-1}$. In the extrinsic regime, oxygen diffusion again has a lower activation energy than Mg diffusion, with a predicted value of 81 $kJmol^{-1}$. There is little experimental data against which to compare these predictions. However, Knittle and Jeanloz [1987] have reported the activation energy for the back transformation of perovskite to pyroxene to be 70 ± 20 $kJmol^{-1}$. Knittle and Jeanloz chose to interpret this energy in terms of the jump energy of Mg within the perovskite lattice. It is not obvious that this activation energy can be interpreted in terms of any simple jump process, but if it is, then we would suggest that it is more likely to correspond to a process related to extrinsic oxygen diffusion rather than to the energetically more difficult migration of Mg. Further support for the accuracy of our predicted activation energy for extrinsic oxygen diffusion in perovskites comes from recent studies of oxygen mobility in structurally related superconductors. Routbort et al. [1988] have recently studied the diffusion of oxygen in $La_{2-x}Sr_xCuO_{4-y}$, structurally related to perovskite, and found values for the activation energy in the range 77 to 108 $kJmol^{-1}$. The similarity of these results with those predicted in this study, suggests that our computer model predictions for defect energies are quantitatively reliable.

We also note that our predicted value for the activation volume for oxygen diffusion along $<100>_O$, 1.6 cm^3mol^{-1}, is in the range required by Yuen and Zhang [1987] to explain the magnitude of the topography of the core-mantle boundary (CMB). They suggest that the processes determining lower mantle viscosity must have an activation volume of between 1 and 2 cm^3mol^{-1} in order to produce CMB relief of a few kilometers. Although it is unlikely that O diffusion will be the rate determining process in mantle rheology, it is encouraging to find that such small activation volumes can indeed be associated with any diffusion process in such a densely packed material.

We have not presented simulations of Si diffusion in this paper. There is no simple path for silicon diffusion through the perovskite structure and the determination of the lowest energy route will require the evaluation of the energy of a large number of interstitial positions. However, even though the activation energy for silicon diffusion has not yet been evaluated, it will probably be much higher than that for magnesium or oxygen diffusion, because of the high energies of formation calculated for both Schottky and Frenkel defects involving silicon. Silicon is, therefore, likely to be the slowest diffusing species in perovskite, and silicon mobility will be the rate controlling process in determining the rheological behaviour of $MgSiO_3$ perovskite. The details of silicon diffusion in $MgSiO_3$ perovskite are under investigation.

Acknowledgements. It is a pleasure to thank Steve Parker, Richard Catlow, Mark Doherty and Janet Baker for their help and encouragement. AW gratefully thanks the NERC for the receipt of a research studentship, and GDP acknowledges the receipt of NERC research grant GR3/5993.

References

Catlow, C.R.A., Computer simulation of defects in solids. In: *Defects in solids. Modern techniques* (eds Chadwick, A.V. and Terenzi, M.). Plenum Press, New York, 1986.

Catlow, C.R.A., A.N. Cormack and F. Theobald, Structure prediction of transition-metal oxides using energy minimisation techniques. *Acta Crysallogr., B40*, 195-200, 1984.

Catlow C.R.A. and W.C. Mackrodt, Theory of simulation methods for lattice and defect energy calculations in crystals. In: *Computer simulation in solids* (eds Catlow, C.R.A. and Mackrodt, W.C.). *Lecture notes in physics 166*. Springer-Verlag, Berlin, 1982.

Cochran, W.G., *The dynamics of atoms in crystals*. Arnold, London, 1973.

Harding, J.H., Calculation of defect processes at high temperatures. In: *Defects in solids. Modern techniques* (eds Chadwick, A.V. and Terenzi, M.). Plenum Press, New York, 1986.

Heinz, D.L., and R. Jeanloz, Measurement of the melting curve of $Mg_{0.9}Fe_{0.1}SiO_3$ at lower mantle conditions and its geophysical implications. *J. Geophys. Res., 92*, 11437-11444, 1987.

Horiuchi, H., E. Ito and D.J. Weidner, Perovskite-type $MgSiO_3$ perovskite: Single crystal X-ray diffraction study. *Amer. Mineral., 72*, 357-360, 1987.

Knittle, E., and R. Jeanloz, The activation energy of the back transformation of perovskite to enstatite. In: *High pressure research in mineral physics* (eds Manghnani, M.H., and Syono, Y.). *Geophysical monograph 39*. A.G.U., Washington, 1987.

Kudoh, Y., E. Ito and H. Takeda, Effect of pressure on the crystal structure of perovskite-type $MgSiO_3$. *Phys. Chem. Minerals, 14*, 350-354, 1987.

Lasaga, A.C., The atomistic basis of kinetics: defects in minerals. In: *Kinetics of geochemical processes* (eds Lasaga, A.C., and Kirkpatrick, R.J.). *Reviews in mineralogy 8*. Min. Soc. America, Washington, 1981.

Lewis, G.V., and C.R.A. Catlow, Defect studies of doped and undoped $BaTiO_3$ using computer simulation techniques. *J. Phys. Chem. Solids, 47*, 89-97, 1986.

Liebermann, R.C., L.E.A. Jones and A.E. Ringwood, Elasticity of aluminate, titanate, stannate and germanate compounds with the perovskite structure. *Phys. Earth Planet. Interiors, 14*,165-178, 1977.

Matsui, M., and W.R. Busing, Computational modelling of the structure and elastic constants of the olivine and spinel forms of Mg_2SiO_4. *Phys. Chem. Minerals, 11*, 55-59, 1984.

Matsui, M., M. Akaogi and T. Matsumoto, Computational model of the structural and elastic properties of the ilmenite and perovskite phases of $MgSiO_3$. *Phys. Chem. Minerals, 14*, 101-106, 1987.

Mott, N.F., and M.J. Littleton, *Conduction in polar crystals. I: Electrolyte conduction in solid salts. Trans. Farad. Soc., 34*, 485-499, 1938.

Parker, S.C., Prediction of mineral crystal structures. *Solid State Ionics, 8*, 179-186, 1983a.

Parker, S.C., *Computer simulation of minerals*. PhD Thesis, University of London, 1983b.

Poirier, J.P., *Creep of crystals*. Cambridge University Press, Cambridge, 1985.

Price, G.D., and S.C. Parker, Computer simulations of the structural and physical properties of the olivine and spinel polymorphs of Mg_2SiO_4. *Phys. Chem. Minerals, 10*, 209-216, 1984.

Price, G.D., S.C. Parker and M. Leslie, The lattice dynamics of forsterite. *Min. Mag., 51*, 157-170, 1987a.

Price, G.D., S.C. Parker and M. Leslie, The lattice dynamics and thermodynamics of the Mg_2SiO_4 polymorphs. *Phys. Chem. Minerals, 15*, 181-190, 1987b.

Rahman, A., Molecular dynamics studies of superionic conductors. In: *Fast Ion transport in Solids* (eds Vashishta, P., Mundy, J.N., and Shenoy, G.K.). Elsevier North Holland Inc., Amsterdam, 1979.

Routbort, J.L., S.J. Rothman, B.K. Flandermeyer, L.J. Nowicki and J.E. Baker, Oxygen diffusion in $La_{2-x}Sr_xCuO_{4-y}$. *J. Mat. Res., 3*, 116-121, 1988.

Walker, A.B., M. Dixon and M.J. Gillan, Computer simulation of superionic lead fluoride. *Solid State Ionics, 5*, 601-604, 1981.

Wall, A., and G.D. Price, Computer simulation of the structure, lattice dynamics and thermodynamics of ilmenite-type $MgSiO_3$. *Amer. Mineral., 73*, 224-231, 1988a.

Wall, A., and G.D. Price, A molecular dynamics simulation of $MgSiO_3$ perovskite. *In preparation*, 1988b.

Wall, A., G.D. Price and S.C. Parker, A computer simulation of the structure and elastic properties of $MgSiO_3$ perovskite. *Min. Mag., 50*, 693-707, 1986.

Yuen, D.A., and S. Zhang, Geodynamics of the lower mantle with depth-dependent equation of state parameters. *EOS 68*, 1488, 1987.

ELECTRONIC STRUCTURE AND TOTAL ENERGY CALCULATIONS FOR OXIDE PEROVSKITES AND SUPERCONDUCTORS

Ronald E. Cohen, Larry L. Boyer, Michael J. Mehl, Warren E. Pickett,

Condensed Matter Physics Branch, Naval Research Laboratory, Washington, D.C. 20375-5000

and

Henry Krakauer,

Department of Physics, College of William and Mary, Williamsburg, VA 23815

Proceedings of the Chapman Conference on Perovskites, Bisbee, Arizona, 1987, to be published by the American Geophysical Union,

Abstract. $MgSiO_3$, $BaTiO_3$, La_2CuO_4, and $YBa_2Cu_3O_7$ are studied using the Linearized Augmented Plane Wave (LAPW) method and the Potential Induced Breathing (PIB) model. $MgSiO_3$ is an important geophysical material and prototypical dense silicate, $BaTiO_3$ is an important ferroelectric material, and La_2CuO_4 and $YBa_2Cu_3O_7$ are examples of the new high-temperature superconductors. $MgSiO_3$ is found to be quite ionic, whereas $BaTiO_3$ has significant covalent character, and the high-temperature superconductors have an unusual combination of ionic, metallic, and covalent character. The equation-of-state of orthorhombic $MgSiO_3$ is in good agreement with experiment, and the elastic constants, which were calculated and published before experimental data were available, are in close agreement with experiment. The ferroelectric phase transition in $BaTiO_3$ from cubic to tetragonal is not found using highly accurate LAPW calculations. This lends support to order-disorder models for the phase transition. Also, contrary to many current models, the charge distortion in $BaTiO_3$ centers around the Ti rather than the O ions. The PIB model gives the correct tetragonal to orthorhombic transition in La_2CuO_4. Lattice dynamics in the high-temperature superconductors shows numerous double-well type modes. A stable oxygen breathing mode is found using PIB, in agreement with experiment. A combination of elaborate LAPW and simple model calculations is shown to be a fruitful approach for understanding the bonding, total energies, lattice dynamics, and elasticity of oxide perovskites.

Introduction

Computational solid state physics has now advanced to the point that complex crystals such as perovskites and the new high temperature superconductors can be studied using non-empirical methods. This is partly because of recent accessibility of supercomputers with large memories, partly because of developments in self-consistent electronic structure methods, and partly because of the development of ab initio models, which though not self-consistent, are often quite accurate. The appeal of non-empirical methods is that one can be sure that artifacts are not introduced into the physics by fitting a parameterized theory to experiment, and that results can be obtained for systems or regimes where experimental data are not yet available. Although theory has become quite accurate recently, it is a mistake to expect approximate theories to have the same accuracy as experiment. On the other hand, there are regimes in which experiments are extremely difficult or impossible, for example, measurement of phonon dispersion at high pressures, determination of phonon eigenvectors in complex crystals, or measurements of elasticity at ultra-high pressures, and in such cases theoretical calculations give the only information that can be obtained.

All of the calculations presented here are performed within the local density approximation (LDA). The LDA is based on the Hohenberg-Kohn density functional theory. Hohenberg and Kohn [1964] and Kohn and Sham [1965] proved that obtaining the ground state properties of the many electron problem can be reduced exactly to solution of a set of single particle Schrödinger-like equations for an effective potential. The LDA assumes a local form for this potential. The LDA has been found in the past eight years to generally give quite accurate total energies and occupied state energies. The exceptions to this general success lie in the areas of magnetic crystals. Even for the more problematic systems, LDA typically gives fairly good total energies.

The LDA can be applied at various levels of approximation. At the most accurate end of the scale is the Lineari-

TABLE 1. Calculated energies for cubic $MgSiO_3$ using the LAPW method.

V(Å^3)	E (Ryd)
30.37784	-1426.2349
32.00787	-1426.3362
33.34153	-1426.4027
35.56430	-1426.4820
38.84570	-1426.5448
42.60248	-1426.5619

zed Augmented Plane Wave (LAPW) method [Wei and Krakauer, 1985]. In the LAPW method, no essential approximations (other than the LDA) are made for either the potential or the charge density. In LAPW there is a dual representation for the charge density and the potential: each is expanded as a sum of plane waves in the "interstitial" region between spherical "muffin tins". Inside the muffin tins, the electron orbitals are represented as radial solutions to Schrödinger's equation for a given energy parameter E for each state, the E derivatives of the radial wave functions, and spherical harmonics for the non-spherical contributions. See Mehl et al. [1988] for a more extended discussion of the LAPW method. Although it is highly accurate, the disadvantage of the LAPW method is that it is extremely computationally intensive. A converged calculation for a single volume for a cubic perovskite requires on the order of 10-20 Cray-XMP hours. For more complex crystals over 100 Cray hours are often required. The memory requirements grow as the square of the system size and the time requirements grow as the cube. Ab initio models have the advantages of self-consistent methods in that they are non-empirical, but are up to six orders of magnitude faster than LAPW.

In an ab initio model, a physical model is chosen for the crystal charge density. The total energy and its derivatives can then be calculated directly from the model charge density using the LDA. The success of an ab initio model depends on the accuracy of the model charge density and on the accuracy of the density functionals that are used. Typically one uses the Thomas-Fermi form for the kinetic energy in an ab initio model, whereas the more accurate Kohn-Sham kinetic energy, which cannot easily be obtained in a non-self-consistent method, is used in self-consistent applications of density functional theory.

We have developed an ab initio model for oxides that has shown considerable success in the cubic alkaline earth oxides [Mehl et al., 1986, Cohen et al., 1987a], bromellite (BeO) [Jephcoat et al., 1988], corundum, stishovite [Cohen 1987a], $MgSiO_3$ perovskite [Cohen 1987b], and copper-oxide superconductors [Cohen et al., 1988b,c]. The Potential Induced Breathing (PIB) model is a derivative of the Gordon-Kim [1972] model. In the latter, the crystal charge density is approximated by overlapping rigid ion charge densities. A difficulty with oxides is that O^{2-} is not stable in the free state. In the PIB model, the oxygen ion is stabilized by including a charged Watson sphere in the quantum mechanical calculation of the ionic charge density. The radius of the Watson sphere is chosen to give the Madelung potential at the site in the crystal. This gives many-body forces that lead to, for example, the observed trends in the Cauchy violations in the cubic alkaline earth oxides. There are three contributions to the total energy in PIB: the Madelung, or electrostatic, energy between ions, the self-energy, or energy of each separated ion, and the overlap energy, or energy change when ionic charge densities are overlapped. More details on the PIB model are given in the above papers and in Boyer et al., [1985a,b, 1986] and Cohen et al., [1987b].

Here we discuss results for $MgSiO_3$, $BaTiO_3$, La_2CuO_4, and $YBa_2Cu_3O_7$. These oxides encompass a wide variety of important phenomena. $MgSiO_3$ is an important geophysical mineral which is believed to be the major mineral in the Earth's lower mantle. The behavior of $MgSiO_3$ perovskite is crucial input to geophysical models. $BaTiO_3$ is a most important member of the ferroelectric materials, which are interesting both scientifically and technologically. La_2CuO_4 and $YBa_2Cu_3O_7$ are of course high temperature superconductors, whose fascinating properties are leading to a resurgence in the study of superconductivity, and may someday lead to a revolution in technology.

$MgSiO_3$

LAPW Calculations for Cubic $MgSiO_3$

In order to examine the bonding in $MgSiO_3$ and to test the PIB model for a dense silicate, a series of LAPW calculations were performed for the cubic perovskite. Cubic $MgSiO_3$ has not been observed in the laboratory. This also exemplifies another advantage of theory--calculations can be performed for crystals that are not stable in the laboratory. [Sometimes such crystals are predicted to have desirable properties and then a way can be found to stabilize and grow them.] In this case we know from PIB calculations that at low pressures cubic $MgSiO_3$ is only slightly unstable with respect to orthorhombic Pbnm perovskite. Furthermore, the bonding in orthorhombic $MgSiO_3$ is probably not qualitatively different from the bonding in cubic $MgSiO_3$. An LAPW calculation for orthorhombic $MgSiO_3$ would require 64 times more time, and is thus impracticable with current supercomputers.

Calculations were performed for 6 volumes. The resulting energies are given in Table 1 (1 Ryd = 13.6 eV). A third-order Birch [1978] fit to the energies gives a=3.48 Å, K_0=279 GPa, and K_0' =3.41 with an root mean squared error of 0.7 mRyd.

The calculations are highly converged in an absolute sense. Two energy windows were used to obtain maximum accuracy for the calculation. The lower (semicore) window included the Mg 2p and O 2s states and the upper (valence) window included the states with primarily O 2p character. The core states were calculated fully relativistically, and the band states were calculated semi-relativistically (all relativistic effects except spin-orbit shifts, which are negligible for these atoms) [Koelling and Harmon, 1977]. Bloch functions were included out to a radius K_{max}=10.8 Å^{-1} in k-space for the valence window and out to K_{max}=9.4 Å^{-1} in the semicore window. This gives 430 LAPW basis functions in

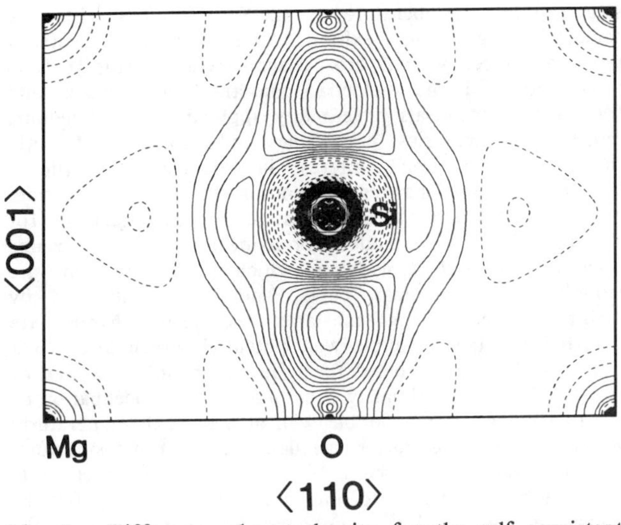

Fig. 1. Difference charge density for the self-consistent LAPW charge density for cubic $MgSiO_3$ perovskite compared with PIB for the (110) plane. The contour interval is five milli-electrons per cubic Bohr (1 Bohr = 0.529 Å). The pressure obtained from the equation of state is 155 GPa and the lattice constant a=5.896 Bohr. The silicon is in the center, oxygen at the top and bottom, and magnesium at the corners.

the lower window and 650 in the valence window at a volume of 30.4 $Å^3$ for each k-point. Four special k-points in the irreducible wedge (64 points in the whole Brillouin zone) were used for the k-point integration. Tests indicate that eigenvalues are converged to better than 1 mRyd and the total energy is absolutely converged to better than 3 mRyd.

Figure 1 shows a contour plot of the difference between the self-consistent LAPW charge density and the overlapping ion charge density of the PIB model for a pressure of 150 GPa. [Although we use self-interaction corrected [Perdew and Zunger, 1981] charge densities in PIB, we must use LDA charge densities for the overlapping ions when comparing PIB and LAPW, since no self-interaction corrections have been included in the LAPW calculations.] The differences are fairly small. A small buildup of charge in the Si-O bond is evidence of some covalency, but the number of electrons involved is small (<<0.1). The self-consistent charge density is quite ionic. At lower pressures, the bonding is expected to be even more ionic since there is decreased overlap between the ions at larger volumes. The high ionicity of $MgSiO_3$ is responsible for the success of the fully ionic PIB model, discussed next.

PIB

The difference charge density in Figure 1 shows a small Si-O bond charge and evidence of a small amount of covalency. The question immediately arises: even though these differences are small, could they significantly affect calculated bulk properties such as the equation of state? Figure 2 shows the static equation of state for cubic $MgSiO_3$ using LAPW and PIB. The PIB calculations were over five orders of magnitude faster than the LAPW calculations, yet the equations of state are in almost perfect agreement. A third order Birch fit to the PIB energies for the cubic structure gives a=3.46 Å, K_0=280 GPa, and K'_0=3.86 compared with a=3.48 Å, K_0=279, and K'_0=3.41 calculated using PIB. Agreement in other systems will in general not be this good, but clearly one can see that the ionic description gives an accurate equation of state for $MgSiO_3$ compared with the self-consistent result. The small amount of covalency seen in Figure 1 evidently does not appreciably affect bulk properties such as the equation of state in this material.

Cohen [1987b] calculated the elastic constants, equation of state, and thermal properties of orthorhombic (space group Pbnm) $MgSiO_3$. The calculated elastic constants, which were calculated before experimental data were available, are in quite good agreement with experiment [see Yeganeh-Haeri et al., 1988]. Figure 3 shows the calculated room temperature equation of state for Pbnm perovskite. Also shown are the experimental data of Knittle and Jeanloz [1987] and Yagi et al. [1982] and the theoretical results of Hemley et al. [1987] and Wolf and Bukowinski [1987]. One sees increasing agreement with experiment as approximations in the theory are removed. Table 2 shows the elastic constants calculated in Cohen [1987b] as well as the measured elastic constants.

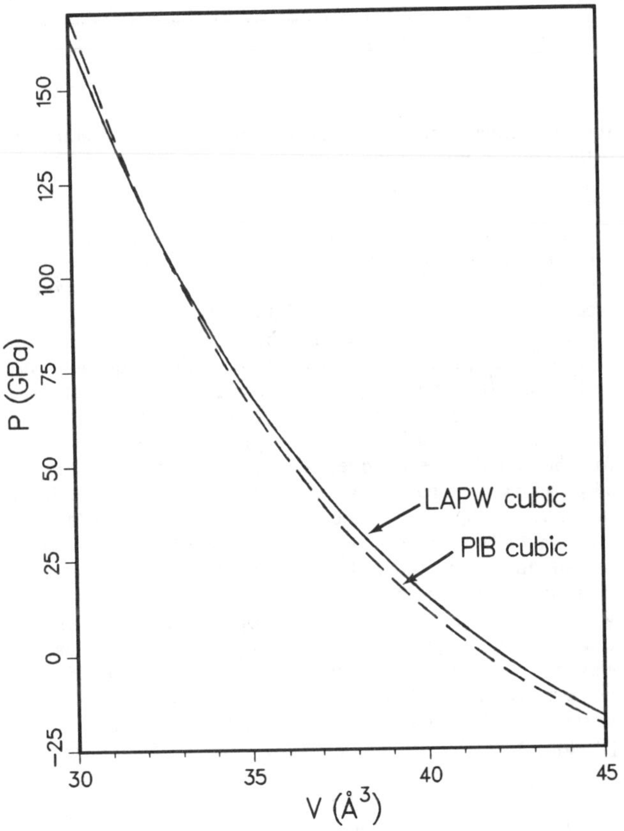

Fig. 2. Equation of state for cubic $MgSiO_3$ perovskite in PIB and LAPW.

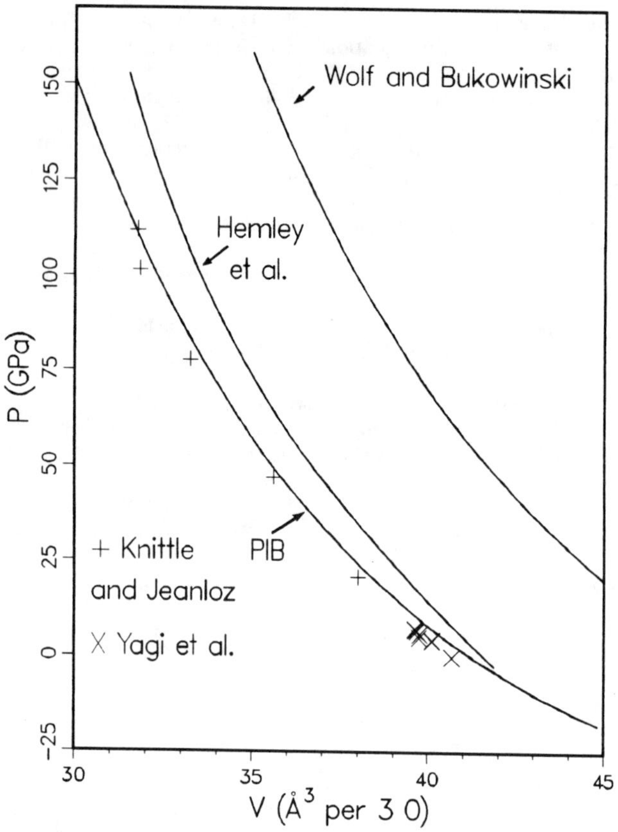

Fig. 3. Experimental and theoretical equations of state for orthorhombic Pbnm perovskite.

The thermal properties of $MgSiO_3$ calculated using PIB are problematic because of the occurrence of some imaginary frequencies in PIB for complex structures. These imaginary frequencies are present only if the Thomas-Fermi kinetic energy functional is used in the self-energy; when the Kohn-Sham kinetic energy is used all of the frequencies are real. In PIB, there are effective charge terms that arise from the first derivative of the self-energy with respect to the Madelung potential at the site. The Thomas-Fermi form for the kinetic energy overestimates this term by a factor of up to two compared with the self-energy calculated using the exact Kohn-Sham kinetic energy. This leads to a large error in LO type modes in complex crystals. Unfortunately, use of the Kohn-Sham kinetic energy in the self-energy gives inaccurate equations of state when used in conjunction with SIC charge densities; this problem is under investigation. Fortunately, zone center properties, such as the elasticity and TO-type Raman and infrared frequencies are not greatly affected by the effective charges. The thermal expansivity is discussed in more detail in Hemley et al. [1988].

$BaTiO_3$

$BaTiO_3$ has the cubic perovskite structure at high temperatures and transforms to a ferroelectric tetragonal structure at a temperature of 120°C. Crystals with ferroelectric transitions exhibit a spontaneous electrical polarization below some critical temperature. This leads to large values of the dielectric constant, for instance, and ferroelectric materials have many applications as detectors, modulators, memories, etc. [Lines and Glass, 1977]. Although there is much interest in ferroelectrics, little is known about the changes in bonding during the transition, and there is much controversy about the driving force for the transition. Most models for oxide ferroelectrics assume a parameterized shell model, in which the oxygen atoms are modeled as mechanical "cores" and "shells" attached by springs. The spring constants and shell charges are adjusted to give a double-well potential which leads to a transition from a distorted form, in which only one well of the double well is primarily occupied, to equal occupation of the two halves of the double well at a critical temperature. We have calculated the charge density, electronic structure, and total energy for the cubic and distorted structure to test whether oxygen distortion is the driving force for the transition. We find that the distortion centers around the Ti in the structure, rather than the oxygen, which calls into question many current models for ferroelectric transitions in oxides.

Strictly speaking the presence of a macroscopic electric field invalidates Bloch's theorem, on which electronic structure calculations for crystals are based, since a macroscopic field implies an electrostatic potential gradient, so that the electrostatic potential differs from one unit cell to the next. This means that a completely rigorous first principles calculation must somehow account for surface depolarization effects, e.g. by treating the solid as a thin slab, for which Bloch's theorem holds in two dimensions. Macroscopic electric field effects obviously greatly complicate a rigorous first principles electronic structure calculation, and for this reason workers in the field have avoided calculations for ferroelectrics. On the other hand, it is possible that the driving mechanism for the transition is

TABLE 2. Elastic constants (in GPa) at zero pressure for $MgSiO_3$ perovskite.

	PIB		Exp*
	Static	298K	
C_{11}	548	531	520
C_{12}	54	44	114
C_{22}	551	531	510
C_{13}	153	143	118
C_{23}	175	166	139
C_{33}	441	425	437
C_{44}	241	237	181
C_{55}	253	249	202
C_{66}	139	136	176
K	256	249	245
μ	196	192	184

*Yeganeh-Haeri et al., 1988

TABLE 3. Calculated energies for $BaTiO_3$.

	a(bohr)	V(Å^3)	E (Ryd)
cubic			
K_{max} = 8.9Å^{-1}			
	7.0	50.8273	-18419.3614
	7.2	55.3096	-18419.4438
	7.4	60.0479	-18419.4743
	7.5	62.5154	-18419.4743
K_{max} = 7.8Å^{-1} semicore, 8.9Å^{-1} valence			
	7.5	62.5154	-18419.4608
ferroelectric			
K_{max} = 7.8Å^{-1} semicore, 8.9Å^{-1} valence			
	7.5	62.5154	-18419.4468

essentially independent of the macroscopic field, in which case restricting the potential to be periodic would not drastically alter the forces driving the transition. By applying the LAPW method, which is based on Bloch's theorem, to ferroelectric $BaTiO_3$ without surfaces, we are imposing periodic boundary conditions on each unit cell. Whether or not this constraint would drastically alter the energetics of the ferroelectric distortion was a fundamental question we wanted to address.

The LAPW program was used to calculate the band structure, total energy and charge density of $BaTiO_3$ in the cubic structure as a function of volume as well as in the experimental ferroelectric structure. We used a xenon core for barium, an argon core for the titanium, and a helium core for the oxygen; atomic-like states were calculated in each iteration. The O 2s states, Ba 5s and 5p states and the Ti 3s and 3p states, as well as the valence band states (predominately oxygen 2p character) were treated as bands. In the cubic case the LAPW matrix is real, but in the ferroelectric structure it is complex due to the lack of an inversion center in the space group. The core states were calculated fully relativistically, and the band states were calculated semi-relativistically. We found that four special k-points were sufficient to converge the integration for the cubic structure, and six were necessary to converge the ferroelectric structure.

The tetragonal ferroelectric structure used was that of Harada et al. [1970] with Ba displaced by 0.015 (in lattice coordinates), Ti displaced by 0.028, and O(1) displaced by-0.0077 along z. The c/a ratio was 1.01 for the ferroelectric structure. For the cubic perovskite, plane waves up to K_{max}=8.9 Å^{-1} were used, corresponding to about 750 basis functions at zero pressure. The ferroelectric calculation was done at lower convergence for the lower window, with a maximum K_{max}=7.8 Å^{-1}, corresponding to about 500 basis functions at zero pressure. Therefore, the cubic calculation was also done at this convergence in order to compare total energies. Convergence tests have not yet been performed for the ferroelectric calculation, so that the ferroelectric energy should be regarded as preliminary.

Table 3 shows the energies (in Rydbergs per unit cell) calculated for the cubic and ferroelectric structures. A second-order Birch fit to the energy values gives a static value of the zero pressure cubic lattice constant of 3.94 Å and a bulk modulus K_0 of 209 GPa. A third-order fit gives a=3.94 Å, K_0=194 GPa, and K_0'=4.79. The experimental values are a=4.01 Å and K_0=200 GPa [Evans, 1961]. Table 3 also shows the calculated energy for the cubic and ferroelectric structures.

Figure 4 shows the differences between LAPW and PIB

TABLE 4. Minimum energy structures for La_2CuO_4 and $YBa_2Cu_3O_7$.

	La_2CuO_4 Tetragonal I4/mmm		La_2CuO_4 Orthorhombic Abma			$YBa_2Cu_3O_7$ Orthorhombic Pmmm	
	PIB	Exp*	PIB	Exp**		PIB	Exp***
V (Å^3)	99.19	94.88	205.76	189.76	V	184.57	173.45
c/a	3.01	3.49	2.14	2.43	c/a	3.143	3.056
b/a	(1.0)	(1.0)	1.016	0.992	b/a	1.071	1.017
La(X)	(0.0)	(0.0)	-0.023	0.007	Ba(Z)	0.202	0.185
La(Z)	0.366	0.362	0.350	0.362	Cu(Z)	0.351	0.355
O_{xy}(Z)	(0.0)	(0.0)	0.062	0.007	O_2(Z)	0.389	0.378
O_z(X)	(0.0)	(0.0)	-0.151	-.031	O_3(Z)	0.383	0.378
O_z(Z)	0.193	0.182	0.183	0.187	O_4(Z)	0.174	0.158

* J. M. Longo and P. M. Raccah [1973]
** V. B. Grande et al. [1977]
*** M. A. Beno et al. [1987]

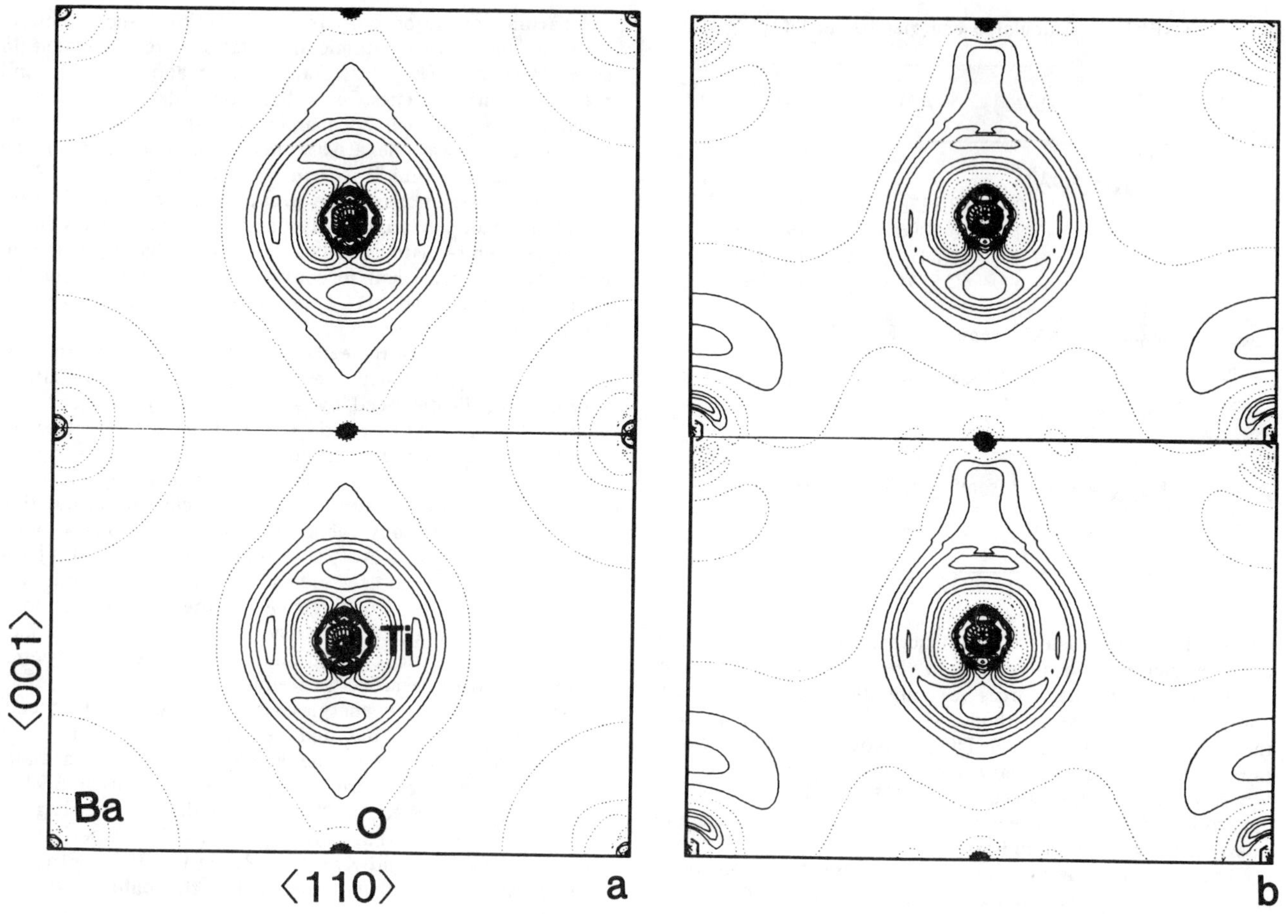

Fig. 4. Difference charge densities between LAPW and PIB for $BaTiO_3$ at zero pressure (a=7.5 Bohr) for the (110) plane. Two unit cells are shown to clarify the differences for the oxygen ion. a. cubic b. tetragonal.

charge densities for cubic and tetragonal $BaTiO_3$. Two unit cells are shown in each figure, in order to bring out the environment of the oxygen atom, which is in the center of each figure. The barium is at the edges of each figure, and the titanium is at the center of each unit cell. The charge densities are shown for the (110) plane. The oxygen and barium are modeled quite well as O^{2-} and Ba^{2+}. However, the differences at the Ti are quite complex. For the ferroelectric structure, the distortion is centered around the Ti atom rather than the oxygen.

PIB gave an energy increase of about 28 mRyd for the ferroelectric distortion [Boyer et al., 1985]. Using LAPW, the ferroelectric structure was found to have a higher energy by 14 mRyd (0.2 eV). One possible explanation is that surfaces must be included in order to obtain a lower energy for the ferroelectric structure, for the reasons discussed above. This is not easily tested, since slab calculations for ferroelectrics are beyond the current capabilities of both current computer codes and current supercomputers. Another possibility is that the ferroelectric structure used in the calculation does not represent the proper local structure, but represents an average structure. If ferroelectric $BaTiO_3$ is disordered, with Ti atoms locally displaced along the cube diagonals rather than the four fold axis, it might be expected that a calculation for the average tetragonal structure would give a higher energy. There has been a great deal of discussion in the literature on this point, and some "split-atom" structure refinements have been performed for $BaTiO_3$ [Comes et al. 1968, 1970]. Further work remains to fully characterize ferroelectric $BaTiO_3$.

Copper Oxide Superconductors

High temperature superconductivity in metallic oxides came as quite a surprise to the physics community. There is now an intense amount of work being done to understand these materials. All of the real progress so far has been firmly rooted in experiment, but the hope is that a theoretical understanding of these materials will lead to the design of materials with higher critical temperatures, T_c, higher critical fields, and better material properties. We have studied La_2CuO_4 (alloying with Sr leads to T_c=40K) and $YBa_2Cu_3O_7$ (T_c=90K) using both the PIB model and the LAPW method. The calculations using PIB were first done

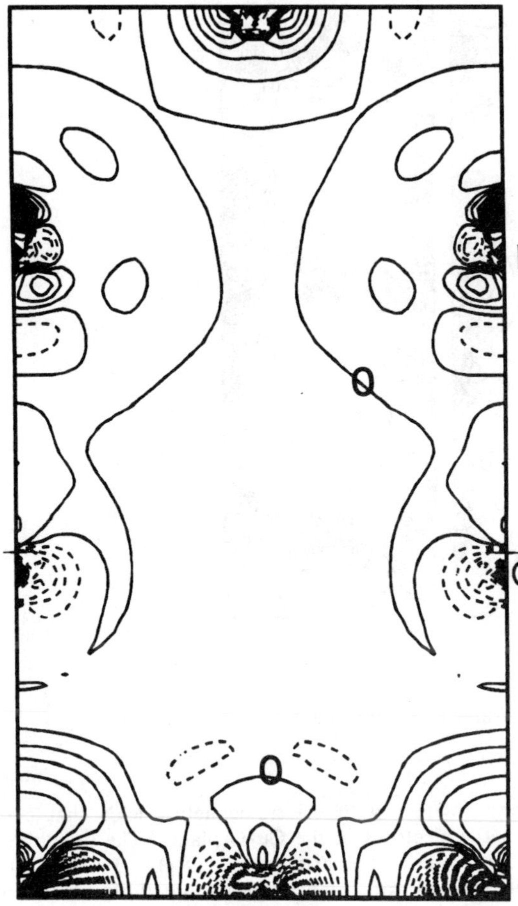

Fig. 5. Difference charge density for the (100) plane in La_2CuO_4 between LAPW and PIB.

for La_2CuO_4 with no expectations that an ionic model would be able to make even qualitative predictions about metallic oxides. We found that PIB even gives semiquantitative agreement for the structure and lattice dynamics of La_2CuO_4. After some initial confusion in the literature, it has been found that undoped La_2CuO_4 is indeed not a metal, but is an insulator. This makes the agreement somewhat less surprising, though there has been no experimental evidence that there are gross changes in structure or lattice dynamics between the insulating and metallic (i.e. doped) samples of La_2CuO_4. We have also found it illuminating to compare charge densities and densities of states between PIB and those obtained self-consistently using LAPW.

Table 4 shows the structures calculated using PIB and the experimental structures for La_2CuO_4 and $YBa_2Cu_3O_7$. The PIB calculations use O^{2-} and Cu^{2+} for the La_2CuO_4 calculations and O^{2-} and $Cu^{2.33+}$ for $YBa_2Cu_3O_7$ (the ionicity in these materials is discussed in more detail below.) Table 4 shows that PIB gives the observed orthorhombic distortion in La_2CuO_4; this is the first (and so far only) calculation that gives this distortion. Complicated many body forces [Anderson, 1987] are not required in order to give this "tilt" distortion. Also, note that the ionic description gives a longer Cu-O_z distance than the Cu-O_{xy} distance as is observed: ionic forces account for about half of the distortion of the Cu-O octahedron when spherical ions are used. The Jahn-Teller effect, which is not included in PIB, presumably accounts for the rest of the observed distortion. The structure of $YBa_2Cu_3O_7$ also comes out quite good, except that the order of the Cu-O bond lengths is predicted incorrectly.

Rather than use the Thomas-Fermi kinetic energy functional, we have used the Kohn-Sham kinetic energy for the self-energy here since it gives a better agreement for the volume. The Thomas-Fermi kinetic energy gives too large a volume due to the non-spherical, open-shell nature of the copper. The Kohn-Sham kinetic energy gives too small a volume, generally, when used with SIC charge densities. These errors fortuitously cancel here. This was the only empirical input to these calculations. Use of the Kohn-Sham kinetic energy in the self-energy also gives more reliable lattice dynamics, as discussed above for $MgSiO_3$.

Figure 5 shows the difference in PIB and LAPW total charge densities (LAPW minus PIB). The O_z is found to be quite ionic. The main differences occur in the copper-oxygen plane. Charge moves out of the overlap regions between the copper and the oxygen. There is only a very small covalent bond charge between the copper and the oxygen. This type of bonding behavior is probably quite general in oxides with open-shelled cations. For example, this is the general pattern observed in FeO [Mehl et al., 1987].

We have found that the copper oxide superconductors have a significant amount of ionic character. In order to determine the ionicity of the ions in $YBa_2Cu_3O_7$ we have compared densities of states and amounts of charge in each muffin tin between self-consistent LAPW results and PIB [Krakauer et al., 1988]. Table 5 shows the differences in muffin tin charges for three different charge configurations. We find that the best fit is obtained with a copper ionicity of +1.62 and an oxygen ionicity of -1.7. Better fits for the Y could be obtained by also varying the Watson sphere radii for the oxygen ion calculation in PIB. We also examined a

TABLE 5. Differences between LAPW and PIB muffin tin charges for different charge configurations in $YBa_2Cu_3O_7$.

	$Cu^{2.333+}$	Cu^{2+}	$Cu^{1.62+}$
Y	0.17	0.24	0.31
Ba	0.00	0.01	0.03
Cu(1)	0.34	0.18	0.05
Cu(2)	0.40	0.20	0.02
O(1)	-0.05	-0.02	0.01
O(2)	-0.13	-0.10	-0.08
O(3)	-0.12	-0.10	-0.06
O(4)	-0.02	0.00	0.02
Interstitial	-0.72	-0.43	-0.19

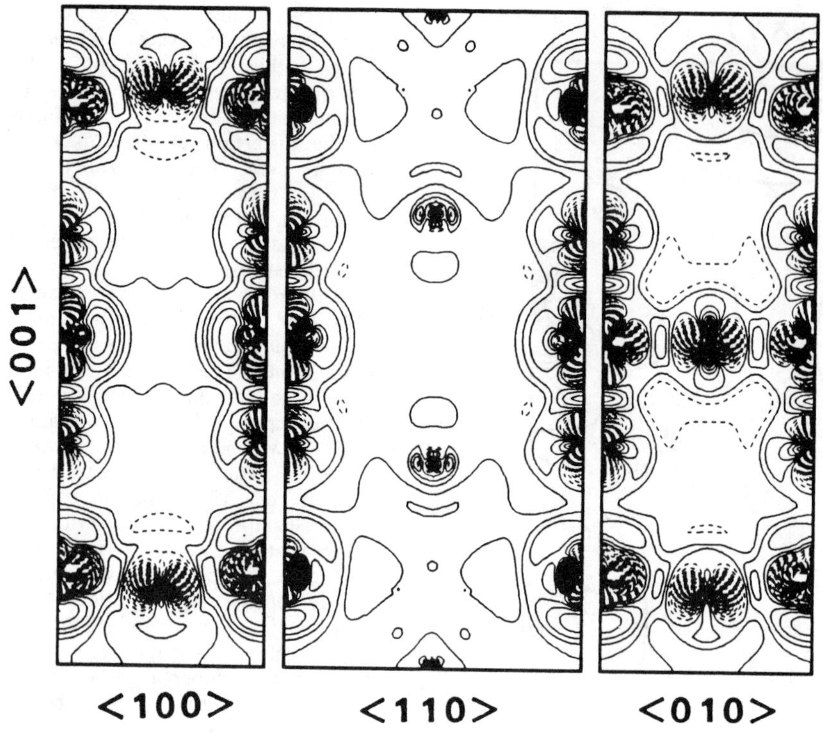

Fig. 6. Difference charge density for $YBa_2Cu_3O_7$ between LAPW and PIB using the charge configuration with Cu $^{+1.62}$.

configuration with $Cu(1)^{3+}$ and $Cu(2)^{2+}$; this gave a much worse charge density. There is no evidence from our calculations to suggest any trivalent copper, or any "mixed valence" character between 2+ and 3+. The oxygen ion is also much closer to -2 than to -1. Thus, it appears that the bonding is only slightly less ionic than insulating oxides.

Figure 6 shows the differences in the PIB and LAPW charge densities for $YBa_2Cu_3O_7$ with $Cu^{1.62+}$. Similar features are observed as for La_2CuO_4, except that $YBa_2Cu_3O_7$ does not have any spherical ionic oxygen like the O_z in La_2CuO_4. The non-spherical and covalent character, consistent with the metallic character of the material, is significantly greater. Note that no differences are observed in the empty oxygen site.

Almost all known superconductivity is driven by the electron-phonon interaction. Thus, it is of considerable interest to examine the lattice dynamics of the copper oxide superconductors. Figure 7 shows the calculated dispersion curves for La_2CuO_4 in both the tetragonal and orthorhombic (minimum energy) structures. A number of unstable (imaginary) harmonic frequencies are calculated, indicating double well, anharmonic potentials. The most unstable mode is an X_2 mode in which the O_{xy} atoms rotate in the Cu-O plane. The next most unstable is an X_4 mode which is the tilt mode corresponding to the orthorhombic to tetragonal transition. The shape of this branch is found to agree very well with inelastic neutron scattering measurements. [Birgeneau et al., 1987] The branch from $\Gamma_{5'}$ to $X_{3'}$ is very flat, and represents sliding motions of the O_z atom in the x-y plane. These motions are infrared active and thus generate and interact with electromagnetic fields that can couple with the electrons in the Cu-O plane. The breathing mode is the highest frequency mode (X_1). In contrast, Weber [1987] found an unstable oxygen breathing mode, in disagreement with experiment, probably at least partly due to the neglect of long-range ionic contributions to the bare phonons. The breathing mode does not involve only O_{xy} motions, but also includes in-phase O_z motions in PIB. There are four modes with X_1 symmetry which can mix. Thus, the "frozen phonon" calculations of Fu and Freeman [1987] may not actually correspond to a normal mode of the system. Many of the unstable branches become stable for the orthorhombic structure. However, there are still two very flat branches which correspond to the sliding motions of O_z in the x-y plane. There are also two other soft branches which suggest a lower symmetry ground state, consistent with the experimental indications of a monoclinic ground state [Moss et al., 1987]. Cohen et al. [1988b,c] give more details.

Figure 8 shows the calculated phonon dispersion curves using PIB for $YBa_2Cu_3O_7$ with the Cu $^{2.333+}$ configuration. As in La_2CuO_4, unstable, flat, double well phonon branches are observed. The most unstable branch (-279 cm^{-1} at Γ) corresponds to movement of the chain oxygens (O(1)) in the x-direction and out-of-chain oxygens (O(4)) in the -x direction. The next most unstable branch (-138 at Γ) corresponds to motions of O(2) and O(3) in opposite directions along z. Both of these modes are infrared active.

An important feature of these materials is that they are metallic in two dimensions and ionic in the third dimension.

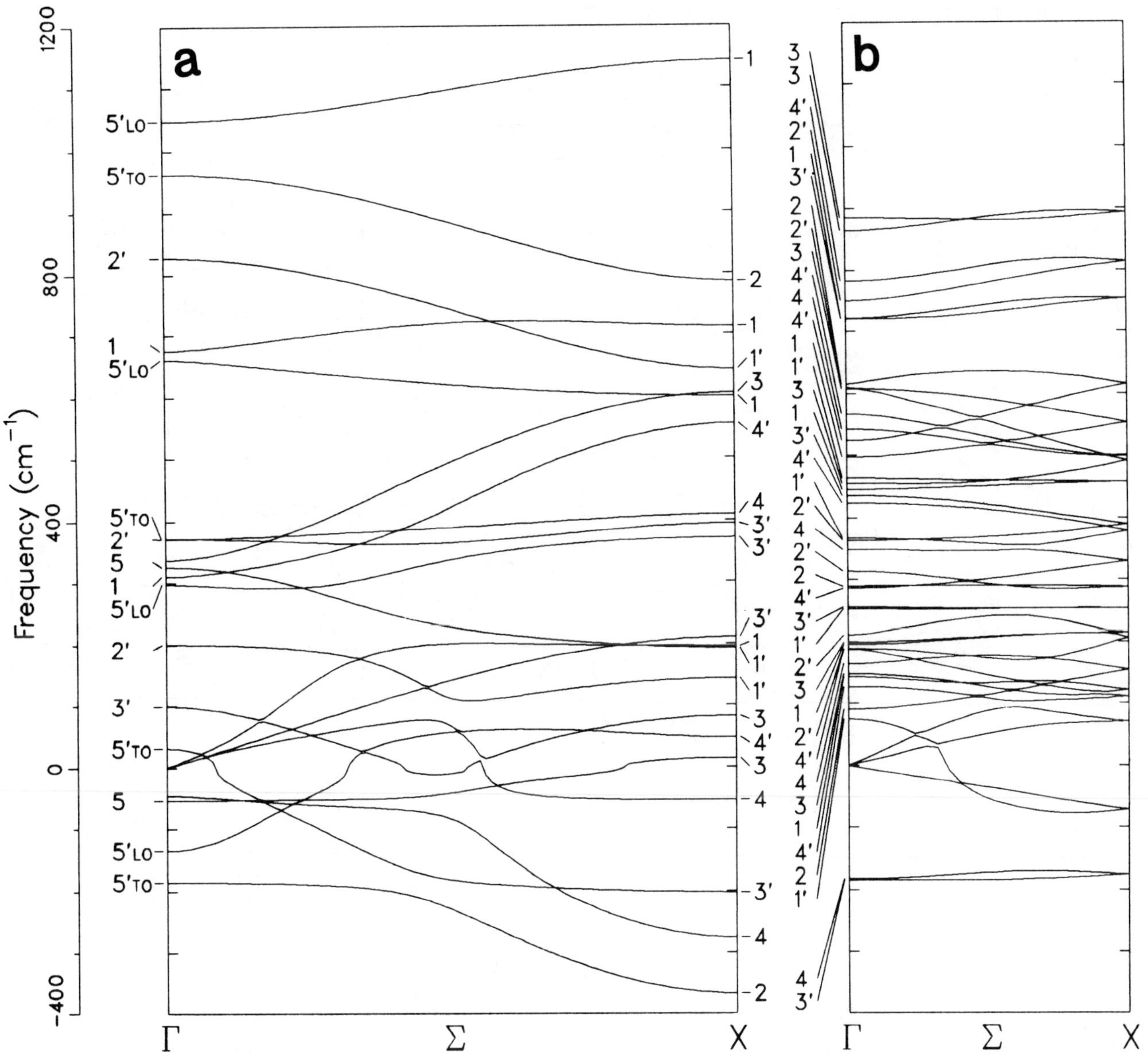

Fig. 7. Calculated phonon dispersion curves for (a) tetragonal and (b) orthorhombic La_2CuO_4 using the PIB model.

This allows unscreened LO phonons to coexist with metallic conduction. It also makes traditional analysis of the electron-phonon interaction in these materials very difficult, since three-dimensional superconductivity in these materials may depend intimately on the small coupling between planes. The intrinsic instability of the O^{2-} ion and the ionic contributions to the electron-phonon matrix elements should greatly enhance the electron-phonon coupling and may contribute to the high T_c. Ionic materials in general have extremely large electron-phonon interactions which lead to polaron formation in insulators. The effect of huge electron-phonon interactions in metals has not been given much attention in the past, due to the fact that the screening in normal metals greatly reduces the electron-phonon interaction relative to a material such as NaCl. The coexistence of metallic conductivity with unscreened ionicity may be the key to high temperature superconductivity.

All of the quantitative theoretical studies of electron-phonon superconductivity in these materials have been based on a rigid atom (or ion) model, such as the rigid muffin tin approximation. In the rigid muffin tin approximation, the electron-phonon interaction is determined by the potential change when the spherical part of the potential for an atom is rigidly displaced [Gaspari and Gyorffy, 1972], and is

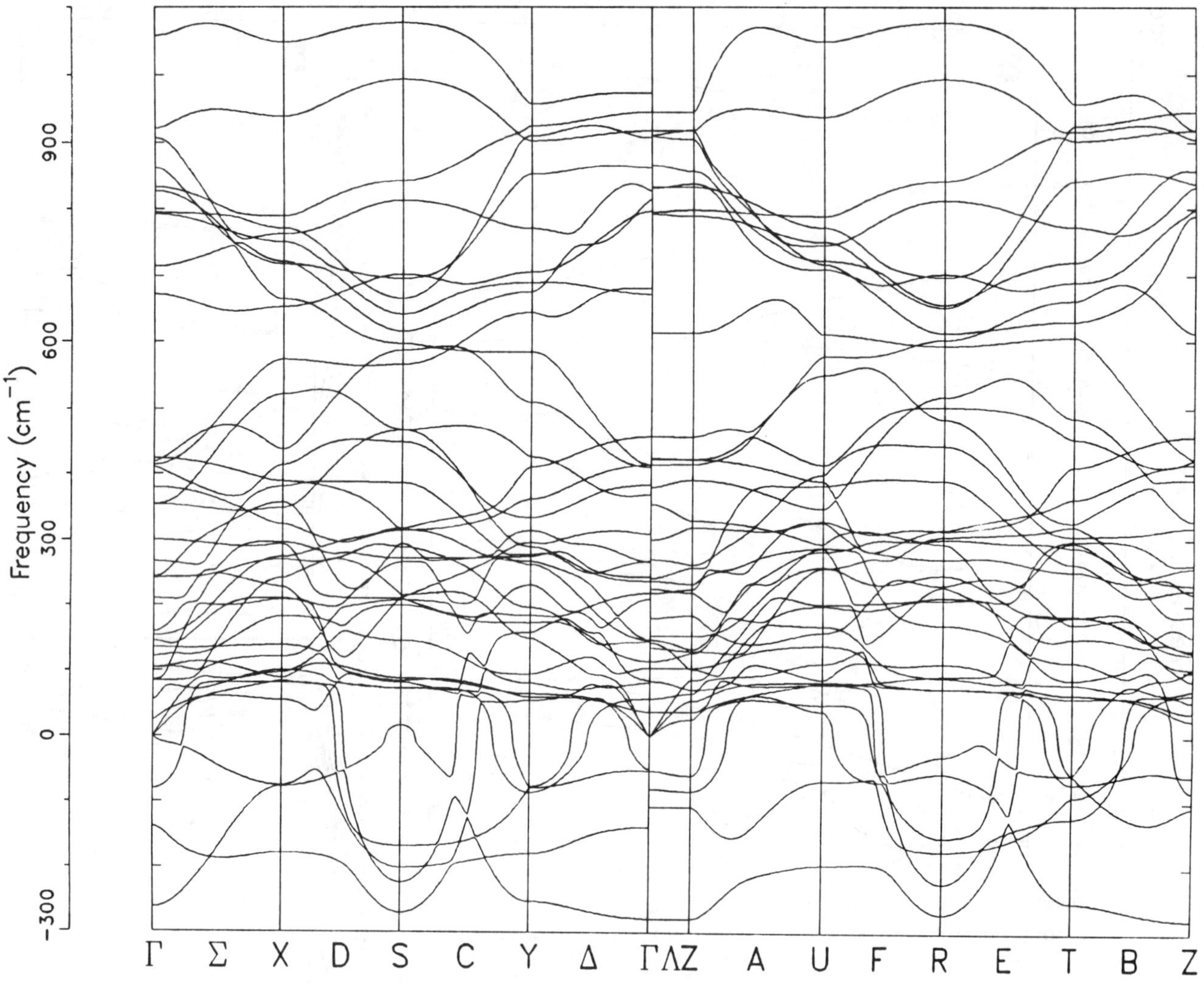

Fig. 8. Calculated phonon dispersion curves for $YBa_2Cu_3O_7$ with $Cu^{+2.333}$.

presumed to be screened within half the nearest neighbor distance or less. In an ionic metal, there can be large changes in the potential at one atom when another ion is displaced, and we find such effects in La_2CuO_4 [Cohen et al., 1988c]. We are now investigating whether these contributions can be large enough to induce T_c's of 90K, or whether another mechanism is necessary in addition to the electron-phonon interaction.

Conclusions

Electronic structure, elasticity, equations of state, and lattice dynamics can be calculated with varying degrees of reliability for oxide perovskites and perovskite-like crystals. These calculations have only recently become powerful enough to study relatively complex crystals, such as perovskites. These methods will continually be improved, and in some cases reliability approaching that of experiment is found. The primary goal of these calculations, however, is to better understand why the materials behave as they do, and to give some guidance to the experimentalist.

The elastic properties and room temperature equation of state of $MgSiO_3$ are given quite well by the PIB model. Some improvements to the PIB model are necessary for better calculations of lattice dynamical and thermal properties. LAPW calculations for $MgSiO_3$ show a quite ionic form of bonding, which is consistent with the high quality of predictions of elasticity and equations of state using PIB. The calculation of the single crystal elastic constants for PIB was done before experimental measurement, and has turned out to be quite accurate. The ferroelectric transformation in $BaTiO_3$ is not well understood. Though various models can be devised that give a ferroelectric transition, it is not clear which, if any, current model is correct. It appears that the distortion in $BaTiO_3$ centers around the Ti, rather than the O, as is assumed in most current models for ferroelectricity in $BaTiO_3$.

Many properties of the copper oxide superconductors can be predicted surprisingly well using an ionic description. The superconductors may behave more like ionic perovskites than metals, at least in terms of lattice dynamical and structural properties.

Acknowledgements. We wish to thank D. J. Singh for much help with the LAPW program, P. Edwardson for pointing out the possible order-disorder character of tetragonal $BaTiO_3$, R. J. Hemley for discussions regarding $MgSiO_3$ perovskite, and J. Serene, D. Papaconstantopolous, J. D. Axe, P. Boni, G. Burns, and J. Hardy for discussions regarding the high-T_c superconductors. B. M. Klein, J. Feldman and K. L. Cohen also contributed helpful discussions. Computations were performed on the Cray-XMP at the Naval Research Laboratory, the IBM 3090 at the Cornell National Supercomputer Facility, and the Cray 2 at the Minnesota Supercomputer Center. This work was supported in part by the Office of Naval Research and by National Science Foundation Grant No. DMR-84-16064.

References

Anderson, P. W., RVB theory of high T_c superconductivity, in Novel Superconductivity, eds. S. A. Wolf and V. Z. Kresin, Plenum, New York, pp. 295-299, 1987.

Beno, M. A., L. Soderholm, D. W. Capone, II, D. G. Hinks, J. D. Jorgensen, J. D. Grace, I. K. Schuller, C. U. Segre, and K. Zhang, Structure of the single-phase high-temperature superconductor $YBa_2Cu_3O_{7-\delta}$, Appl. Phys. Lett., 51, 57-59, 1987.

Birch, F., Finite strain isotherm and velocities for single crystal and polycrystalline NaCl at high pressure and 300 K, J. Geophys. Res., 83, 1257-1268, 1978.

Birgeneau, R. J., C. Y. Chen, D. R. Gabbe, H. P. Jenssen, M. A. Kastner, C. J. Peters, P. J. Picone, Tineke Thio, T. R. Thurston, H. L. Tuller, J. D. Axe, P. Boni, and G. Shirane, Soft-phonon behavior and transport in single crystal La_2CuO_4, Phys. Rev. Lett., 59, 1329-1332, 1987.

Boyer, L. L., M. J. Mehl, J. L. Feldman, J. R. Hardy, J. W. Flocken, and C. Y. Fong, Beyond the rigid ion approximation with spherically symmetric ions, Phys. Rev. Lett., 54, 1940-1943, 1985a; 57, 2331, 1986.

Boyer, L. L., M. J. Mehl, J. W. Flocken, and J. R. Hardy, Ab initio calculations of the effect of non-rigid ions on the static and dynamics of oxides, Jap. J. Appl. Phys., 24 Suppl. 24-2, 204-205, 1985b.

Cohen, R. E., Calculation of elasticity and high pressure instabilities in corundum and stishovite with the potential induced breathing model, Geophys. Res. Lett., 14, 37-40, 1987a.

Cohen, R. E., Elasticity and equation of state of $MgSiO_3$ perovskite, Geophys. Res. Lett., 14, 1053-1056, 1987b.

Cohen, R. E., L. L. Boyer, and M. J. Mehl, Lattice dynamics of the potential-induced breathing model: Phonon dispersion in the alkaline-earth oxides, Phys. Rev. B, 35, 5749-5760, 1987a.

Cohen, R. E., L. L. Boyer, and M. J. Mehl, Theoretical studies of charge relaxation effects on the statics and dynamics of oxides, Phys. Chem. Minerals, 14, 294-302, 1987b.

Cohen, R. E., M. J. Mehl, and L. L. Boyer, Phase transitions and elasticity in zirconia, Physica B, in press, 1988a.

Cohen, R. E., W. E. Pickett, H. Krakauer, and L. L. Boyer, Applications of ionic models to the high-temperature superconductor La_2CuO_4, Physica B, in press, 1988b.

Cohen, R. E., W. E. Pickett, L. L. Boyer, and H. Krakauer, Ionic contributions to lattice instabilities and phonon dispersion in La_2CuO_4, Phys. Rev. Lett., 60, 817-820, 1988c.

Comes, R., M. Lambert, and A. Guinier, The chain structure of $BaTiO_3$ and $KNbO_3$, Solid State Comm., 6, 715-719, 1968.

Comes, R., M. Lambert, and A. Guinier, Desordre Lineaire dans les Cristaux: cas du Silicium, du Quartz, et des Perovskites Ferroelectriques, Acta Cryst., A26, 244-254, 1970.

Evans, H. T., Jr., An x-ray diffraction study of tetragonal barium titanate, Acta Cryst., 14, 1019-1026, 1961.

Fu, C. L. and A. J. Freeman, Optic breathing mode, resonant charge fluctuations, and high-T_c superconductivity in the layered perovskites, Phys. Rev. B, 35, 8861-8864, 1987.

Gordon, R. G. and Y. S. Kim, Theory for the forces between closed-shell atoms and molecules, J. Chem. Phys., 56, 3122-3133, 1972.

Grande, V.B., Hk. Muller-Buschbaum, and M. Schweizer, Zur kristallstruktur von Seltenerdmetalloxocupraten: La_2CuO_4, Gd_2CuO_4, Z. Anorg. Allg. Chem., 428, 120-124, 1977.

Harada, J., J.D. Axe, and G. Shirane, Neutron scattering study of soft modes in cubic BaTiO3, Phys. Rev. B, 4, 155-162, 1971.

Hemley, R. J., M. D. Jackson, and R. G. Gordon, Theoretical study of the structure, lattice dynamics, and equations of state of perovskite-type $MgSiO_3$ and $CaSiO_3$, Phys. Chem. Minerals, 14, 2-12, 1987.

Hemley, R. J., H. K. Mao, A. Yeganeh-Haeri, D. J. Weidner, R. E. Cohen, and E. Ito, Raman spectroscopy and lattice dynamics of $MgSiO_3$ perovskite at high pressure, this volume.

Hohenberg P. and W. Kohn, Inhomogeneous electron gas. Phys. Rev., 136, B 864-871, 1964.

Jephcoat, A.P., R.J. Hemley, H.K. Mao, R.E. Cohen, and M.J. Mehl, Raman spectroscopy and theoretical modeling of BeO at high pressure, Phys. Rev. B, 37, 4727-4734, 1988.

Koelling, D. D. and B. N. Harmon, A technique for relativistic spin-polarized calculations, J. Phys. C, 10, 3107-3114, 1977.

Kohn W. and L. J. Sham, Self-consistent equations including exchange and correlation effects, Phys. Rev., 140, A1133-1140, 1965.

Knittle, E. and R. Jeanloz, Synthesis and equation of state of (Mg,Fe)SiO_3 perovskite to over 100 Gigapascals, Science, 235, 668-670, 1987.

Knittle, E., R. Jeanloz, and G. L. Smith, Thermal expansion of silicate perovskite and stratification of the Earth's mantle, Nature, 319, 214-216, 1986.

Krakauer, H., W. E. Pickett, and R. E. Cohen, Analysis of Electronic Structure and Charge Density of the High-Temperature Superconductor $YBa_2Cu_3O_7$, J. Superconductivity, 1, 111-139, 1988.

Lines, M. E. and A. M. Glass, Principles and Applications of Ferroelectrics and Related Materials, Oxford Univ. Press, New York, 1977.

Longo, J. M. and P. M. Raccah, The structure of La_2CuO_4 and $LaSrVO_4$, J. Solid State Chem., 6, 526-531, 1973.

Mehl, M. J., R. J. Hemley, and L. L. Boyer, Potential induced breathing model for the elastic moduli and high-pressure behavior of the cubic alkaline-earth oxides, Phys. Rev. B, 33, 8685-8696, 1986.

Mehl, M. J., R. E. Cohen, H. Krakauer, and R. Rudd, LAPW study of the high pressure behavior of stoichiometric FeO, EOS, 68, 1455.

Mehl, M. J., R. E. Cohen, and H. Krakauer, LAPW electronic structure calculations for MgO and CaO, J. Geophys. Res., in press, 1988.

Moss, S.C., K. Forster, J. D. Axe, H. You, D. Hohlwein, D. E. Cox, P. H. Hor, R. L. Meng, and C. W. Chu, High-resolution synchrotron x-ray study of the structure of $La_{1.8}Ba_{0.2}CuO_{4-y}$, Phys. Rev. B, 35, 7195-7198, 1987.

Perdew, J. P. and A. Zunger, Self-interaction correction to density-functional approximations for many-electron systems, Phys. Rev. B, 23, 5048-5079, 1981.

Pickett, W. E., H. Krakauer, D. A. Papaconstantopoulos, and L. L. Boyer, Evidence for conventional superconductivity in La-Ba-Cu-O compounds, Phys. Rev. B, 35, 7252-7255, 1987.

Waldman, M. and R. G. Gordon Scaled electron gas approximation for intermolecular forces, J. Chem. Phys., 71, 1325-1338, 1979.

Weber, W., Electron-phonon interaction in the new superconductors $La_{2-x}(Ba,Sr)_xCuO_4$, Phys. Rev. Lett., 58, 1371-1374, 1987.

Wei, S.-H., and H. Krakauer, Local density functional calculation of the pressure induced phase transition and metallization of BaSe and BaTe, Phys. Rev. Lett., 55, 1200-1203, 1985.

Wolf, G. H. and M. S. T. Bukowinski, Theoretical study of the structural and thermoelastic properties of $MgSiO_3$ and $CaSiO_3$ perovskites: Implications for lower mantle composition, in High Pressure Research in Mineral Physics, eds. M. H. Manghnani and Y. Syono, Terra Scientific, Tokyo, 313-331, 1987.

Yagi, T., H.K. Mao, and P.M. Bell, Hydrostatic compression of perovskite-type $MgSiO_3$, in Advances in Physical Geochemistry, Vol. 2, ed. S.K. Saxena, Springer-Verlag, 1982, pp. 317-325.

Yeganeh-Haeri, A., D. J. Weidner, and E. Ito, Single-crystal elastic properties of $MgSiO_3$, this volume.

THERMOCHEMISTRY OF PEROVSKITES

Alexandra Navrotsky

Department of Geological and Geophysical Sciences, Princeton University

Abstract. This paper reviews several aspects of perovskite thermodynamics. The energetics of formation of $A^{2+}B^{4+}O_3$ perovskites are related to lattice energies and crystal chemical parameters. Perovskite-related oxide superconductor phases are reviewed and new calorimetric data are presented for the partial molar enthalpy of oxygen in $YBa_2Cu_3O_x$ ($5.9 < x < 7$). High pressure transitions in silicates and germanates show that perovskite is generally a phase of unfavorable energy, high density, and relatively high entropy. The high entropy leads, in general, to negative P-T slopes for perovskite-forming reactions. The entropy of $MgSiO_3$ perovskite is calculated using Kieffer's vibrational modelling approach and the thermal expansion coefficient is estimated from the observed pressure dependence of vibrational spectra.

Introduction

Oxides with perovskite-related structures form a varied family of materials, ranging from ceramics and superconductors to minerals in the Earth's crust and mantle. To better predict and understand $MgSiO_3$ perovskite as a lower mantle phase, it is useful to examine systematic trends in perovskite stability for a wide range of compositions, temperatures, and pressures. Available data are summarized in this paper for ceramic perovskites, for $YBa_2Cu_3O_x$ oxide superconductor phases, and for high pressure silicate and germanate perovskites. Trends in thermodynamic properties are correlated with structure and bonding in the perovskite structure.

Enthalpies of Formation of $A^{2+}B^{4+}O_3$ Perovskites from the Oxides

Enthalpies of formation of ABO_3 perovskites, corresponding to the reaction:

$$AO + BO_2 = ABO_3 \quad (1)$$

have been measured by high temperature oxide melt calorimetry and are summarized in Table 1 (Takayama-Muromachi and Navrotsky, 1988). They range from large exothermic values (-152 kJ/mol) for $BaTiO_3$ to distinctly positive values (at atmospheric pressure) for the high pressure phases $CdGeO_3$ and $MgSiO_3$. In general, for a given B^{4+} ion, the heat of formation of the perovskite becomes more negative as the size of the A^{2+} ion increases, and for a given A^{2+} ion, the heat of formation becomes more negative in the order Si, Ge, Zr, Ti.

The tolerance factor, a useful parameter related to the stability of perovskite structures, is given by

$$t = (r_A + r_O)/\ 2(r_B + r_O) \quad (2)$$

where r_A, r_B, r_O are empirical radii of the respective ions. By geometry, the ideal cubic structure should have t=1. Navrotsky (1981) has shown that the heat of formation from oxides of perovskite compounds becomes less negative nearly linearly with the absolute value of 1-t. Perovskites ($CdGeO_3$, $CaGeO_3$ and $MgSiO_3$) stable only at high pressure have standard enthalpies of formation from the oxides which are much less negative than the above correlation would suggest (-33 kJ for $CaGeO_3$, +23 kJ for $CdGeO_3$ and an estimated value of +31 ± 20 kJ for $MgSiO_3$). Clearly at 1 atm these perovskites are rather unstable, and qualitatively this instability probably relates to the rather small sizes of Mg^{2+} in the central site and of Si^{4+} and Ge^{4+} in the octahedral sublattice.

The concept of tolerance factor is complicated by the fact that ionic radius depends on coordination number. The numerical value of the tolerance factor therefore depends on whether one considers A^{2+} to be 12-coordinated (as it would be in the ideal cubic structure) or to have a lower coordination number (as low as 8) in a distorted structure.

Because of these questions, Takayama-Muromachi and Navrotsky (1988) took a somewhat different approach based on lattice energies. For an ionic crystal, the internal energy can be separated into two terms

$$E = E_M + E_N \quad (3)$$

where E_M is the electrostatic or Madelung energy and E_N contains all the other terms, of which

TABLE 1. Enthalpy of Formation of Perovskites from Oxides (kJ/mol)

Compound	ΔH_f°
$CaTiO_3$	-80.9±2.3[a]
$SrTiO_3$	-135.1±2.2[a]
$BaTiO_3$	-152.3±4.0[a]
$PbTiO_3$	-31.1±4.1[a]
$CdTiO_3$	-22.3±2.4[a]
$CaZrO_3$	-31.3±4.0[a]
$SrZrO_3$	-75.9±4.5[a]
$BaZrO_3$	-123.9±4.1[a]
$PbZrO_3$	+1.7±6.6[a]
$CaGeO_3$	-33.4±3.3[b]
$CdGeO_3$	+22.6±4.3[c]
$MgSiO_3$	+12.2±2.0[d]

a. 1068 K, Takayama-Muromachi and Navrotsky (1988)
b. 973 K, Ross et al. (1986)
c. 973 K, Akaogi et al. (1987)
d. estimated, referred to SiO_2 stishovite

the repulsive interaction is the largest (positive) contribution, but terms related to Van der Waals energy, vibrational energy, and specific directional (covalent) interactions may contribute. Then, the heat of formation of a perovskite compound can be divided into two parts,

$$\Delta H_f \simeq \Delta E = \Delta E_M + \Delta E_N \qquad (4)$$

where ΔE_M is the difference of Madelung energy between product and reactants in reaction (3), while ΔE_N is the difference of repulsion and other energy. If the ΔE_M term is calculated, one can obtain the ΔE_N term using experimental values of ΔH_f according to Eq. (4).

Table 2 shows the Madelung energies of perovskites and values of ΔE_M and ΔE_N calculated by Eqs. (3) and (4) above. The change in Madelung (electrostatic) energy, ΔE_M, is negative for all perovskites studied except $PbTiO_3$. ΔE_M spans a large range (-392 to 58 kJ). The change in nonelectrostatic energy, ΔE_N, for perovskite formation is positive except for $BaTiO_3$ and $PbTiO_3$. ΔE_N also spans a large range (-125 to 395 kJ) for the compounds studied. The generally positive values of ΔE_N suggest that repulsions increase on forming perovskites from binary oxides. If one considers the perovskite structure ($t < 1$) to consist of an octahedral sublattice with tetravalent ions and oxygens in contact, and divalent ions filling the large central site, then the lattice constant for the cubic perovskite is related to the ionic radii:

$$a_{calc} = 2(r_B + r_O) \qquad (5)$$

The calculated lattice constant from this equation is 4.01 Å for titanates, 4.24 Å for zirconates, 3.86 Å for germanates, and 3.60 Å for silicates. The observed lattice constant, a, is usually smaller than a_{calc}. It seems that in the perovskite structure, the octahedral sublattice is somewhat compressed relative to that predicted by "ideal" M^{4+}-O distances calculated from ionic radii. Of course much of this apparent compression is accounted for by tilting of the octahedra in noncubic perovskites rather than by shorter B-O bond lengths. Figure 1 shows both ΔE_M and ΔE_N as functions of Δa, where Δa is the difference between the sum of B^{4+} and O^{2-} radii and the observed cubic lattice constant. ΔE_M for perovskite formation becomes more negative (more favorable) as Δa increases, while ΔE_N becomes more positive (destabilizing). The systematic trend seen in ΔE_N suggests that changes in repulsive energies are indeed the dominant contribution to ΔE_N. The energy of formation of a perovskite from the binary oxides results from a balance of ΔE_M and ΔE_N terms, which depend in opposite fashions on Δa (see Fig. 1).

Fig. 2 shows ΔE_M and ΔE_N plotted against the tolerance factor, t, defined by Eq. (2). Coordination numbers were chosen to be consistent with observed coordination numbers in perovskites, namely 6 for oxygen and the

TABLE 2. The Madelung and Non-Coulombic (repulsion) energy for Perovskite Formation (kJ/mol)

Compound	E_M[a]	ΔE_M[b]	ΔE_N[c]
$CaTiO_3$	-17986[d]	-322	241
$SrTiO_3$	-17615	-225	90
$BaTiO_3$	-17160	-27	-125
$PbTiO_3$	-17275	58	-89
$CdTiO_3$	-18071	-308	286
$CaZrO_3$	-17156	-392	361
$SrZrO_3$	-16771	-281	205
$BaZrO_3$	-16406	-173	49
$PbZrO_3$	-16591	-158	160
$CaGeO_3$	-18478	-237	204
$CdGeO_3$	-18567	-227	250
$MgSiO_3$	-20017	-364	395
MgO	-4612		
CaO	-4038		
SrO	-3674		
BaO	-3507		
CdO	-4137		
PbO	-3707		
TiO_2	-13626		
ZrO_2	-12726		
GeO_2	-14203		
SiO_2	-15041		

a Madelung energy calculated using formal ionic charges
b Madelung energy change for reaction $AO + BO_2 = ABO_3$
c Non-Coulombic contribution to enthalpy of formation, $\Delta E_N = \Delta H_f - \Delta E_M$
d from Takayama-Muromachi and Navrotsky (1988)

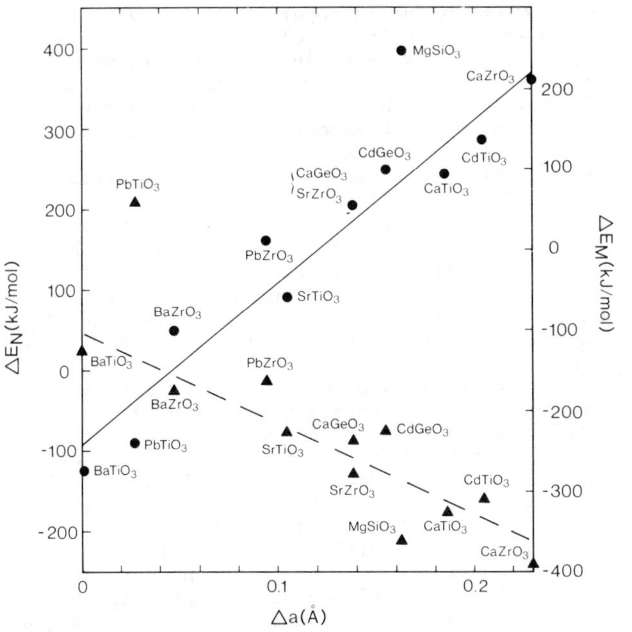

Fig. 1. The triangles, dashed line, and right hand scale show the relation between the change in Madelung energy (ΔE_M) upon perovskite formation from the oxides and the difference between calculated and observed lattice parameters, Δa. The circles, solid line, and left hand scale show the relation between the change in non-electrostatic (largely repulsive) energy and Δa. From Takayama-Muromachi and Navrotsky (1988).

tetravalent ions, 8 for the smaller divalent ions (Mg, Cd, Ca) and 12 for the larger divalent ions (Sr, Pb, Ba). Using this convention, values of t are slightly greater than unity for $SrTiO_3$, $BaTiO_3$ and $PbTiO_3$. Both ΔE_M and ΔE_N correlate with t, the electrostatic energy change becoming less favorable and the repulsive energy change more favorable with increasing t. Qualitatively, the same balance of electrostatic and repulsive energies is seen whether one uses Δa or t as a measure of bond length mismatch in the perovskite structure.

The high pressure germanate and silicate perovskites fall on roughly the same general trends as the titanates and zirconates, though for $MgSiO_3$ the change in nonelectrostatic energy appears some 50 kJ more destabilizing than the trend would predict. However, the heat of formation of that compound is an estimated value which still has large uncertainties.

Reznitskii (1978) pointed out that the entropy of formation of perovskites from the binary oxides is related to the tolerance factor, perovskites with $t < 0.95$ having positive ΔS values, those with $t > 0.95$ having ΔS zero or somewhat negative. This is consistent with positive ΔS^o values for perovskite-forming phase transitions (e.g. ilmenite → perovskite) for phases such as $MgSiO_3$, $CaGeO_3$, $CdGeO_3$ and $CdTiO_3$ (see below). However, phases such as $CaTiO_3$ and $BaTiO_3$ do not have particularly high entropies and the recently discovered $MnTiO_3$ perovskite (Ko and Ross, pers. comm.) also need not necessarily have a positive entropy of formation from lower pressure structures (ilmenite and lithium niobate).

The perovskite structure is not restricted to oxides of the $A^{2+}B^{4+}O_3$ charge type. Oxide perovskites of the type $A^{3+}B^{3+}O_3$ and fluoride perovskites of the type $A^{+}B^{2+}F_3$ form extensive families, and many other complex perovskites are known. The presently available thermochemical data are insufficient to test whether systematics similar to those observed here apply to the other charge types. The perovskite-forming reaction is different for other charge-types because some of the binary oxides and fluorides have other structures. Additional systematic thermochemical studies are needed.

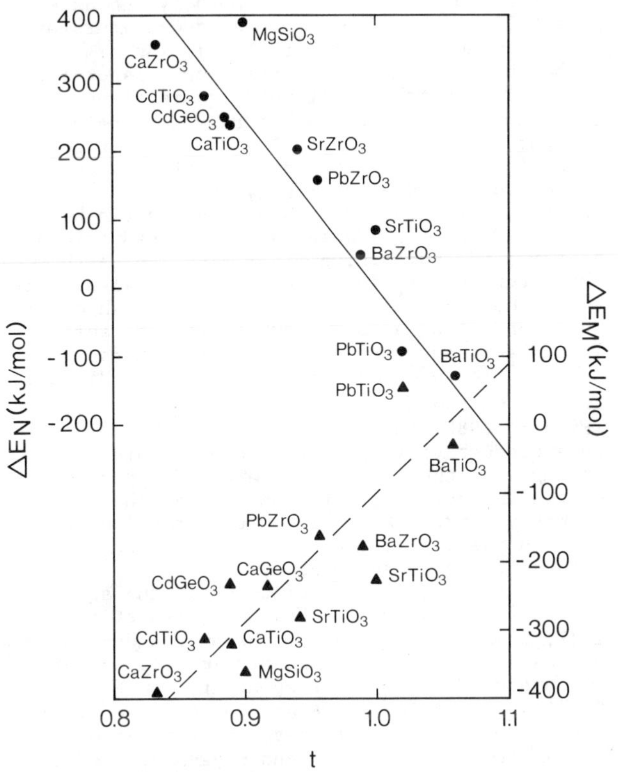

Fig. 2. The triangles, dashed line, and right hand scale show the relation between the change in Madelung energy (ΔE_M) upon perovskite formation from the oxides and the tolerance factor, t. The circles, solid line and left-hand scale show the relation between the change in non-electrostatic (largely repulsive) energy and t. From Takayama-Muromachi and Navrotsky (1988).

Oxide Superconductors

Since late 1986, the crescendo of excitement and blizzard of publications in high T_c oxide superconductors have placed solid state chemistry in a new perspective. To date, three families of superconducting compounds, all containing perovskite-related structural elements, have been found. The first is a series of doped materials based on the K_2NiF_4 structure (Michel and Reveau, 1984; Bednorz and Muller, 1986; Hazen, 1988). This structure contains layers of the perovskite structure alternating with rocksalt-like layers (see Fig. 3), leading to retention of octahedral (perovskite-like framework) coordination for the smaller cation but to a coordination number of 9 for the larger cations. Starting with the composition La_2CuO_4, doping with a large alkaline earth results in a series of compounds $La_{2-x}Sr_xCuO_{4-t}$. The substitution of Sr^{2+} for La^{3+} is charge- compensated by the conversion of Cu^{2+} to Cu^{3+}, at least in terms of formal oxidation state. However, loss of some oxygen converts part of the Cu^{3+} back to Cu^{2+}. The amount of Cu^{3+} is controlled both by the extent of alkaline earth doping and by temperature and oxygen fugacity (Gallagher, 1987). An orthorhombic-tetragonal transition, distortion of the CuO_6 octahedra (toward square planar coordination), copper polyhedra with missing oxygens and possible ordering of these oxygen vacancies are structural complications which affect both crystal chemistry and superconductivity. Many other series (in both these and the 1:2:3 materials discussed below) of compounds containing other alkaline earth substitutions or almost any other trivalent rare earth show similar structures, many of which are superconducting at some compositions (Fisher, 1988; Tarascon et al., 1987a). In contrast, substitution of other divalent cations (Ni, Co, Zn) for Cu rapidly destroys superconductivity (Fisher, 1988; Tarascon et al., 1987b). Thus superconductivity appears to depend more strongly on the oxidation state of copper, on oxygen vacancy concentration, and on the geometry and linkages of the copper polyhedra than on overall lattice dimensions or on the nature of the large cations.

The second group of compounds is the so called 1:2:3 materials, $YBa_2Cu_3O_x$ (Wu et al., 1987; Roth et al., 1987). The oxygen content x varies between limits of approximately six and seven (see Fig. 4), corresponding, nominally, to $2Cu^{2+}$ and $1Cu^{+}$ for x = 6 and to $2Cu^{2+}$ and $1Cu^{3+}$ for x = 7. The structure again is related to perovskite (see Fig. 3), and oxygen content is controlled by temperature and oxygen fugacity. Systematic vacancy formation appears essential to the structure and defines two sets of oxygen sites, one which contains mainly or entirely Cu^{2+} and one which is quite variable in oxygen coordination and seems to accomodate the changing average copper oxidation state. For superconductivity, a high oxygen content and an orthorhombic structure appear essential. The role of separated copper-containing planes and of Cu-O-Cu chains in superconductivity has been proposed, but it is not clear what structural features are absolutely necessary. To what extent is superconductivity governed by gradual changes in the number of electrons and the band structure; to what extent is it dictated by specific local crystal chemical factors? Indeed, it is not clear whether it is useful to think of Cu^{2+}, Cu^{3+}, and Cu^{+} as distinct entities, to think of a dynamic equilibrium between these ions and itinerant electron holes, or to consider most of the electron transfer to occur not at the cations but in the oxygen "p" band. A common "chemist's" and "physicist's" language for these complex structures has yet to emerge.

The third class of materials, having the highest critical temperatures (>120 K) of any discovered thus far, are complex oxides of copper with alkaline earths (Ca, Ba) and large ions other than rare earths, namely Bi and Tl (Fisher, 1988; Hazen, 1988). They are layer structures (see Fig. 3) analogous to previously known phases (Aurivillius, 1949), with superconductivity related to the number of copper planes sandwiched between perovskite-like layers (see Fig. 3). Interestingly, they do not contain copper-oxygen chains. These materials can also be described in terms of perovskite-related building blocks.

Rather little is known yet about the thermodynamic properties of these new materials. Thermodynamic stability involves at least three questions - namely stability with respect to other phases in the multicomponent (typically 3-5 component) oxide systems to which these phases belong, stability with respect to reactions with water and carbon dioxide (a major problem for Ba-rich phases) and stability with respect to oxygen content within a homogeneous solid solution. The first two questions have received some attention in the Y-Ba-Cu-O system. Studies relating oxygen fugacity, temperature, and oxygen content, x, in $YBa_2Cu_3O_x$, have been reported, and the tetragonal-orthorhombic transition has been mapped as a function of T, x and fO_2 (Gallagher, 1987). It is clear that the highest values of x (and highest T_c) are achieved by low-temperature anneals (see Fig. 5), but it is not certain whether these oxygen-rich phases are stable or metastable with respect to decomposition to other phase assemblages at 300-500 ^{o}C. Heats of formation and of some chemical reactions involving $YBa_2Cu_3O_x$ and La_2CuO_4 have recently been measured by acid solution calorimetry (Morss et al., 1988) (see Table 3).

Calorimetric studies have recently been completed on the enthalpy of oxidation as a function of composition, x, in the system $YBa_2Cu_3O_x$ ($5.9 \le x \le 7$) (Lampert, 1988; Lampert and Navrotsky, in preparation). Three sets of samples, prepared and analyzed in different laboratories, were studied. The calorimetric experiments consisted of the following. Each composition, encapsulated but not sealed in Pt, was dropped from room temperature into an empty Pt crucible in a Calvet-type microcalorimeter (Navrotsky, 1977) operating in air at 784 ^{o}C.

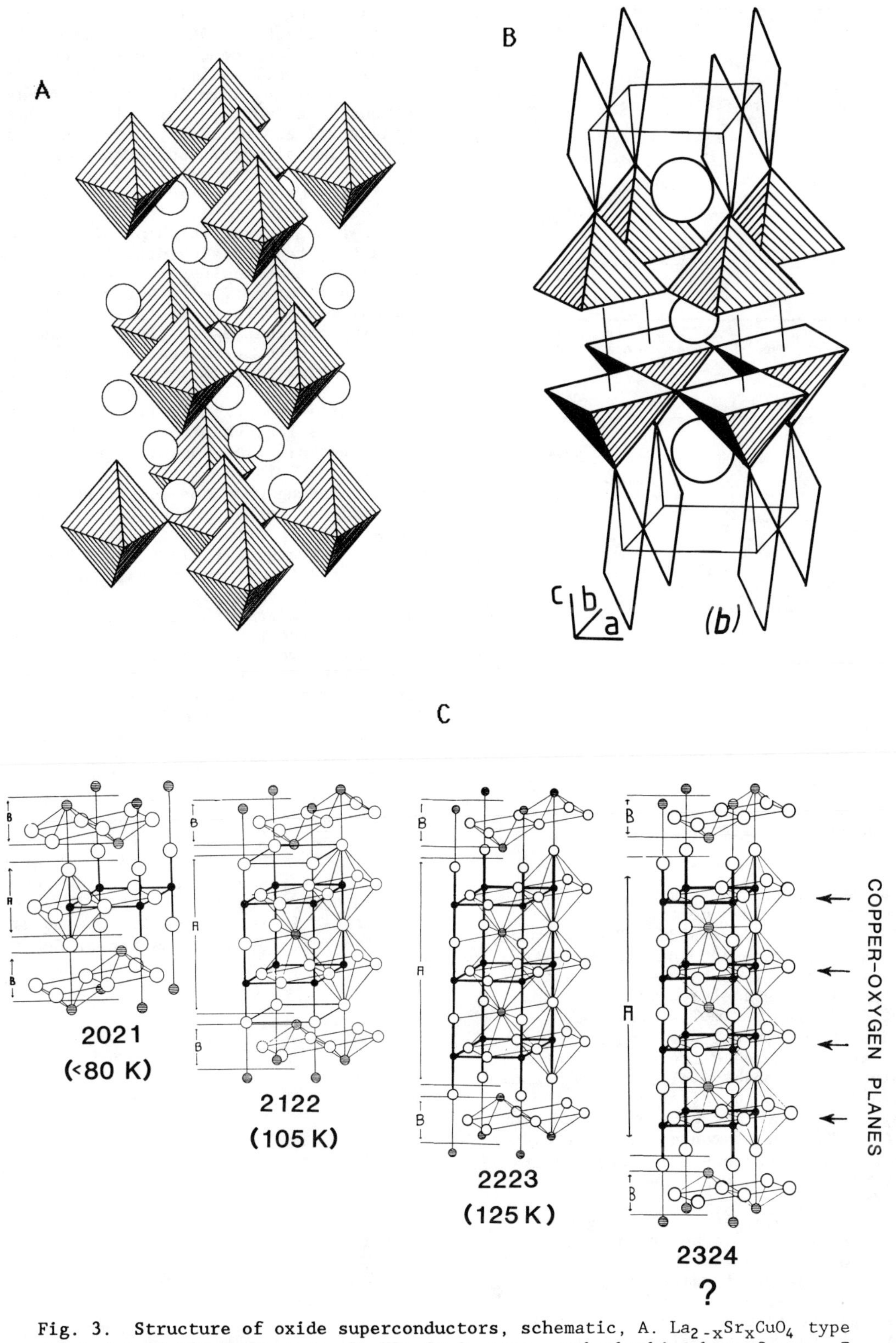

Fig. 3. Structure of oxide superconductors, schematic, A. $La_{2-x}Sr_xCuO_4$ type related to K_2NiF_4 structure, B. $YBa_2Cu_3O_x$ type, orthorhombic phase for x = 7, C. Layer types found in Bi and Tl phases related to Aurivillius phase. From Hazen (1988 and pers. comm.).

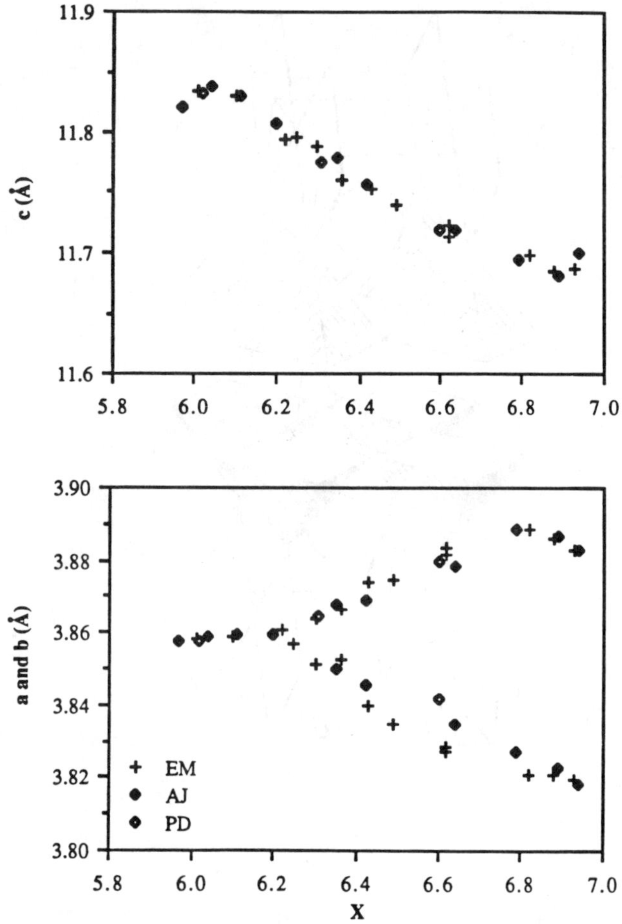

Fig. 4. Lattice parameters versus composition for $YBa_2Cu_3O_x$ phases. From Lampert (1988).

Under those conditions, the samples rapidly equilibrated to the oxygen content characteristic of 784 ^oC, x=6.36. Thus samples with x < 6.36 were oxidized, while those with x > 6.36 were reduced. The measured heat effects (see Fig. 6) were then analyzed to obtain the enthalpy of oxidation per mole $YBa_2Cu_3O_x$.

The figure shows several striking features. First, the data for three different sets of samples coincide reasonably well, although there may be a small systematic offset between those prepared by fast quench and by slow cooling in sealed tubes. This might represent a difference in structural state (vacancy ordering) but it more likely represents a small systematic difference (of x = 0.02 to 0.03) in oxygen analysis. On the whole, it is the agreement among different samples, rather than their differences, which is noteworthy.

Second, the enthalpy of oxidation is best represented by a straight line. The slope of that line gives the partial molar enthalpy of oxygen (per mole O) reacting with the structure. Twice the slope gives -241 ± 5 kJ/mol as the partial molar enthalpy of O_2, in good agreement with values and -200 to -221 kJ/mol (Gallagher, 1987) estimated from the temperature dependence of oxygen content and of -190 ± 20 kJ/mol estimated from calorimetric cycles involving solution of several samples in aqueous acid (Morss et al., 1988). Because the present values are measured by a more direct thermochemical cycle and because they cover a large number of compositions, these data can be considered more detailed and accurate than previous work.

The linearity of enthalpy with composition means that the enthalpy of oxidation, and the partial molar enthalpy of oxygen, is constant throughout the entire rather wide range of stoichiometry. There are neither steps in the enthalpy nor changes in slope at the orthorhombic-to-tetragonal transition which, for these samples, see lattice parameters in Fig. 4, is near x = 6.3. Thus, if the orthorhombic-tetragonal transition is first order, its enthalpy is less than the resolution of our measurements, namely about ±4 kJ per mole of $YBa_2Cu_3O_x$. This small enthalpy of transition is not surprising when one considers the small magnitude of enthalpies of distortional transitions in other perovskite-related components, e.g. 0.1 kJ/mol in $BaTiO_3$ (Gallagher et al., 1987). If the transition from orthorhombic to tetragonal were higher-order, rather than a step in the enthalpy of oxidation, one might expect a change in slope in the enthalpy curve. Once more, to a resolution of about ±4 kJ/mol, no change is seen, and the partial molar enthalpy of oxygen dissolving in the two phases is essentially the same.

Equally striking is the lack of any discontinuity or change in slope near x = 6.5, the composition at which all the copper is nominally divalent. For x < 6.5, oxidation represents, formally, the reaction $Cu^+ = Cu^{2+}$, for x > 6.5, oxidation represents the reaction $Cu^{2+} = Cu^{3+}$. The calorimetric data strongly suggest that the enthalpy of oxidation is the same on both sides of x = 6.5. This could be interpreted in several ways. One possibility is that the enthalpy of oxidation of Cu^+ to Cu^{2+} is

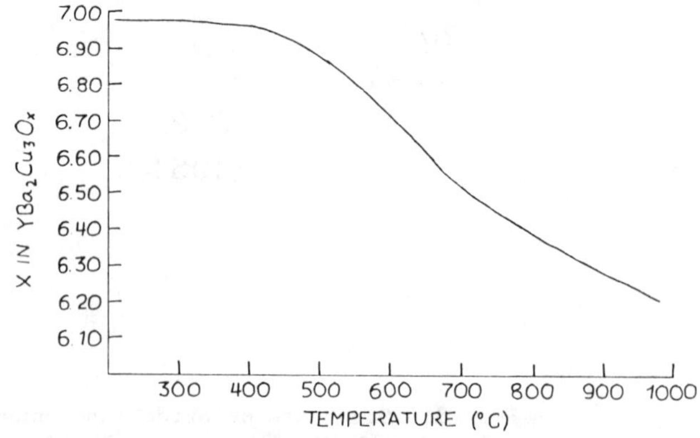

Fig. 5. Temperature dependence of x in $YBa_2Cu_3O_x$ in air. From Gallagher (1987).

TABLE 3. Thermochemical Data for Oxide Superconductors and Related Phases

Reaction	ΔH^o_{298} (kJ/mol)	ΔG^o_{298} (kJ/mol)
Formation Reactions		
$La_2O_3 + CuO = La_2CuO_4$	-28[a]	
$\frac{1}{2}Y_2O_3 + 2BaO + 3CuO = YBa_2Cu_3O_{6.5}$	-143[a]	
$Y + 2Ba + 3Cu + 3.465O_2 = YBa_2Cu_3O_{6.93}$	-2713±17[a]	
Decomposition Reactions		
$La_2CuO_4 + (x+0.5)H_2O + 3.5CO_2 = La_2(CO_3)_3 \cdot xH_2O + 0.5CuCO_3 \cdot Cu(OH)_2$		-1870[a]
$YBa_2Cu_3O_{6.93} + 1.5H_2O + 5CO_2 = 0.215O_2 + 0.5Y_2(CO_3)_3 + 2BaCO_3 + 1.5CuCO_3 \cdot Cu(OH)_2$		-355[a]
Partial Molar Enthalpy of Oxygen		
O_2(gas) = O_2(dissolved in $YBa_2Cu_3O_x$)	-190±20[a] -241±5[b] -200 to -221[c]	

a. Morss et al. (1988), acid solution calorimetry
b. Present work, oxidation calorimetry
c. Gallagher (1987) from temperature dependence of oxygen content

identical to that of Cu^{2+} to Cu^{3+}, but that the oxidation occurs stepwise with Cu^{2+} the only species present at x = 6.5. This seems an unlikely coincidence, especially since Cu^{+} and Cu^{2+} are common oxidation states while Cu^{3+} occurs only under special conditions in other oxides. A second possibility is that the disproportionation reaction

$$2Cu^{2+} = Cu^{+} + Cu^{3+} \qquad (6)$$

occurs significantly even at the stoichiometric composition (x = 6.5). Thus that composition does not represent a unique state with only one valence state but is simply one point in a gradual progression where, with increasing x, the amount of Cu^{+} decreases and the amount of Cu^{3+} increases. A defect model has been proposed by Su et al. (1988) which gives an equilibrium constant for reaction (6) K_d = ~0.40 (independent of T,x), and for

$$Cu^{+} + \tfrac{1}{2}O_2(gas) = O_o^{=} + Cu^{3+}, \quad \ln K_{ox} = -10.08 + 10885/T \text{ (independent of x)} \qquad (7)$$

Therefore, for reaction (7), ΔG^o = -90498 + 83.81 T (J/mol) and ΔH^o = -90500 J/mol, ΔS^o = -83.8 J/mol·K.

The oxidation can also be represented by the reaction:

$$YBa_2Cu_2^{2+}Cu^{+}O_6 + \tfrac{1}{2}O_2 = YBa_2Cu_2^{2+}Cu^{3+}O_7 \qquad (8)$$

From the calorimetric measurements, ΔH^o for reaction (8) is -120 kJ/mol, in reasonable agreement with the -90.5 kJ above. The entropy of oxidation, -83.8 J/mol·K per mole O, is in the range commonly seen for many oxidation reactions, -92±17 J/mol·K as discussed by Navrotsky, (1974).

Can one reconcile an equilibrium between Cu^{+}, Cu^{2+}, and Cu^{3+} and the observed linear variation of enthalpy with composition? Because of the small temperature dependence of K_d seen by Su et al. (1988), and the relatively small range of temperatures over which it was studied (650-850 K), one cannot reliably separate enthalpy and entropy terms for the disproportionation reaction. Were K_d truly temperature-independent, then reaction (6) would be

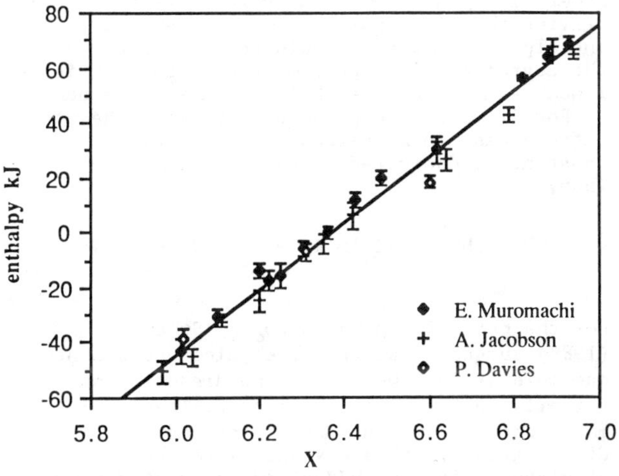

Fig. 6. Enthalpy of oxidation or reduction for reaction $YBa_2Cu_3O_x + (\frac{x-6.36}{2})O_2 = YBaCu_3O_{6.36}$ at 298 K as measured by high temperature calorimetry. From Lampert (1988). Muromachi, Jacobson, and Davies refer to sources of samples.

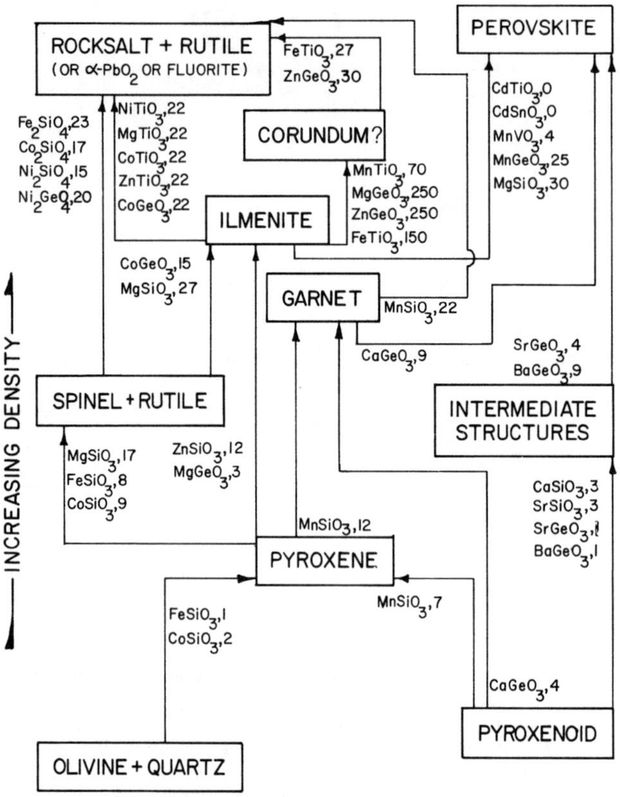

Fig. 7. Phase transitions in compounds of ABO_3 stoichiometry. Numbers indicate approximate transition pressures (in GPa) at 1273 K. From Navrotsky (1981, 1987).

athermal, with $\Delta H = 0$ and entropy effects alone driving the disproportionation. Then the enthalpy of oxidation would not be influenced by the degree of disproportionation, and a strictly linear variation of enthalpy with x would occur.

For illustrative purposes, consider the other extreme case, namely that the equilibrium constant is dominated by the enthalpy term. Then,

$$\Delta G^o \sim \Delta H^o = -RT \ln K_d = -750 \text{ R} \ln 0.40 = +5.7 \text{ kJ/mol} \quad (9)$$

For the composition $YBa_2Cu_3O_{6.5}$, Su et al. (1988) suggest that the site which would contain one mole of Cu^{2+} were there no dissociation actually contains 0.4 moles of Cu^{2+} and 0.3 moles each of Cu^+ and Cu^{3+}. The degree of dissociation is thus 60% and, were the enthalpy of dissociation 5.7 kJ/mol as calculated above, an enthalpy of 0.4 x 5.7 = 2.3 kJ/mol would be subtracted from the observed enthalpy of oxidation to account for the back-reaction to form Cu^{2+}. This would be at the level of resolution of the calorimetric data and would not produce definitely detectable curvature in the plot of enthalpy of oxidation versus x. At other values of x, this correction terms would be even smaller. Thus one can conclude that the defect model suggested by Su et al. (1988) is consistent with these calorimetric results.

Rather than distinguishable Cu^{2+}, Cu^+, and Cu^{3+} related by a chemical equilibrium, all cationic species may be equivalent with changes in oxidation state accomodated by the oxygen p-band. This would also result in a very gradual, and perhaps linear, variation of energetics with x, with no sharp changes at or near x = 0.5. Thus the calorimetric results can not be used to distinguish the defect structure proposed by Su et al. (1988) and Morss et al. (1988) from models containing extra electrons in the oxygen p-band. However, the calorimetric results show conclusively that the composition x = 6.5 is not energetically unique, and is not a titration equivalence point separating two regions of strikingly different defect chemistry.

The calorimetric results also show no change in slope of the enthalpy of oxidation curve near the limits of homogeneity of the phase, namely x = 5.9 and x = 7.0. This might suggest that these limits of stability arise, not from any intrinsic destabilization of the 1:2:3 phase by too high or too low oxygen content, but by a balance of free energy between that phase and other phases in the $BaO\text{-}Y_2O_3\text{-}Cu\text{-}O$ system.

High Pressure Phase Transitions Involving Perovskite: Systematics in Silicates and Germanates

Fig. 7 shows reported phase transitions and their pressures (in GPa) for phases of ABO_3 stoichiometry. A wealth of structures is seen: pyroxenoid, pyroxene, garnet, ilmenite, perovskite, as well as two phase mixtures (olivine plus quartz, spinel plus stishovite, or rocksalt plus stishovite). Clearly very closely balanced enthalpy, entropy and volume factors determine which phases occur at a given composition, pressure, and temperature. Table 4 lists available thermochemical data and Fig. 8 compares the enthalpies, relative to the phase stable at ambient conditions, of the various polymorphs. In the germanates, the wollastonite and garnet phases are very similar in energy; the garnet is a high pressure phase not because of significantly unfavorable energetics but because of its lower entropy and volume. The perovskite, when it occurs, appears to always be a phase of rather unfavorable energy in silicates and germanates.

Phase diagrams for $CaGeO_3$, $CdGeO_3$ and $MgSiO_3$ are shown in Fig. 9. Transitions to form perovskite phases (γ + st → pv, gar → pv, il → pv) have positive ΔS^o values and negative P-T slopes as initially pointed out by Navrotsky (1980). Qualitatively, such behavior can be rationalized as follows. In a phase such as $CaGeO_3$ or $MgSiO_3$ perovskite, the large central site is occupied by a relatively small cation which has rather large M-O distances, relatively weak bonding, and, one surmises, rather easy vibration involving it and neighboring oxygens. In addition, the octahedral Si-O or Ge-O bond is longer and has a lower vibrational frequency than the strong covalent tetrahedral Si-O or

TABLE 4. Thermochemical Parameters for Phase Transitions in Silicates and Germanates of ABO_3 Stoichiometry (at 298 K, 1 atm)

	$\Delta H^o (J \cdot mol^{-1})$	$\Delta S (J \cdot mol^{-1} \cdot K^{-1})$	$\Delta V^o (cm^3 \cdot mol^{-1})$
pyroxene or pyroxenoid → garnet			
$CaGeO_3$ [a]	-4900 ± 4200	-5.9 ± 1.5	-5.97
$CdGeO_3$ [a]	+500 ± 2700	-8.4 ± 2.0	-5.30
$MnSiO_3$ [a]	+34600 ± 2500	-6.7 ± 2.1	-4.00
$MgSiO_3$ [a]	+35660 ± 3000	-1.9 ± 2.0	-2.83
garnet → perovskite			
$CaGeO_3$ [a]	+43300 ± 5000	+10.9 ± 3.8	-5.35
$CdGeO_3$ [a]	+43100 ± 5000	-1.7 ± 3.0	-4.88
$MgSiO_3$ [a]	+63960 ± 5000	-3.9 ± 2.0	-4.00
pyroxene → ilmenite			
$MgGeO_3$ [a,b]	7500 ± 600	-6.3 ± 3.0	-5.11
$MgSiO_3$ [c]	55370 ± 5000	-9.1 ± 2.0	-4.98
ilmenite → perovskite			
$CdTiO_3$ [a]	15000 ± 800	+14.2 ± 2.0	-2.94
$CdGeO_3$ [a]	34300 ± 4000	+2.6 ± 2.0	-3.00
$MgSiO_3$ [a,c]	50000 ± 2000	+3.3 ± 3.0	-1.91

a. Navrotsky (1987)
b. Ross and Navrotsky (1988)
c. Ashida et al. (1988)

Ge-O bond. Indeed while the tetrahedral Si-O or Ge-O stretching vibrations can be identified relatively unambiguously in the vibrational spectrum, those of octahedrally coordinated Si or Ge occur amidst many other lattice vibrations and probably are not simple isolated modes. Nor can M-O vibrations be unambiguously identified in the vibrational spectra. The factors above lead to greater excitation of vibrations at lower temperatures and a higher entropy for the perovskite. To illustrate these structural changes, bond lengths for cation-oxygen bonds in some low and high pressure phases are summarized in Table 5.

To quantify such reasoning, one needs either complete knowledge of the vibrational density of states (generally unavailable for high pressure phases) or models which approximate this density of states in detail sufficient for thermodynamic calculations. Some lattice dynamical calculations in $MgSiO_3$ perovskite are discussed by Hemley et al. (this volume). In addition, the approach pioneered by Kieffer (1979, 1980) offers a useful middle ground between theoretical rigor and practical use of available data. In it, the vibrational density of states is considered to consist of three contributions (see Fig. 10). Three acoustic modes, assumed to be Debye-like with cutoff frequencies calculated from elastic data, contribute at low frequency (usually below 150 cm^{-1}). The remaining 3n-3 vibrational modes for a primitive unit cell containing n atoms are optic modes. Some of them can be assigned fairly unambiguously to specific vibrations (tetrahedral Si-O stretching modes, O-H stretching modes in hydrous phases) generally seen at high frequency (> ~900 cm^{-1}) in infrared and Raman spectra. These are treated as "Einstein oscillators", spikes in the vibrational density of states, each at a single frequency. The number of modes assigned to each Einstein oscillator is inferred from the crystal

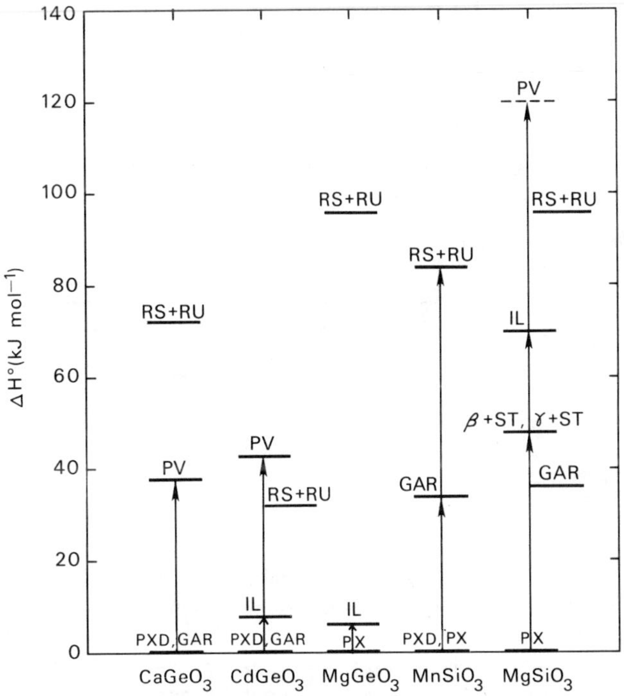

Fig. 8. Enthalpies (kJ/mol) of ABO_3 polymorphs relative to enthalpy of phase stable at atmospheric pressure. From Navrotsky (1987).

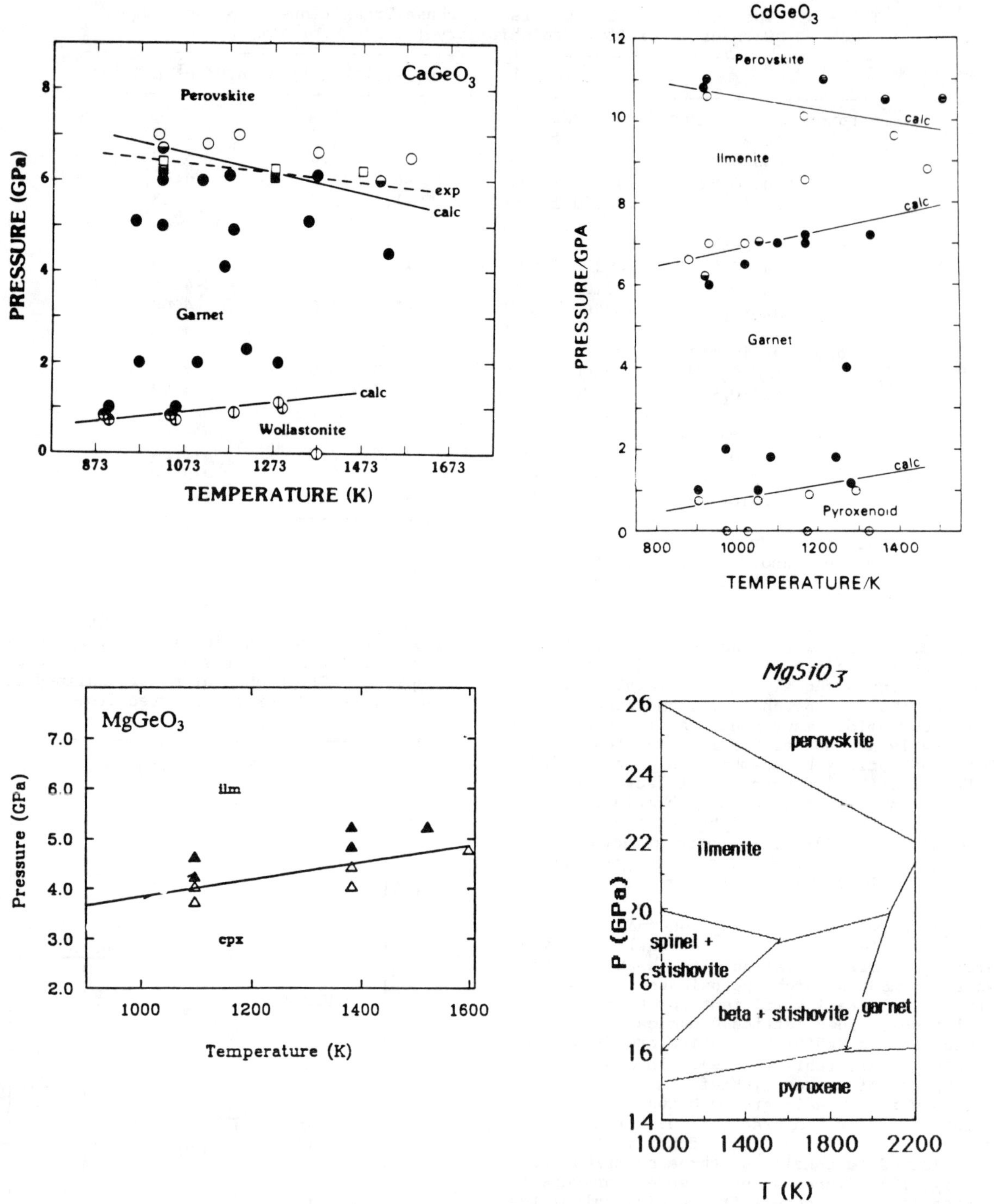

Fig. 9. Phase relations in $CaGeO_3$, $CdGeO_3$, $MgGeO_3$ and $MgSiO_3$. From Ross et al. (1986), Akaogi and Navrotsky (1987), and Sawamoto (1987).

structure. The remaining modes (in the mid-infrared) generally cannot be assigned readily; they correspond to various stretches, bends, and lattice deformations. Kieffer's model treats them as an "optic continuum", a constant density of states between a low and high frequency cutoff. These cutoffs are inferred from the infrared and Raman spectra of

TABLE 5. Bond Lengths in Low and High Pressure Phases Involving Changes in Cation Coordination

	A^{2+}-O (Å)[a]			B^{4+}-O (Å)	
	VI	VIII	XII	IV	VI
$MgSiO_3$					
pyroxene	2.15	—	—	1.64	—
ilmenite	2.08	—	—	—	1.80
perovskite	—	2.20	(2.47)	—	1.79
$CaGeO_3$					
garnet	2.29	2.47	—	1.78	1.92
perovskite	—	—	2.65	—	1.89

[a] A^{2+}-O and B^{4+}-O are average lengths of divalent (Mg,Ca) metal-oxygen and tetravalent (Si,Ge) metal oxygen bonds, Roman numerals refer to coordination numbers.

each phase. Because such spectra give information on the vibrational density of states only at the center of the Brillouin zone, some assumptions about dispersion must also be made. The approximate vibrational density of states can then be integrated to yield thermodynamic functions. A comparison of S^o for different structures gives the entropy of transition.

This approach is particularly useful for high pressure phases. The input data needed to constrain the model are the space group, molar volume, elastic constants, bulk modulus, thermal expansion, and vibrational (IR and Raman) spectra. All these data are obtainable for ultrahigh pressure phases and indeed are of interest to geophysics for other reasons. The vibrational models work best when comparing phases with quite different crystal structures and vibrational spectra, such as the $MgSiO_3$ polymorphs (pyroxene, garnet, ilmenite, and perovskite). One model for each phase is shown in Table 6. Each model is representative of a family of models, with small variations in parameters for each phase, all of which give consistent thermodynamic parameters for that phase.

The model for $MgSiO_3$ (opx) is similar to that used by Kieffer (1979) except that the high frequency modes are described as a second optic continuum rather than as several Einstein oscillators. This makes almost no difference to the results. The entropy and heat capacity are in good agreement with measured values.

The model for $MgSiO_3$ garnet is constrained by the following. Acoustic velocities and thermal expansion are assumed the same as for pyrope. The bulk modulus is from Akaogi et al. (1987). An infrared spectrum of tetragonal garnet is reported by Kato and Kumazawa (1985). It is broadly similar to that of pyrope, and the model is constructed to be consistent with that spectrum and with models constructed for pyrope (Kieffer, 1980). The number of modes at frequencies above 650 cm^{-1} is smaller for garnet than for pyroxene, and the total number below 650 cm^{-1} is correspondingly greater. This reflects the fact that in $MgSiO_3$ garnet only half the silicon is in tetrahedral coordination, while the other half is octahedrally coordinated. The low frequency cutoffs for the optic continua appear to be similar in pyroxene and garnet. The result of these differences in the vibrational density of states, especially of the greater density of states in the lowest optic continuum, is that the calculated entropy of garnet is somewhat greater than would be expected only on the basis of its increased density. In fact garnet and pyroxene are predicted to have very similar entropies and heat capacities. The entropy of the pyroxene-garnet transition is calculated to be $\Delta S^o_{1000} = -2.4\ J \cdot K^{-1} \cdot mol^{-1}$. This supports the virtually horizontal P-T slope reported by Kato and Kumazawa (1985).

Vibrational models of ilmenite were discussed by McMillan and Ross (1987). One of these is shown in Table 6. Its two striking features are that the low frequency cutoff is at substantially higher frequency than for the other $MgSiO_3$ polymorphs and that there are no high frequency modes (>900 cm^{-1}). The reason for the former may be related to the regularity edge-sharing, and strong interaction of the MgO_6 octahedra and for the latter is the absence of tetrahedral silicon. The result is a compression of the vibrational spectrum into a much smaller frequency range. The absence of low frequency modes causes ilmenite to have the lowest entropy of all the $MgSiO_3$ polymorphs. The calculated heat capacities are in excellent agreement with those measured by Watanabe (1982) and Ashida et al. (1988). The calculated value of S^o_{298} is 51.7 $J \cdot K^{-1} \cdot mol^{-1}$. The values of ΔS^o for px = il calculated from the vibrational model are in reasonable agreement with those inferred from high pressure phase relations (Ashida et al., 1988).

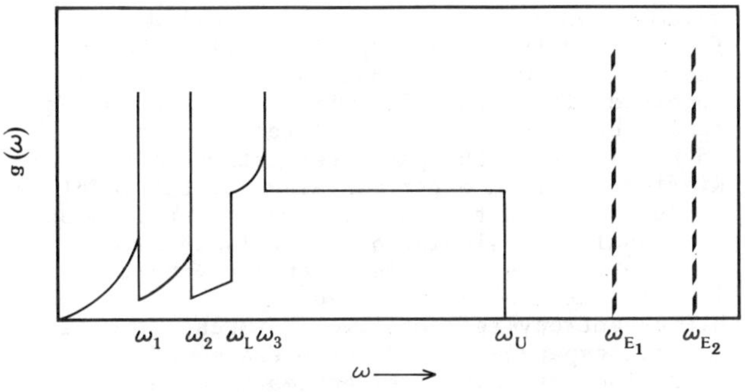

Fig. 10. Schematic representation of vibrational density of states (VDOS) in the Kieffer model (1979, 1980). ν_1, ν_2, ν_3 are acoustic modes. Optic continuum spans ν_ℓ to ν_u. Einstein oscillators are at ν_{E1}, ν_{E2}. These can be replaced by a second optic continuum. Total number of modes, $\int g(\nu)d\nu$, is 3n. In this sketch, acoustic modes overlap optic continuum; this is not always the case.

TABLE 6. Vibrational Calculations for $MgSiO_3$ Phases

	Pyroxene	Garnet	Ilmenite	Perovskite 1 atm	Perovskite 10 GPa
Volume (cm^3/mol)	31.40	28.35	26.41	24.50	23.64
Bulk Modulus (GPa)	108	154	211	262	302
Thermal Expansivity x10^5 (K^{-1})	2.70	2.57	2.60	a	a
Acoustic Velocities (km/s)	4.67	3.58	5.30	5.75	5.75
	5.09	5.05	5.80	6.30	6.30
	8.07	8.98	9.80	10.40	10.40
Lower Optic continuum					
Lower cutoff (cm^{-1})	240	150	288	225	255
Upper cutoff (cm^{-1})	545	630	800	900	930
Fraction of modes	0.6775	0.7875	0.90	0.95	0.95
Upper Optic continua					
Lower cutoff (cm^{-1})	650 760	800	-	-	-
Upper cutoff (cm^{-1})	760 1045	1050	-	-	-
Fraction of modes	0.19 0.12	0.200	-	-	-
Entropy (J/K·mol)					
298 K	65.60	63.20	51.72	48.88	44.42
1000 K	193.43	190.99	176.95	180.92	175.28
1500 K	243.48	241.22	225.82	234.62	230.60

a $\alpha = (3+0.037T)x10^{-5}$ at 100-1000 K, $\alpha = 4x10^{-5}$ above 1000 K, to fit data in Fig. 12 but not diverge at higher T

The bulk modulus, elasticity, and thermal expansivity of perovskite now are known provisionally (Yagi et al., 1982; Kudoh et al., 1987; Weidner and Ito, 1987; Knittle et al., 1986; Ross and Hazen, 1988). Its heat capacity is not. Raman and infrared spectra have recently been reported by Williams et al. (1987) and Hemley et al. (this volume). A vibrational model is constructed consistent with those spectra. Its optic continuum extends from 225 to 900 cm^{-1}. The calculated spectra based on lattice dynamical models (Hemley et al., this volume) contain modes at higher and lower frequencies than seen experimentally. However, because agreement with observed modes is only moderate, it is not clear if these other modes really exist near the calculated frequencies. In keeping with the parameterization of the Kieffer model on experimental data, the model is chosen consistent with observed spectral bonds. The resulting calculations (see Table 6) confirm that perovskite is a higher entropy phase than ilmenite at T > ~800 K, (see Fig. 11). This higher entropy reflects two features: the large thermal expansivity and the presence of vibrations at lower frequencies than in ilmenite. The latter may be related to the large central Mg-site. Indeed, a thermal expansivity greater than about $3x10^{-5}$ at T > 500 K appears necessary to ensure a positive ΔS^o for ilmenite = perovskite in these model calculations. The vibrational calculations generally support the hypothesis, consistent with phase equilibria, that perovskite is a phase of relatively high entropy at high temperature and that perovskite-forming reactions have negative P-T slopes.

Williams et al. (1987) and Hemley et al. (this volume) report pressure shifts for Raman and infrared modes in $MgSiO_3$ perovskite. All observed modes shift upward with slopes of 1.7

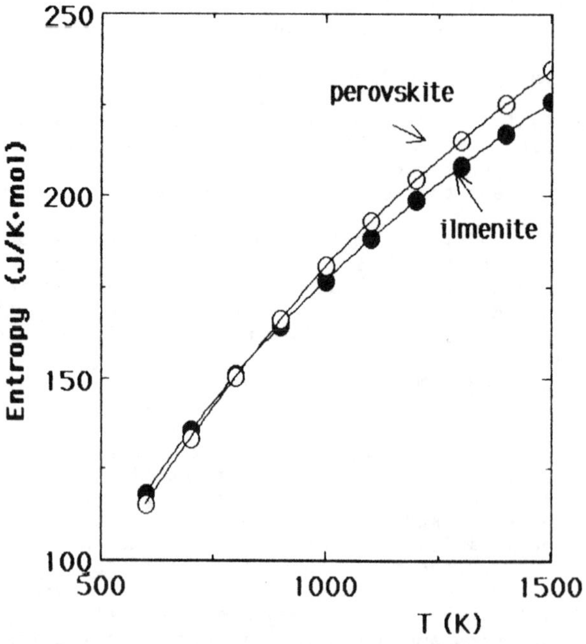

Fig. 11. Calculated entropies of $MgSiO_3$ pyroxene, garnet, ilmenite and perovskite from vibrational models in Table 6.

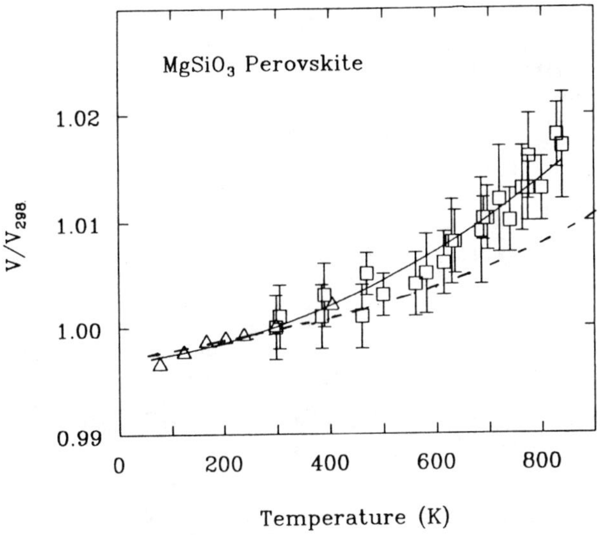

Fig. 12. Relative volume, V_T/V_{298}, of $MgSiO_3$ perovskite at various temperatures. Triangles represent data of Ross and Hazen (1988), squares data of Knittle et al. (1986). Solid curve is "best fit" by Ross and Hazen (1988), dashed curve is calculated from $(dS/dP)_T$ (present work).

to 4.2 cm^{-1}/GPa, with no indication of soft modes or phase transitions up to 2.5 GPa at room temperature. Within the spirit of the Kieffer model, one can calculate $(dS/dP)_T$ by performing the vibrational calculation using spectra, unit cell volume, and bulk modulus at different pressures. The simplest model consistent with the spectral shifts seen would move the whole optic continuum upward in frequency at a rate given by an average spectral shift of about 3 cm^{-1}/GPa. Parameters and results for the calculations at 1 atm and 10 GPa are shown in Table 6.

Since $(dS/dP)_T = -(dV/dT)_P$, the above calculation constrains the thermal expansivity, α. One should note that although the calculated value of C_P and of S^o depends on the value chosen for α, this dependence cancels out of $(dS/dP)_T$ and thus the calculated value of α does not depend on the value of α initially put into the heat capacity calculation. The results give $\alpha = 1.8 \times 10^{-5}$ K^{-1} at 300 K and 2.4×10^{-5} K^{-1} at 800 K. The results compare reasonably well to those measured (see Fig. 12) although the rather high thermal expansion (~4×10^{-5} K^{-1} at 500-800 K) suggested by Knittle et al. (1986) is not observed in this calculation. However the disagreement between calculated and observed V/V_0 at high temperature is barely outside the quoted experimental uncertainty. It is also noteworthy that the mode averaging scheme (a flat optic continuum) adopted in this application of the Kieffer approach gives thermal expansion coefficients quite comparable to these calculated by far more complex lattice vibrational models (see Ross and Hazen, 1988, for a summary).

Several further checks on this approach were performed. First, the location of the optic continuum was varied. Based on the earlier data of Williams et al. (1987), the optic continuum was placed between 250 and 800 cm^{-1} and its pressure shift at 2 cm^{-1} GPa at the low end and at 3 cm^{-1}/GPa at the high end. This gave $\alpha = 1.4 \times 10^{-5}$ at 300 K and 2.3×10^{-4} at 800 K. Second, the bulk modulus was varied by ±10%. This had little effect. Third, similar calculations were done for Mg_2SiO_4 olivine and β-phase, where the pressure dependence of some spectral bands has been measured. Once more, very reasonable values of thermal expansivity were obtained. Therefore one can conclude that the Kieffer model works quite adequately, not only to model C_P and S^o of silicate phases but, given appropriate spectral shifts, to model $S(T,P)$, $(dS/dP)_T$, and thermal expansion.

Acknowledgments. Much of the work reviewed here resulted from an ongoing collaboration between Navrotsky's group in the U.S. and a number of Japanese colleagues. I thank S. Akimoto, E. Ito, S. Kume, M. Akaogi, E. Takayama-Muromachi, T. Ashida and many other friends in Japan for their help. I also thank S. Kieffer, P. McMillan, N.L. Ross, and M.E. Lampert in the U.S.A. for their contributions. R. Hazen provided the drawings for Fig. 3. This work received financial support from the National Science Foundation (Grants DMR 8610816 and INT 8515337).

References

Akaogi, M., and A. Navrotsky, Calorimetric study of high pressure phase transitions among the $CdGeO_3$ polymorphs (pyroxenoid, garnet, ilmenite, and perovskite structures), Phys. Chem. Min., 14, 435-440, 1987.

Akaogi, M., A. Navrotsky, T. Yagi, and S. Akimoto, Pyroxene-garnet transition: thermochemistry and elasticity of garnet solid solutions, and application to mantle models, High Pressure Research in Mineral Physics, edited by M.H. Manghnani and Y. Syono, Terra Publications, Tokyo, Japan, 1987.

Ashida, T., S. Kume, E. Ito, and A. Navrotsky, $MgSiO_3$ ilmenite: heat capacity, thermal expansivity, and enthalpy of transformation, Phys. Chem. Min., in press, 1988.

Aurivillius, J.B., Mixed bismuth oxides with layer lattices 1. The structure type of $CaNb_2Bi_2O_9$, Arkiv. Kemi., 1, 463-512, 1949.

Bednorz, J.G., and K.A. Müller, Possible high T_c superconductivity in the Ba-La-Cu-O system, Z. Phys. B: Condens. Matter, 64, 189-193, 1986.

Fisher, G., Superconductivity mysteries unravel as developments proceed, Ceram. Bull., 67, 725-735, 1988.

Gallagher, P.K., Characterization of $Ba_2YCu_3O_x$ as a function of oxygen partial pressure, Part I: thermoanalytical measurements, Adv. Ceram. Mater., 2, 632-639, 1987.

Hazen, R.M., Perovskites, Scientific American, June, 74-81, 1988.

Kato, T., and M. Kumazawa, Garnet phase of $MgSiO_3$ filling the pyroxene-ilmenite gap at very high temperatures, Nature, 316, 803-804, 1985.

Kieffer, S.W., Thermodynamics and lattice vibrations of minerals: 3. Lattice dynamics and approximation for minerals with application to simple substances and framework silicates, Rev. Geophys. Space Phys., 17, 35-59, 1979.

Kieffer, S.W., Thermodynaimcs and lattice vibrations of minerals: 4. Application to chain and sheet silicates and orthosilicates, Rev. Geophys. Space Phys., 18, 862-886, 1980.

Knittle, E., R. Jeanloz, and G.L. Smith, The thermal expansion of silicate perovskite and stratification of the earth's mantle, Nature, 319, 214-216, 1986.

Kudoh, Y., E. Ito, and H. Takeda, Effect of pressure on the crystal structure of perovskite-type $MgSiO_3$, Phys. Chem. Min., 14, 350-354, 1987.

Lampert, M.E., Thermodynamic study of the $YBa_2Cu_3O_x$ system, Senior Thesis, Physics Dept., Princeton University, Princeton, NJ, 1988.

McMillan, P.F., and N.L. Ross, Heat capacity calculations for Al_2O_3 corundum and $MgSiO_3$ ilmenite, Phys. Chem. Min., 14, 225-234, 1987.

Michel, C., and B. Raveau, Oxygen intercalation in mixed valence oxides related to the perovskites, Rev. Chim. Min., 21, 407-425, 1984.

Morss, L., C. Sonnenberger, and P.J. Thorn, Thermochemistry of rare earth-alkaline earth-copper oxide superconductors, Inorg. Chem., in press, 1988.

Navrotsky, A., Thermodynamics of binary and ternary transition metal oxides in the solid state, MTP Reviews of Science, Inorganic Chemistry, Series 2, edited by D.W.A. Sharp, Butterworths-University Park Press, Baltimore, MD, 29-70, 1974.

Navrotsky, A., Recent progress and new directions in high temperature calorimetry, Phys. Chem. Min., 2, 89-104, 1977.

Navrotsky, A., Lower mantle phase transitions may generally have negative pressure-temperature slopes, Geophys. Res. Lett., 7, 709-711, 1980.

Navrotsky, A., High pressure transitions in silicates and related compounds, Prog. Solid State Chem., 17, 53-86, 1987.

Navrotsky, A., Energetics of phase transitions in AX, ABO_3, and AB_2O_4 compounds, Structure and Bonding in Crystals, Vol. II, edited by M. O'Keeffe and A. Navrotsky, Academic Press, New York, 71-93, 1981.

Reznitskii, L.A., Tolerance factor and entropy of perovskites (in Russian), Neorgan. Materialy, 14, 2127-2128, 1978.

Ross, N.L., and R.M. Hazen, Single crystal xray diffraction study of $MgSiO_3$ perovskite from 77 to 400 K. Phys. Chem. Min., submitted, 1988.

Ross, N.L., A. Akaogi, A. Navrotsky, and P. McMillan, Phase transitions among the $CaGeO_3$ polymorphs (wollastonite, garnet, and perovskite structures): studies by high pressure synthesis, high temperature calorimetry, and vibrational spectroscopy and calculation, Jour. Geophys. Res., 91, 4685-4698, 1986.

Ross, N.L., and A. Navrotsky, Study of the $MgGeO_3$ polymorphs (orthopyroxene, clinopyroxene, and ilmenite structures) by calorimetry, spectroscopy, and phase equilibria, Am. Min., submitted, 1988.

Roth, R.S., K.L. Davis, and J.R. Dennis, Phase equilibria and crystal chemistry in the system Ba-Y-Cu-O, Adv. Ceram. Mater., 2, 3B, 303-312, 1987.

Sawamoto, H., Phase diagram of $MgSiO_3$ at pressures up to 24 GPa and temperatures up to 2200°C. Phase stability and properties of tetragonal garnet, High Pressure Research in Mineral Physics, edited by M.H. Manghnani and Y. Syono, Terra Publishing Co., Tokyo, Japan, 209-220, 1987.

Su, H.-Y., S.E. Dorris, and T.O. Mason, Defect model and transport at high temperature in $YBa_2Cu_3O_{6+y}$, Jour. Solid State Chem., submitted, 1988.

Takayama-Muromachi, E., and A. Navrotsky, Energetics of compounds ($A^{3+}B^{4+}O_3$) with the perovskite structure, Jour. Solid State Chem., 72, 244-256, 1988.

Tarascon, J.M., L.H. Greene, W.R. McKinnon, and G.W. Hall, Superconductivity in rare earth-doped oxygen-defect perovskites $La_{2-x-y}Ln_ySr_xCuO_{4-z}$, Solid State Comm., 63, 6, 499-505, 1987a.

Tarascon, J.M., L.H. Greene, W.R. McKinnon, G.W. Hull, T.P. Orlando, K.A. Delin, S. Foner, and E.J. McNiff, Jr., 3d-metal doping of the high-temperature superconducting perovskites La-Sr-Cu-O and Y-Ba-Cu-O, Phys. Rev. B: Condens. Matter, 36, 16, 1987b.

Watanabe, H., Thermochemical properties of synthetic high-pressure compounds relevant to the earth's mantle, High-Pressure Research in Geophysics, edited by S. Akimoto and M.H. Manghnani, 411-464, Cent. Acad. Pub. Japan, Japan, 1982.

Weidner, D.J., and E. Ito, Elasticity of $MgSiO_3$ in the ilmenite phase, Phys. Earth Planet. Inter., 40, 65-70, 1985.

Williams, Q., R. Jeanloz, and P. McMillan, Vibrational spectrum of $MgSiO_3$ perovskite: Zero-pressure Raman and mid-infrared spectra to 27 GPa, J. Geophys. Res., 92, 8116-8128, 1987.

Wu, M.K., J.R. Ashburn, C.J. Torng, P.H. Hor, R.L. Meng, L. Gao, Z.L. Huang, Q. Wang, and C.W. Chu, Superconductivity at 93 K in a new mixed phase Y-Ba-Cu-O compound system at ambient pressure, Phys. Rev. Lett., 58, 908, 1987.

Yagi, T., H.K. Mao, and P.M. Bell, Hydrostatic compression of perovskite type $MgSiO_3$, Advances in Physical Geochemistry, Vol. 2, edited by S.K. Saxena, Springer-Verlag, New York, 317-326, 1982.

MOSSBAUER SPECTRA OF ^{57}FE IN RARE EARTH PEROVSKITES: APPLICATIONS TO THE ELECTRONIC STATES OF IRON IN THE MANTLE

Roger G. Burns

Department of Earth, Atmospheric and Planetary Sciences,
Massachusetts Institute of Technology, Cambridge, Massachusetts 02139

Abstract. The influence of temperature and effects of next-nearest cation interactions on electronic states of iron in mantle silicate perovskites have been deduced from Mossbauer spectral data for ^{57}Fe-bearing perovskites containing rare earth (*A*) and alkaline earth (*R*) elements. Results for $ACoO_3$ perovskites are critically examined because Co(III), being isoelectronic with Fe(II), may herald changes of electronic states of iron at high temperatures in the mantle. The identification of octahedral Fe^{2+} and Fe^{3+} replacing Ti^{4+} in $CaTiO_3$ perovskite indicates that iron cations could also substitute for Si^{4+} in silicate perovskites. Thermally-induced spin-unpairing of low-spin (LS) Co^{III} to high-spin (HS) Co^{3+} is observed in ^{57}Co-implanted $ACoO_3$ perovskites. The spin-state transition temperatures in these Co(III) perovskites increase with rising atomic number along the rare earth series, indicating that smaller cations in adjacent *A*-sites stabilize LS Co^{III} in octahedral *B*-sites at higher temperatures. At elevated temperatures, the intermediate spin-state (IS) Co^{iii}, $[(t_{2g})^5(e_g)^1]$, becomes stabilized and progressive delocalization of the e_g electrons imparts metallic properties to $LaCoO_3$, as well as to LS Ni^{III} in $LaNiO_3$ and HS Fe^{4+} in stoichiometric and oxygen-deficient $(A,R)FeO_3$ perovskites. Therefore, the presence of appreciable amounts of relatively large Ca^{2+} ions in mantle silicate perovskites should facilitate entry of Fe^{2+} into octahedral *B*-sites, stabilize the IS Fe^{ii} spin-state, and induce metallic properties in iron-rich $CaSiO_3$-$MgSiO_3$ perovskites at high temperatures if they occur in the lower mantle. In $CaFeO_3$ perovskite, when electrons become localized in e_g orbitals at low temperatures, disproportionation of HS Fe^{4+} to Fe^{5+} and HS Fe^{3+} occurs. By analogy, localized e_g electrons in IS Fe^{ii} could induce disproportionation to LS Fe^{III} and HS Fe^{1+} which would enter *B*-sites and *A*-sites, respectively, in silicate perovskites in the lower mantle.

Introduction

The role of spin-pairing electronic transitions in iron-bearing minerals in the lower mantle has been discussed extensively [Ohnishi, 1978; Ohnishi and Sugano, 1981; Burns, 1985; Sherman, 1988]. Central to most calculations of high-spin to low-spin cross-over points in Fe^{2+} ions has been the effect of pressure on the relative energies of the t_{2g} and e_g groups of 3d orbitals, since compression of an iron-oxygen bond has a profound effect on the energy separation between these orbitals. However, the effects of temperature and the influence of next-nearest neighbor interactions on spin-state transitions in ferrous iron have not received as much attention as pressure and are the focus of the present paper which examines results for Co(III) perovskites.

Trivalent cobalt is isoelectronic with ferrous iron, and there are a number of observations that rising temperatures influence spin-unpairing transitions in Co(III) oxides possessing structure-types relevant to the mantle, including Co_2O_3 corundum [Chenavas et al., 1971], Co_3O_4 spinel [O'Neill, 1985] and $LaCoO_3$ perovskite [Raccah and Goodenough, 1967] phases. Mossbauer spectra of synthetic ^{57}Co-bearing rare earth perovskites were used to study temperature-induced low-spin to high-spin transitions in cobalt, first in $LaCoO_3$ [Bhide et al., 1972], and later in a variety of Co(III) perovskites [e.g. Rajoria et al., 1974; Jadhao et al., 1975; Rao et al., 1975], results of which figured in recent calculations of strain interaction effects on spin-pairing in transition metal compounds [Ohnishi and Sugano, 1981]. However, interpretations of these early Mossbauer spectra are flawed for several reasons; first, they were biased by subsequently unconfirmed structural data for $LaCoO_3$ [Raccah and Goodenough, 1967]; second, complications due to oxygen vacancies and non-stoichiometry of the Co perovskites were not considered; and third, complexities originating from electron capture decay of ^{57}Co-implanted phases were not fully appreciated. Therefore, this paper critically evaluates Mossbauer spectral data for Co(III) perovskites, as well as results for ^{57}Fe in other rare earth - transition metal perovskites, highlights information demonstrating that temperature and nearest-neighbor cations influence magnetic coupling and electronic states of transition metal ions, and draws attention to thermally-populated delocalized and intermediate spin-states which might be important if iron-rich silicate perovskites occur in the lower mantle.

Crystal Chemistry

Background

The ternary ABO_3 metal oxides with perovskite structure-types described here comprise lanthanide series elements, *A* (where *A* = La, Pr, Nd, Sm, Eu, Gd, Tb, Dy, Ho, Er, Tm, Yb and Lu, plus Y), and first series transition metals, *B* (where *B* = Ti, V, Cr, Mn, Fe, Co, Ni and Cu). All of the cations are trivalent in stoichiometric ABO_3 phases, and the transition metal cations are octahedrally coordinated by oxygen in the *B* sites such that each BO_6 octahedron shares corners with six other BO_6 octahedra. All of the rare earth - transition metal perovskites, including most of the Co(III) perovskites, are orthorhombic (space group *Pnma*); $LaCoO_3$ alone is rhombohedral ($R\bar{3}c$). In the orthorhombic structure, the four octahedra in the unit cell are tilted in different directions, the extent of tilting being measured by non-linearity of *B*-O-*B* bond angles. In $AFeO_3$ perovskites, Fe-O-Fe bond angles decrease with diminishing ionic radius of the rare earth cation in the *A* sites. As a result, each *B*-site Fe^{3+} cation has eight nearest *A*-site cations with next-nearest neighbor *A*-*B* distances ranging from ~3.04 to ~3.70 Å, the spread of interatomic distances being larger for $LuFeO_3$ than for $LaFeO_3$ [Marezio et al., 1970]. Such structural distortions, which occur also in the orthorhombic $ACoO_3$

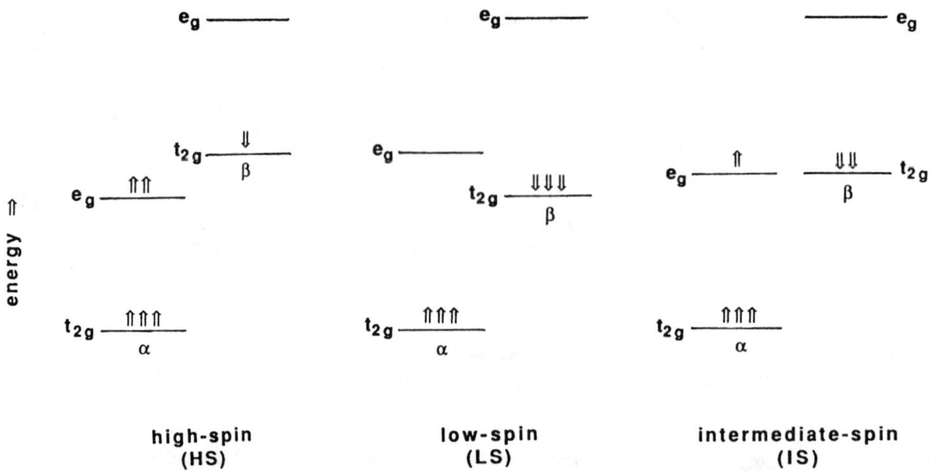

Figure 1. Electronic configurations of $3d^6$ transition metal ions, Co(III) and Fe(II), in octahedral coordination. The schematic energy levels for high-spin (HS), low-spin (LS) and intermediate-spin (IS) states are shown. Clockwise and anticlockwise electron spins are designated by α and β.

perovskites [Demazeau et al., 1974], influence magnetic ordering and spin-state transitions discussed later.

The ABO_3 perovskites exhibit variability of electronic and magnetic properties with temperature, depending on the 3d electronic configurations of the transition metal cations. In octahedral coordination, cations with configurations $3d^N$ (where N = 4, 5, 6 or 7, corresponding to Mn(III), Fe(III), Co(III) and Ni(III), respectively), may exist in either a spin-paired ground state when a large crystal field splitting of t_{2g} and e_g orbitals is predominant, or in states with maximum unpaired electrons when Hund's rules are obeyed. Where the balance between the crystal field splitting and the intraionic exchange interactions is fine, low-spin (LS) and high-spin (HS) states may coexist in proportions that vary with temperature and pressure. In some cations, particularly Co(III) discussed later, intermediate spin-states (IS) may exist. These three spin-states for the $3d^6$ cations Co(III) and Fe(II) are depicted in Figure 1, and Table 1 lists the electronic configurations of different spin-states of Fe, Co and Ni. The 3d electrons can occupy either localized (discrete, trapped) or delocalized (itinerant, collective) states, depending on the transition metal ion. Such differences are exhibited by the series of transition metal-bearing lanthanum perovskites. The $LaFeO_3$ and $LaMnO_3$ perovskites, like $LaCrO_3$ and $LaVO_3$, are magnetically ordered semiconductors, reflecting localized 3d electrons in high-spin states; $LiNiO_3$ contains low-spin Ni^{III} and, by

Table 1. Electronic Configurations of Selected Fe, Co and Ni Cations

Oxidation State	Number of 3d Electrons	Spin State	Electronic Configuration	Unpaired Electrons	Design-ation	After Decay of ^{57}Co
Fe(II)	6	HS	$(t_{2g})^4(e_g)^2$	4	HS Fe^{2+}	
		LS	$(t_{2g})^6$	0	LS Fe^{II}	
		IS	$(t_{2g})^5(e_g)^1$	2	IS Fe^{ii}	
Fe(III)	5	HS	$(t_{2g})^3(e_g)^2$	5	HS Fe^{3+}	
		LS	$(t_{2g})^5$	1	LS Fe^{III}	
		IS	$(t_{2g})^4(e_g)^1$	3	IS Fe^{iii}	
Fe(IV)	4	HS	$(t_{2g})^3(e_g)^1$	4	HS Fe^{4+}	
		LS	$(t_{2g})^4$	2	LS Fe^{IV}	
Fe(V)	3	HS=LS	$(t_{2g})^3$	3	Fe^{5+}	
Co(II)	7	HS	$(t_{2g})^5(e_g)^2$	3	HS Co^{2+}	HS Fe^{2+}
		LS	$(t_{2g})^6(e_g)^1$	1	LS Co^{II}	LS Fe^{II}
Co(III)	6	HS	$(t_{2g})^4(e_g)^2$	4	HS Co^{3+}	HS Fe^{3+}
		LS	$(t_{2g})^6$	0	LS Co^{III}	LS Fe^{III}
		IS	$(t_{2g})^5(e_g)^1$	2	IS Co^{iii}	LS Fe^{III}
Co(IV)	5	HS	$(t_{2g})^3(e_g)^2$	5	HS Co^{4+}	HS Fe^{4+}
		LS	$(t_{2g})^5$	1	LS Co^{IV}	LS Fe^{IV}
		IS	$(t_{2g})^4(e_g)^1$	3	IS Co^{iv}	LS Fe^{IV}
Ni(III)	7	HS	$(t_{2g})^5(e_g)^2$	3	HS Ni^{3+}	
		LS	$(t_{2g})^6(e_g)^1$	1	LS Ni^{III}	

analogy with $LaCuO_3$ (and $LaTiO_3$), shows metallic conductivity and Pauli paramagnetism as a result of delocalization of itinerant 3d electrons in e_g orbitals (or t_{2g} orbitals for $LaTiO_3$). The $LaCoO_3$ perovskite is intermediate between these two extremes; the 3d electrons exhibit localized to itinerant behavior at elevated temperatures, as well as thermally-induced spin-unpairing transitions.

Properties of $LaCoO_3$ Perovskite

In addition to possessing rhombohedral symmetry, which is unique among the Co(III) perovskites [Demazeau et al., 1974], $LaCoO_3$ has unusual magnetic properties [Raccah and Goodenough, 1967] which are manifested in its magnetic susceptibility, χ_g or χ_M. Plots of the temperature (T) variation of the inverse magnetic susceptibility, $1/\chi_g$, as well $\chi_g T$ versus T plots, show three distinct regions: (1) a low temperature region (0-400 K) where $1/\chi_g$ is linear with temperature; (2) an intermediate temperature region (400-650 K) where $1/\chi_g$ is almost independent of temperature; and (3) a high-temperature region (600-1100 K) where $1/\chi_g$ is essentially linear again. Such a plateau between the low and high temperature regions is unusual and was interpreted by Raccah and Goodenough [1967] to signal a rapid change in the relative proportions of two spin-states of Co(III), designated by them as Co^{III} and Co^{3+}, which undergo short-range ordering. At low temperatures, diamagnetic LS Co^{III} is present in $LaCoO_3$, but in the interval 0 to 650 K progressive conversion of LS Co^{III} to HS Co^{3+} occurs. The high-spin state was estimated to be <0.08 eV higher in energy than the low-spin state [Raccah and Goodenough, 1967], so that in the temperature interval 0-400 K thermal population of localized electrons in e_g orbitals occurs such that by about 675 K equal concentrations of LS Co^{III} and HS Co^{3+} exist. Raccah and Goodenough [1967] proposed that between 400 and 650 K, short-range cation ordering occurs in the $LaCoO_3$ structure resulting in alternating (111) planes of larger HS Co^{3+} and smaller LS Co^{III} ions. From an analysis of x-ray powder diffraction patterns as a function of temperature, Raccah and Goodenough [1967] suggested that a change of space group occurs above 650 K from $R\bar{3}c$ to $R\bar{3}$, and postulated that long-range cation ordering occurs in $LaCoO_3$ in the temperature interval 650-1215 K; in space group $R\bar{3}$ there are two distinguishable cobalt sub-lattices, Co_I and Co_{II}, which were assumed to accommodate LS Co^{III} and HS Co^{3+}, respectively. In order to rationalize the insignificant differences of Co-O distances between the two sites, Raccah and Goodenough [1967] further proposed that transfer of an e_g electron from HS Co^{3+} to LS Co^{III} created (smaller) LS Co^{IV} and (larger) HS Co^{2+} ions on sub-lattices Co_{II} and Co_I, respectively. They correlated this electron transfer process with a maximum observed in the electrical conductivity of $LaCoO_3$ above 650 K. Differential thermal analysis (dta) data were reported by Raccah and Goodenough [1967] to support their proposed cation ordering model for $LaCoO_3$. They also stated that the calorimetric data revealed a first-order endothermic transition at 1210 K, which was interpreted as a semiconductor to metal transition in $LaCoO_3$ resulting from delocalized e_g electrons in (antibonding) σ^* orbitals. The x-ray diffraction data also appeared to support the calorimetric evidence for this first-order phase transition [Raccah and Goodenough, 1967].

Optical absorption spectra confirmed the presence of LS Co^{III} in $LaCoO_3$ [Marx and Happ, 1975] and demonstrated that at room temperature a minority of thermally activated HS Co^{3+} ions exists among a majority of LS Co^{III} ions. Energy level calculations based on these crystal field spectra [Marx and Happ, 1976; Marx, 1980], as well as results from X-ray photoelectron spectra [Main et al., 1979; Lam et al., 1980], confirmed the small energy gap between the spin-states of Co(III) in $LaCoO_3$ as well as the low concentrations of HS Co^{3+} at room temperature. The XPS measurements at elevated temperatures [Main et al., 1979] also did not identify any Co(II) or Co(IV) ions which were suggested by Raccah and Goodenough [1967], and subsequently by Bhide et al. [1972, 1975], to be formed by electron transfer from HS Co^{3+} to LS Co^{III} ions. Furthermore, recent dta, electrical conductivity and neutron diffraction measurements [Thornton et al., 1982] did not confirm the evidence for a first-order localized to itinerant electron phase transition in $LaCoO_3$ near 1210 K. A re-examination of the $LaCoO_3$ structure by powder neutron diffraction measurements in the temperature range 4.2-1248 K [Thornton et al., 1982, 1986] revealed that only at 668 K, alone, is there any evidence for the rhombohedral $R\bar{3}$ space group. Above and below this temperature, $R\bar{3}c$ symmetry is preferred including the proposed first-order transition point at 1210 K. These findings led Thornton et al. [1982, 1986] to suggest that a more gradual semiconductor to metallic transition occurs in $LaCoO_3$ in the region 520-750 K brought about by the stabilization of intermediate spin-state Coiii, $[(t_{2g})^5(e_g)^1]$, which undergoes a smooth transition of progressive delocalization of the e_g electrons into partially filled σ^* molecular orbitals. The high temperature electron conductivity and magnetic susceptibility of $LaCoO_3$ thus resemble $LaNiO_3$ and $LaCuO_3$ which exhibit itinerant electron behavior.

Magnetic susceptibility measurements of other Co(III) perovskites have also provided evidence of spin-state transitions. In plots of $1/\chi_g$ versus T, plateaus analogous to that for $LaCoO_3$ were observed for $NdCoO_3$ [Rajoria et al., 1974], $EuCoO_3$ [Jadhou et al., 1976] and solid solutions of $La_{1-x}Nd_xCoO_3$ with x = 0.1 [Madhusaden et al., 1980]. However, since unpaired 4f electrons in rare earth cations also contribute to magnetic susceptibilities and complicate interpretations of Co(III) spin-states, cobalt-only values were estimated from differences of molar magnetic susceptibilities ($\Delta\chi_M$) between $ACoO_3$ and corresponding aluminate $AAlO_3$ perovskite phases (where A = Pr, Nd, Gd, Td, Dy, Ho and Yb) [Madhusadan et al., 1980]. Maxima in plots of $\Delta\chi_M^{-1}$ versus T data shown in Figure 2, as well as inflexion points in $\chi_M T$ (or $\chi_g T$) versus T plots [Rajoria et al., 1974; Jadhao et al., 1975; Bahadur, 1976; Main et al., 1979], indicate the onset of spin-unpairing transitions at different temperatures in LS Co^{III} in a variety of Co(III) perovskites. Results illustrated in Figure 2 show that Co(III) spin-unpairing transition

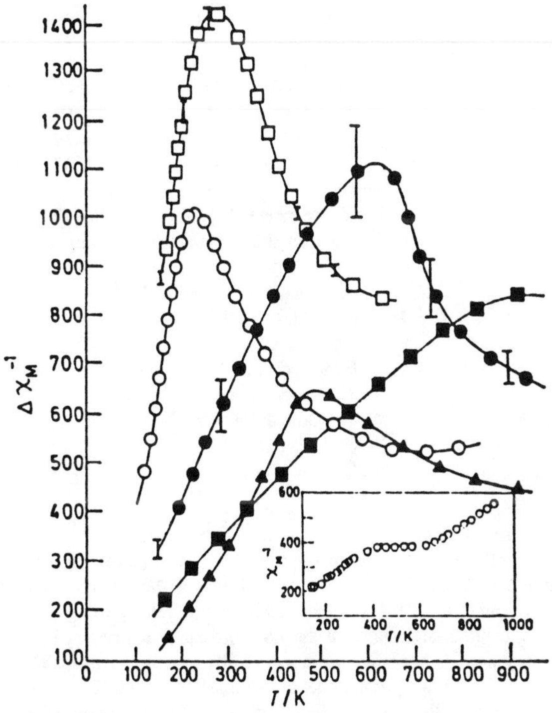

Figure 2. Plots of cobalt contributions to the inverse magnetic susceptibility of $ACoO_3$ perovskites against temperature for A = Pr (O), Nd (U), Tb (▲), Dy (●) and Yb (■). The χ_M^{-1} versus T curve of $LaCoO_3$ is shown in the inset for comparison [after Madhusudan et al., 1980].

Table 2. Mossbauer Parameters for ^{57}Fe in Perovskites.

Composition	T[a]	δ[b]	Δ[c]	H[d]	Fe Species[e]	Ref[f]
$LaFeO_3$	RT	0.39	-0.02	521	HS Fe^{3+}	1
	LN	0.46	-0.02	562	HS Fe^{3+}	1
	LH			564	HS Fe^{3+}	1
$CaTiO_3$	RT	0.35	0	-	HS Fe^{3+}	2
	RT	1.05	2.15	-	HS Fe^{2+}	2
$CaFeO_3$	RT	0.20	0	-	av Fe^{+4}	3
	LH	0.34	0	416	HS Fe^{3+}	3
	LH	0.00	0	279	HS Fe^{5+}	3
$SrFeO_3$	RT	0.05	0	-	HS Fe^{4+}	3,4
	LH	0.15	0	331	HS Fe^{4+}	3,4
$La_{1.5}Sr_{0.5}Li_{0.5}Fe_{0.5}O_3$	RT	-0.19	1.10	-	HS Fe^{4+}	5
$LaLi_{0.5}Fe_{0.5}O_3$	RT	-0.41	0	-	HS Fe^{5+}	5
$La_{0.5}Sr_{0.5}FeO_3$	RT	0.19	0.18	-	av $Fe^{+3.5}$	3,6
	LH	0.44	0	505	HS Fe^{3+}	3,6
		0.00	0	262	mixed	3,6
		0.10	0	298	mixed	3,6
$La_{0.3}Sr_{0.7}FeO_3$	LH	0.36	0	460	HS Fe^{3+}	3,6
		-0.05	0	269	HS Fe^{5+}	3,6
$La_{0.5}Ca_{0.5}FeO_3$	RT	0.20	0	-	av $Fe^{+3.5}$	7
	LH	0.46	0	~500	HS Fe^{3+}	7
		-0.07	0	~270	HS Fe^{5+}	7
$GdBa_2(Cu_{0.94}$	RT	0.23	1.97	-	HS Fe^{3+} planar	8
$Fe_{0.06})_3O_{9-\delta}$		0.18	1.05	-	HS Fe^{3+} pyramid	8
$LaCoO_3$	RT	0.36	0	-	HS Fe^{3+}	9
		0.01	0	-	LS Fe^{III}	9
	LN	0.47	0	-	HS Fe^{3+}	9
		0.02	0	-	LS Fe^{III}	9
$SrCoO_{3-x}$	RT	0.0	0	-	HS Fe^{4+}	9
$La_{0.5}Sr_{0.5}CoO_3$	RT	0.21	0	-	av $Fe^{+3.5}$	10
	LN	0.10	0	-	av $Fe^{+3.5}$	10
$Dy_{0.5}Ba_{0.5}CoO_3$	RT	0.29	0	-	HS Fe^{3+}	10
	RT	-0.07	0	-	LS Fe^{IV}	10
	LN	0.39	0	-	HS Fe^{3+}	10
	LN	0.02	0	-	LS Fe^{IV}	10

a T = Temperature: RT, ~300 K; LN, ~77 K; LH, ~4.2 K.
b δ = Isomer shift, mm/sec, relative to Fe foil calibration.
c Δ = Quadrupole splitting, mm/sec.
d H = Magnetic hyperfine field, kOe.
e Fe spin-states, see Table 1.
f Key to references:
1 Eibschutz et al. [1967]
2 This work
3 Takano and Takeda [1983]
4 Gallagher et al. [1964]
5 Demazeau et al. [1983]
6 Takano et al. [1981]
7 Grenier et al [1983]
8 Tang et al. [1987]
9 Bhide et al. [1972]
10 Chakrabarty et al. [1983].

temperatures increase with rising atomic number along the rare earth series, indicating that diminishing ionic radius of the rare earth cation coupled with decreasing unit cell volume of the $ACoO_3$ perovskites [Demazeau et al., 1974] produce larger crystal field splittings between t_{2g} and e_g orbitals and stabilize the LS Co^{III} state at progressively higher temperatures. Other magnetic susceptibility measurements of $YCoO_3$ and $LuCoO_3$ containing diamagnetic Y^{3+} and Lu^{3+} ions [Demazeau et al., 1974] support this trend. However, discrepancies between magnetic susceptibility data for $YCoO_3$ [Jadhao et al., 1975; Demazeau et al., 1974], as well as for $LuCoO_3$ [Bahadur, 1976; Demazeau et al., 1974], were attributed [Main et al., 1979] to contamination by Co^{2+} ions and non-stoichiometry of the $YCoO_3$ perovskite phase measured by Jadhao et al. [1975]. The evidence that Mossbauer spectroscopy has brought to bear on the electronic states of $LaCoO_3$ and related Co(III) perovskites is described later.

Mossbauer Spectra of Perovskites

Background

Because the Mossbauer-active ^{57}Co (half-life = 270 days) and daughter ^{57}Fe (natural abundance = 2.2%) isotopes are particularly sensitive to their crystalline environments, changes of oxidation, magnetic and spin-states of Fe and Co cations in the perovskite structure have been determined from Mossbauer spectral measurements of several iron-bearing ABO_3 phases over ranges of temperatures. Two methods of measuring perovskite Mossbauer absorption spectra have been exploited. In the more conventional Mossbauer experiment, an iron-bearing perovskite serves as the absorber and ^{57}Fe isotopes in it experience resonance absorption of velocity-modulated 14.41 keV nuclear gamma rays when they are emitted from a near-by source such as ^{57}Co diffused into a metallic foil of Rh, Pd, etc. The ^{57}Co atoms in the source decay by electron capture to ^{57}Fe retained in the metallic foil. The daughter ^{57}Fe then emits Mossbauer gamma rays which are resonantly absorbed by ^{57}Fe isotopes contained in the perovskite sample under investigation. Each crystallographically distinct Fe cation in the perovskite absorber contributes a singlet, quadrupole doublet or magnetic sextet to a Mossbauer spectrum, enabling the coordination symmetry, oxidation state, spin-state, magnetic state and relative proportion of an Fe cation to be determined.

In the less conventional Mossbauer emission spectroscopy (MES) experiment, the source becomes ^{57}Co implanted into the perovskite phase of interest, and the absorber is chosen to be a standard compound such as potassium ferrocyanide, $K_4Fe(CN)_6.3H_2O$, or stainless steel which both give single-line spectra. Chemical shifts of this line and/or the appearance of multiple lines in the Mossbauer spectra of the stainless steel or potassium ferrocyanide standards may be indicative of ^{57}Co species having different valences, spin-states, and crystallographic site occupancies in the host perovskite prior to the decay to ^{57}Fe cations which are retained at the cobalt sites in the mineral emitter. The technique of MES, therefore, appears to be a powerful method for determining the crystal chemistry of Co in host perovskites. However, there are complex chemical after-effects during the decay of ^{57}Co to ^{57}Fe [Spencer and Schroeer, 1974; Smith et al., 1978] that may influence the electronic structures and valences of daughter iron cations in the perovskite sources and seriously affect interpretations of MES data. ^{57}Co decays by electron capture to ^{57}Fe by 91% capture from the K electron shell (1s orbital) and 9% L shell capture (2s orbital). Electrons in outer shells, including 3d electrons in the M shell, may cascade into holes in the K or L shells (emitting X-rays), or undergo Auger transitions and be ejected from the cation thereby increasing the charge-state of the daughter ^{57}Fe. A variety of Fe valence and spin-states may be produced, some of which are included in Table 1. These chemical after-effects are very sensitive to the local environment about ^{57}Co (now ^{57}Fe) in oxide structures due to intrinsic crystal defects [Olivella et al., 1983], including those in oxygen-deficient perovskites. Chemical after-effects in MES experiments have been monitored by carrying out parallel measurements on ^{57}Fe-implanted host oxide phases (e.g. Spencer and Schroeer, 1974; Smith et al., 1978].

In the following sections Mossbauer spectral data summarized in Table 2 for a variety of iron-bearing perovskites are described. Note that the isomer shift data listed there are all expressed relative to reference zero velocity for the metallic α-Fe calibration standard. The parameters in Table 2 set the stage later for a critical evaluation of Mossbauer spectral data for ^{57}Co-doped perovskites.

Spectra of $AFeO_3$ Perovskites

Pure end-member $AFeO_3$ perovskites (where A = La through Lu), sometimes termed orthoferrites to distinguish them from cubic spinel ferrites, produce relatively simple six-line magnetic hyperfine Mossbauer spectra at ambient temperatures [Eibschutz et al., 1967], as shown in Figure 3. Magnetic ordering of HS Fe^{3+} cations in the octahedral B sites is essentially antiferromagnetic with a weak ferromagnetic component, and results from strong Fe(B-site) - Fe(B-site) coupling between corner-sharing [FeO_6] octahedra. Mossbauer spectra at elevated temperatures have enabled antiferromagnetic disordering temperatures, T_N, to be resolved to within ±1 K. This is demonstrated by the spectra of $DyFeO_3$ in Figure 3 which show collapse of the magnetic hyperfine sextet to a singlet for paramagnetic HS Fe^{3+} by 650 K. Values of T_N for the $AFeO_3$ orthoferrites, which are plotted in Figure 4, decrease periodically from 740 K for $LaFeO_3$ to 623 K for $LuFeO_3$. Such a drop of the T_N values indicates that the strong antiferromagnetic coupling interactions between Fe^{3+} (B-site) cations are influenced by differences of A-site occupancies resulting from decreasing ionic radii, contraction of the unit cell, departures from linearity of Fe-O-Fe bond angles in corner-shared FeO_6 octahedra and increased covalent bond character with rising atomic number along the lanthanide series. Thus, as the A-O bonds become more covalent, the Fe-O bonds become more ionic. With incrreasing ionicity of the Fe-O bonds, the Fe-O-Fe superexchange interactions decrease, thereby lowering the Neel temperatures.

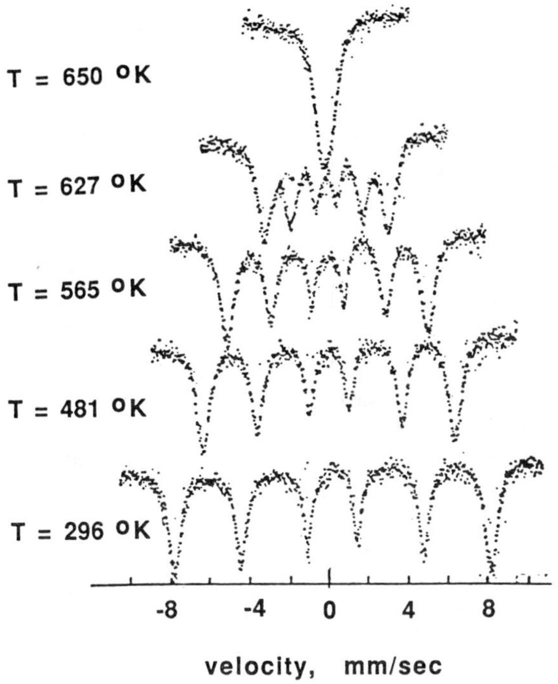

Figure 3. Mossbauer spectra of $DyFeO_3$ perovskite between room temperature and the Neel temperature, T_N ~ 650 K. The spectra are calibrated against ^{57}Co in Cu which is 0.225 mm/sec higher than the α-Fe standard [after Eibschutz et al., 1967].

Magnetic exchange interactions also vary in perovskite solid solutions. For example atomic substitution of Co for Fe in $EuFe_{1-x}Co_xO_3$ solid solutions causes the magnetic ordering temperatures to decrease drastically from 662 K in pure $EuFeO_3$ to <85 K for x = 0.6Co [Gibb, 1983a], due to the replacement of neighboring HS Fe^{3+} ions by diamagnetic LS Co^{III}. Furthermore, the single six-line magnetic hyperfine spectrum of pure $EuFeO_3$ shows additional superhyperfine patterns in Co-substituted phases originating from different next-nearest neighbor cation interactions [Gibb, 1983a]. The composition $EuFe_{0.9}Co_{0.1}O_3$, for example, consists of three component patterns, which were attributed [Gibb, 1983a] to a statistical distribution of Fe^{3+} sites with $6Fe^{3+}$ neighbors (53%), $5Fe^{3+}$ and $1Co^{III}$ neighbors (35%), and $4Fe^{3+}$ and $2Co^{III}$ neighbors (10%). Similar nearest-neighbor cation interactions appear in Mossbauer spectra of $TbFe_{1-x}Cr_xO_3$ [Nishihara, 1975], $LaNi_{1-x}Fe_xO_3$ [Asai and Sekizawa, 1980] and $EuFe_{1-x}Cr_xO_3$ [Gibb, 1983b] solid-solutions. These results demonstrate that next nearest-neighbor cation interactions influence magnetic coupling between iron atoms in the perovskite structure.

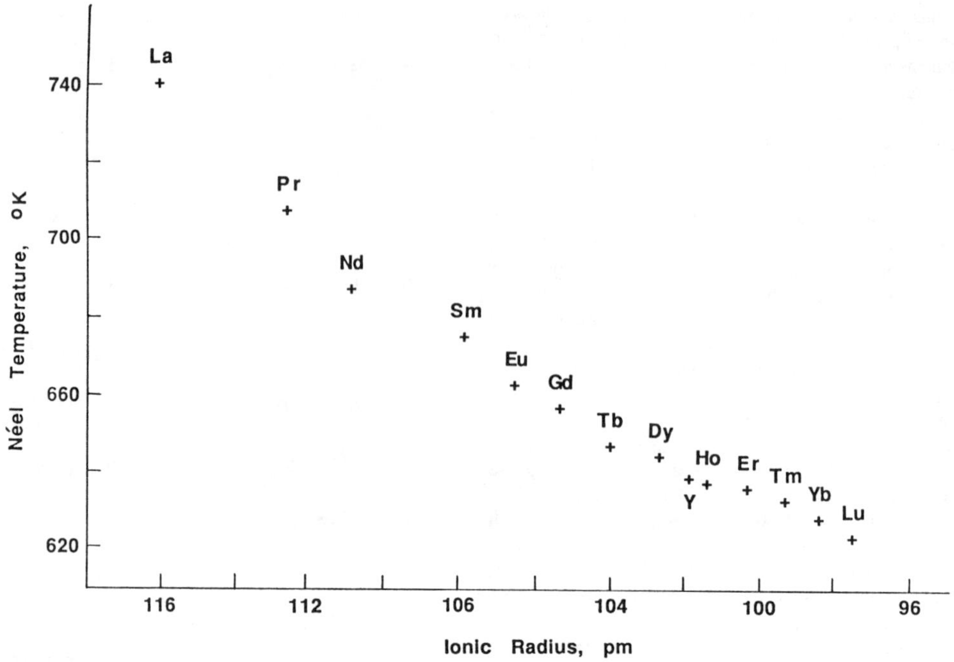

Figure 4. Compositional variations of the Neel temperatures for rare earth (*A*) perovskites $AFeO_3$. Magnetic ordering of Fe^{3+} in *B*-sites is influenced by the rare earth cation in adjacent *A*-sites [data from Eibschutz et al., 1967].

Spectra of Alkaline Earth Perovskites

Natural perovskites, $CaTiO_3$, usually contain small amounts of iron, the concentrations of which are generally higher in rare earth and niobium-rich specimens [Deer et al., 1962, Table 10]. Room temperature Mossbauer spectra of such a perovskite mineral acquired for this study and possessing the formula

$$(Ca_{0.954}Sr_{0.007}REE_{0.046})(Ti_{0.911}Nb_{0.025}Al_{0.018}Fe_{0.052})O_3$$

consists of a single peak at 0.35 mm/sec, the isomer shift of which is indicative of HS Fe^{3+}, and a weak doublet with parameters (δ = 1.05 mm/sec; Δ = 2.15 mm/sec) diagnostic of octahedral HS Fe^{2+} ions. The absence of quadrupole splitting for HS Fe^{3+} is consistent with the presence of ferric ions in regular octahedral sites in the cubic $CaTiO_3$ structure. The detection of iron cations replacing Ti^{4+} in these sites is indicative that Fe^{2+} could replace octahedral Si^{4+} in silicate perovskites in the lower mantle [Jeanloz et al., 1988].

The Mossbauer spectrum of cubic $SrFeO_3$ at room temperature also consists of a single peak occurring at 0.05 mm/sec, and a single magnetic sextet at 4.2 K [Takano and Takeda, 1983] suggestive of just one oxidation state, Fe(IV). Stoichiometric orthorhombic $CaFeO_3$ synthesized under high oxygen pressures gives a single peak, too, at 0.07 mm/sec in the room temperature spectrum, but at 4.2 K two magnetic sextets are present attributable to cations in two valence states, Fe^{5+} and HS Fe^{3+} [Takano and Takeda, 1983]. These authors suggested that Fe(IV) in $SrTiO_3$ exists as HS Fe^{4+}, $(t_{2g})^3(e_g)^1$, with the e_g electrons being delocalized in the σ^* band by analogy with LS Ni^{IV}. The delocalized HS Fe^{4+} state also exists in $CaFeO_3$ at room temperature, but at lower temperatures localization of electrons into e_g orbitals causes the Fe^{4+} ions to disproportionate to equal amounts of Fe^{3+} and Fe^{5+} ions at 4.2 K after passing through temperature-dependent intermediate valence states [Takano and Takeda, 1983].

Substitution of Sr into $CaFeO_3$ produces a single peak around 0.06 mm/sec in the room temperature Mossbauer spectra for all compositions of $Ca_{1-x}Sr_xFeO_3$ ($0 < x < 0.75$) [Takano and Takeda, 1983]. At lower temperatures two sextets appear, the parameters of which vary linearly between those of the sextets for Fe^{5+} and HS Fe^{3+} in $CaFeO_3$ and the values for the single sextet for HS Fe^{4+} in $SrFeO_3$. These results indicate that HS Fe^{4+} ions disproportionate into equal amounts of Fe ions having indiscrete continuous valences higher and lower than +4 [Takano and Takeda, 1983].

The spectrum of stoichiometric $La_{1.5}Sr_{0.5}Li_{0.5}Fe_{0.5}O_3$ consists of a doublet [Demazeau et al., 1983], the large quadrupole splitting of which (Table 2) is indicative of tetravalent LS Fe^{IV}, $[(t_{2g})^5]$, with electrons localized in the t_{2g} orbitals. Stoichiometric $LaLi_{0.5}Fe_{0.5}O_3$, on the other hand, gives a singlet spectrum at room temperature with an isomer shift of -0.41 mm/sec [Demazeau et al., 1983], indicative of pentavalent Fe with the spherical electronic configuration $[(t_{2g})^3]$.

Spectra of oxygen-deficient perovskites show evidence of averaged valence states resulting from electron hopping between adjacent Fe cations at elevated temperatures. Thus, room temperature spectra of $La_{1-x}Sr_xFeO_{3-x}$ and $La_{1-x}Ca_xFeO_{3-x}$ both consist of singlets. The isomer shift of the former varies linearly between the values of HS Fe^{4+} in $SrFeO_3$ and HS Fe^{3+} in $LaFeO_3$ [Takano et al., 1981], whereas the isomer shift of the latter remains at 0.20 mm/sec when La replaces Ca [Grenier et al., 1983; Takano et al., 1981]. These results indicate that the larger more ionic Sr^{2+} cation stabilizes HS Fe^{4+} even at room temperature, whereas smaller Ca and La cations produce the Fe^{3+} and Fe^{5+} states at low temperatures which undergo electron hopping leading to average-valence Fe(IV) species at elevated temperatures. Similar average-valence Fe species occur in a variety of oxygen deficient Sr, La-Sr and La-Ca iron-bearing perovskites [Takano et al., 1981; Grenier et al., 1982, 1983; Takano and Takeda, 1983].

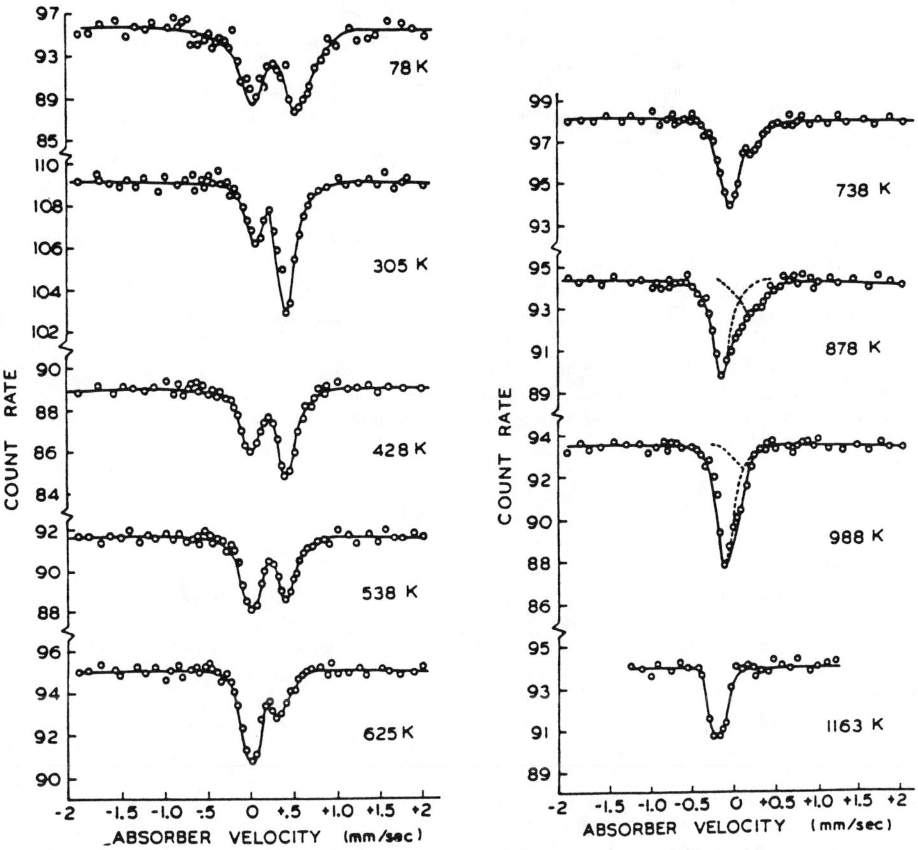

Figure 5. Temperature variations of the Mossbauer emission spectra of $LaCoO_3$ implanted with ^{57}Co [after Bhide et al., 1972].

Spectra of Superconducting Perovskites

Recent Mossbauer spectral measurements of the Fe-bearing superconducting cuprate perovskite, $GdBa_2(Cu_{2.94}Fe_{0.06})_3O_{9-\delta}$ [Tang et al., 1987], yielded two ferric doublets having an approximate intensity ratio of 2:1, which were assigned to HS Fe^{3+} in square planar (Cu_I position, site multiplicity of one) and square pyramidal (Cu_{II} position, site multiplicity of two) coordinations (Table 2). Tang et al. [1987] believed that the larger electric field gradient of the Cu_I site compared to the Cu_{II} site correlated with a higher quadrupole splitting expected for four-coordinated square planar Fe^{3+} relative to five-coordinated square pyramidal Fe^{3+}. If this assignment is correct, the relative peak intensities show that Fe^{3+} ions have a site preference for the Cu_I positions. An alternative assignment of the two doublets is possible on the basis of the isomer shift parameter, which generally is smaller for HS Fe^{3+} in lower coordination number environments. Thus, five-coordinated Fe^{3+} ions in the more abundant square pyramidal Cu_{II} sites should have a larger isomer shift than four-coordinated Fe^{3+} ions in the less abundant square planar sites. This interpretation of the Mossbauer spectra would indicate a more random distribution of Fe^{3+} ions in the superconducting Gd-Ba-Cu perovskite phase than inferred from the assignment of Tang et al. [1987].

Summary.

The conclusion to be drawn from these Mossbauer spectral measurements is that defect lanthanide transition metal perovskites, $(A,R)FeO_{3-x}$, induced by the substitution of trivalent lanthanide cations, *A*, by divalent alkaline earth cations, *R* (where A = Ca, Sr, etc) result in unusual oxidation states, spin-states and coordination symmetries of Fe cations in the *B*-sites. Similar defects may also affect spin-states and valences of Co-bearing perovskites discussed later.

Mossbauer Emission Spectra of ^{57}Co-Doped $ACoO_3$ Perovskites

Pursuant to interpretations of spin-state transitions in $LaCoO_3$ based on the dta, X-ray powder diffraction, magnetic susceptibility and electrical conductivity data assembled by Raccah and Goodenough [1967], a number of Mossbauer emission spectral measurements were made of ^{57}Co in Co(III) perovskites, principally by Indian scientists during the period 1972-1976 [Bhide et al., 1972, 1973, 1975; Rajoria et al., 1974; Jadhou et al., 1975, 1976; Rao et al., 1975; Bahadur, 1976; Chakrabarty et al., 1983], which were aimed at determining variations of the proportions of LS Co^{III} and HS Co^{3+} induced by increasing temperatures and different rare earth cations. Temperature variations of the MES spectra of $LaCoO_3$ [Bhide et al., 1972] are illustrated in Figure 5. The two peaks at ~0.05 mm/sec and ~0.40 mm/sec (referred to potassium ferrocyanide calibration, which is +0.035 mm/sec higher than that for α-Fe) observed in ambient and lower temperature spectra were assigned to LS Fe^{III} and HS Fe^{3+}, respectively, derived from electron capture decay of $^{57}Co^{III}$ and $^{57}Co^{3+}$ in the host $LaCoO_3$ perovskite. The increased intensity of the high velocity peak relative to the low velocity peak above 78 K appeared to demonstrate that LS Co^{III} is converted to HS Co^{3+} with rising temperatures. However,

Figure 4 shows that the high velocity peak decreases in relative intensity at elevated temperatures and disappears completely by 1210 K. Influenced by the earlier suggestion [Raccah and Goodenough, 1967] that HS Co^{3+} cations transfer e_g electrons to some of the remaining LS Co^{III} ions to form Co(II) and Co(IV) ions pairs, Bhide et al. [1972] initially proposed that LS Co^{II}, $[(t_{2g})^6(e_g)^1]$, and HS Co^{4+}, $[(t_{2g})^3(e_g)^2]$ ion pairs are formed which, upon electron capture decay, produce LS Fe^{II}, $[(t_{2g})^6]$, and HS Fe^{4+}, $[(t_{2g})^4(e_g)^1]$, the isomer shifts of each of which were assumed to lie near zero mm/sec and to overlap the LS Fe^{III} peak originating from LS Co^{III} not involved in electron transfer. Later, Bhide et al. [1975] proposed that the ion pairs involved the intermediate spin-state Co^{iv}, $[(t_{2g})^4(e_g)^1]$, and LS Co^{II} cations which, upon electron capture decay, produced LS Fe^{IV}, $[(t_{2g})^4]$, and LS Fe^{II}, respectively. Bhide et al. [1972] also concluded that their Mossbauer spectral data for $LaCoO_3$ were consistent with the cation ordering model proposed by Raccah and Goodenough [1967] in which the Co^{III} and Co^{3+} occupancies of the Co_I and Co_{II} sites led to a change in space group from $R\bar{3}c$ to $R\bar{3}$. However, Bhide et al. [1972] suggested that such cation ordering occurs only after the formation of the Co(IV) and Co(II) lone pairs.

The Mossbauer spectra illustrated in Figure 5 show temperature shifts of the peak centers. The temperature variations of the high velocity peak are consistent with those measured for the isomer shift of HS Fe^{3+} in the $AFeO_3$ perovskites up to ~600 K [Eibschutz et al., 1967]. Bhide et al. [1972] suggested that the sharp decrease of peak positions beyond 650 K, particularly for the low velocity peak, correlated with observations [Raccah and Goodenough, 1967] of structural changes brought about by long-range cation ordering and the resultant change of rhombohedral space group of $LaCoO_3$ which, as noted earlier, was not confirmed by subsequent neutron diffraction studies [Thornton et al., 1982, 1986].

A major problem inherent in the Mossbauer spectra shown in Figure 5 is that the peak intensities indicate that HS Fe^{3+} predominates over LS Fe^{III} at 305 K, suggesting that HS $^{57}Co^{3+}$ was more abundant than LS $^{57}Co^{III}$ in the host $LaCoO_3$ perovskite. However, this result does not correlate with evidence from optical absorption spectra [Marx and Happ, 1975], x-ray photoelectron spectra [Main et al., 1979; Lam et al., 1980], and calculations [Raccah and Goodenough, 1967] which show HS Co^{3+} cations to be the minority species at room temperature. Thus, the high velocity peak originating from HS Fe^{3+} should have the lower intensity. Such a discrepancy points to complications which may be attributable to chemical after-effects in the MES measurements.

The Mossbauer spectra of other ^{57}Co-doped Co(III) perovskites, including $NdCoO_3$ [Rajoria et al., 1974)], $EuCoO_3$ [Jadhao et al., 1976], $GdCoO_3$ [Rajoria et al., 1974], $DyCoO_3$ [Chakrabarty et al., 1983], $HoCoO_3$ [Bhide et al., 1973], $ErCoO_3$ [Jadhao et al., 1975] and $LuCoO_3$ [Bahadur, 1976], as well as $YCoO_3$ [Jadhao et al., 1975], all show evidence of temperature-induced spin-unpairing transitions in Co(III) which occur at progressively higher temperatures from $NdCoO_3$ to $LuCoO_3$. Thus, at ambient temperatures (78-300 K), the ratio of HS Fe^{3+} to LS Fe^{III} (gauged from relative intensities of the peaks at ~0.40 mm/sec and ~zero mm/sec, respectively) decreases from $LaCoO_3$ to $LuCoO_3$. Again, such a variation along the lanthanide series correlates with diminishing ionic radius of the rare earth cation, contraction of unit cell parameters and increased zig-zagging of the corner-shared octahedra in the $ACoO_3$ perovskites, larger crystal field splitting and increased stability of t_{2g} orbitals of LS Co^{III}. It is also significant that the isomer shift attributed to HS Fe^{3+} increases from 0.36 mm/sec to 0.70 mm/sec along the series La<Nd<Gd<(Y)<Ho<Er (Bhide et al., 1973; Jadhao et al., 1975), from which it was concluded that increased ionicity of Co-O bonds in the parent $ACoO_3$ perovskite is associated with increasing covalency along the rare earth series. However, the isomer shifts reported [Jadhao et al., 1975; Bhide et al., 1973] for $ErCoO_3$ (0.70 mm/sec), $YCoO_3$ (0.67 mm/sec) and, perhaps, $HoCoO_3$ (0.57 mm/sec) are anomalously high for octahedral HS Fe^{3+} in oxides, and lie in the range observed for electron delocalized HS (Fe^{2+}-Fe^{3+}) species [Burns, 1981]. The increased stability of LS Co^{III} in Co(III) perovskites of the heavier rare earth elements is also shown by the Mossbauer spectral data for alkaline earth-substituted phases [Bhide et al., 1975; Rao et al., 1975; Chakrabarty et al., 1983}. Thus, $Ba_xDy_{1-x}CoO_3$ phases contain smaller proportions of HS Co^{3+} [Chakrabarty et al., 1983] than do corresponding Sr-La, Sr-Nd and Ba-Nd cobalt perovskite solid-solutions.

As noted earlier for $LaCoO_3$, the Mossbauer spectra at ambient temperatures of other rare earth Co(III) perovskites, too, all show anomalously high peak intensities for HS Fe^{3+} (assumed to originate from parent HS $^{57}Co^{3+}$ cations) than would be expected from magnetic susceptibility and x-ray photoelectron spectroscopy measurements [Main et al., 1979] in which LS Co^{III} ions predominate. Main et al. [1979] attributed the anomalously high magnetic susceptibilities reported for $YCoO_3$ and $ErCoO_3$ [Jadhao et al., 1975] to Co^{2+} impurities in these oxygen deficient perovskites. Decay of $^{57}Co^{2+}$ ions would produce the Fe^{2+} cations involved in Fe^{2+}-Fe^{3+} electron delocalization, thereby accounting for the high isomer shifts reported in the Mossbauer spectra of $ErCoO_3$ and $YCoO_3$. Other discrepancies may be attributable to chemical after-effects in the Mossbauer emission spectral measurements.

Discussion

The magnetic susceptibility and Mossbauer spectral data for rare earth Co(III) perovskites demonstrate that thermally-induced spin-unpairing of electrons occurs in LS Co^{III} and that once initiated, the spin-state transition to HS Co^{3+} proceeds smoothly as the temperature increases. However, the onset of such spin-unpairing transitions in these $ACoO_3$ phases occurs at progressively higher temperatures with rising atomic number of the A^{3+} rare earth cation, indicating that larger more ionic cations (e.g. La^{3+}, Nd^{3+}) in the A-sites facilitate low-spin to high-spin transitions in the B-site cations. As the relative proportions of LS Co^{III} and HS Co^{3+} equalize at high temperatures, electron delocalization is induced and the itinerant electrons produce intermediate spin-states of Co(III), particularly in $LaCoO_3$, $NdCoO_3$ and, perhaps, $EuCoO_3$. The data also suggest that $ACoO_3$ and $(A,R)CoO_3$ perovskites are prone to oxygen vacancies and that non-stoichiometric phases induce electron delocalization and intermediate valence states on B-site cations.

Several inconsistencies exist, however, between the Mossbauer emission spectroscopy data and results obtained from magnetic susceptibility and other measurements. For example, why is the HS Fe^{3+} peak intensity so high in lower temperature spectra of ^{57}Co-doped perovskites, when magnetic susceptibility and x-ray photoelectron spectroscopy data indicate that HS Co^{3+} is the minority cation particularly in heavier rare earth Co(III) perovskites? And, why does the HS Fe^{3+} peak "disappear" and the low velocity peak intensify in higher temperature spectra? The answers to these questions hinge on chemical after-effects accompanying electron capture decay of ^{57}Co and the thermal populations of intermediate spin-states.

Electron capture events during the decay of parent ^{57}Co isotopes are associated with intraionic electronic transitions to a vacated 1s orbital or 2s orbital from higher energy orbitals including the t_{2g} and e_g orbitals which, in turn, may undergo electron rearrangements. Interionic electron transfer accompanying Auger transitions may also redistribute electrons in outer 3d orbitals of neighboring isotopes. These electron rearrangements are further influenced by elevated temperatures, particularly if energy differences between various configurations are small; this appears to be the case for the HS Co^{3+}, IS Co^{iii} and LS Co^{III} spin-states. At lower temperatures, excess HS $^{57}Fe^{3+}$ may result from the electron rearrangements following electron capture decay of LS $^{57}Co^{III}$ isotopes, producing an anomalously intense peak at high velocity in a Mossbauer emission spectrum which is characteristic of HS Fe^{3+}. However, as the IS $^{57}Co^{iii}$ spin state, $[(t_{2g})^5(e_g)^1]$, becomes thermally populated at elevated temperatures, electron capture decay of these isotopes eject electrons from the e_g orbitals producing LS Fe^{III} ions which accentuate the low velocity peak in a Mossbauer emission spectrum.

Applications to Spin States of Iron in the Mantle

Crystal chemical data derived from the Mossbauer spectra of Fe-bearing alkaline earth and rare earth perovskites provide insight into the electronic structures of iron cations that might be expected in silicate perovskites occurring in the lower mantle. In these perovskites, Si^{4+} is the predominant octahedral *B*-site cation. However, evidence from Mossbauer spectroscopy showing that Fe^{2+} and Fe^{3+} replace Ti^{4+} in the parent $CaTiO_3$ perovskite structure support observations [Jeanloz et al., 1988] that some Si^{4+} may be replaced by Fe^{2+} in silicate perovskites. The entry of large Fe^{2+} ions into the octahedral *B* sites would be facilitated by *A*-site occupancy by cations with large ionic radii. Thus, if either extensive atomic substitution of Ca^{2+} for Mg^{2+} occurs in $MgSiO_3$ perovskite, or two-perovskite $CaSiO_3$-$MgSiO_3$ assemblages exist in the lower mantle, significant amounts of Fe^{2+} ions could be present in the calcium-rich perovskite phase.

Octahedral Fe^{2+}, be they in the octahedral sites of the silicate perovskite or in a coexisting magnesiowustite phase, are vulnerable to HS Fe^{2+} to LS Fe^{II} spin-pairing transitions in the lower mantle. Although pressure-induced spin-pairing and temperature-induced unpairing phenomena would offset one another with increasing depth in the lower mantle, the effect of elevated pressures is likely to dominate the influence of rising temperature. While entry of Fe^{2+} in perovskite *B* sites may be facilitated by the presence of Ca^{2+} in neighboring *A* sites, the presence of the larger Ca^{2+} ions there could destabilize LS Fe^{II} compared to *A*-site occupancy by smaller Mg^{2+} ions, by analogy with La^{3+} versus Lu^{3+} in the Co(III) perovskites.

If spin-state transitions are induced in octahedrally coordinated high-spin Fe^{2+}, $[(t_{2g})^4(e_g)^2]$, in perovskite or periclase structures at high pressures in the lower mantle, the influence of high temperatures there could well be to populate the intermediate spin-state Fe^{ii} configuration, $[(t_{2g})^5(e_g)^1]$, as well as the low-spin Fe^{II} state, $[(t_{2g})^6]$, by analogy with IS Co^{iii} which occurs in $LaCoO_3$ and, perhaps, other Co(III) perovskites at elevated temperatures. Delocalized e_g electrons of IS Fe^{ii} in σ^* orbitals could induce metallic properties in host perovskite or periclase phases if they are significantly iron-rich, by analogy with LS Ni^{III} in $LaCoO_3$, IS Co^{iii} in $LaCoO_3$ and HS Fe^{4+} in $SrFeO_3$. Alternatively, localized e_g electrons in IS Fe^{ii} could induce Jahn-Teller distortion of its coordination site, or cause disproportionation to HS Fe^{1+}, $[(t_{2g})^5(e_g)^2]$, and LS Fe^{III}, $[(t_{2g})^5]$, by analogy with the HS Fe^{4+} to Fe^{5+} plus HS Fe^{3+} disproportionation observed in $CaFeO_3$. If such relatively large HS Fe^{1+} cations were to exist in the lower mantle, they would be stabilized in *A*-sites of silicate perovskites and the body-centered cube site of the MgO phase if it transformed to the CsCl structure-type. Certainly, the electronic states of iron cations in the Earth's interior are likely to be vastly different from those observed at ambient temperatures and pressures; isoelectronic cobalt cations in perovskite phases may well be the key to a better understanding of the properties of iron in the lower mantle.

Acknowledgments. I am grateful to Dr. Kenneth B. Schwartz who drew my attention to interesting crystal chemical problems involving rare earth - transition metal perovskites. Helpful discussions were held with Drs. K. B. Schwartz, C. T. Prewitt and Xing Liu. Analyzed specimens of $CaTiO_3$ were kindly provided by Dr. Paul Henderson from the British Museum. I appreciate the comments of Drs. K. B. Schwartz and D. M. Sherman who reviewed the manuscript. The research was supported by the NASA Planetary Geology and Geophysics Program (grant number NSG-7604).

Note Added. In his review of the manuscipt, Dr. David M. Sherman pointed out that the transitions: (1) HS Fe^{2+} to IS Fe^{ii}, and (2) LS Fe^{II} to IS Fe^{ii}, both involve an increase of electronic entropy (0.18*R* and 2.89*R*, respectively) and would be favored by the high temperatures of the lower mantle. He also commented that the disproportionation of IS Fe^{ii} to LS Fe^{III} plus HS Fe^{1+} involves a decrease in electronic entropy (-0.75*R*) and would not be favored by high temperatures.

References

Asai, K., and H. Sekizawa, Magnetization measurements and ^{57}Fe Mossbauer studies of $LaNi_{1-x}Fe_xO_3$ ($0.0 \leq x \leq 0.2$), *J. Phys. Soc. Japan, 49*, 90-98, 1980.

Bahadur, D., Spin-state equilibrium in $LuCoO_3$, *Indian J. Chem., 14a*, 204-205, 1976.

Bhide, V. G., D. S. Rajoria, G. R. Rao, and C. N. R. Rao, Mossbauer studies of the high-spin - low-spin equilibria and the localized-collective electron transition in $LaCoO_3$, *Phys. Rev. B., 6*, 1021-1032, 1972.

Bhide, V. G., D. S. Rajoria, Y. S. Reddy, G. R. Rao, and C. N. R. Rao, Spin-state equilibria in holmium cobaltate, *Phys. Rev. B, 8*, 5028-5034, 1973.

Bhide, V. G., D. S. Rajoria, C. N. R. Rao, G. R. Rao, and V. G. Jadhao, Itinerant-electron ferromagnetism in $La_{1-x}Sr_xCoO_3$; A Mossbauer study, *Phys. Rev. B, 12*, 2832-2843, 1975.

Burns, R. G., Intervalence transitions in mixed-valence minerals of iron and titanium, *Ann. Rev. Earth Planet. Sci., 9*, 345-383, 1981.

Burns, R. G., Thermodynamic data from crystal field spectra, In: S. W. Kieffer and A. Navrotsky, eds, *Microscopic to Macroscopic: Atomic Environments to Mineral Thermodynamics, Rev. Mineralogy, 14*, 277-316, 1985.

Chakrabarty, D. K., A. Bandyopadhyay, S. B. Patil, and S. N. Shringi, Magnetic and Mossbauer studies of some mixed cobaltites of the lanthanides and alkali metals, *phys. stat. sol. (a), 79*, 213-222, 1983.

Chenavas, J., J. C. Joubert and M. Marezio, Low spin - high spin state transition in high pressure cobalt sesquioxide, *Solid State Comm., 9*, 1057-1060, 1971.

Deer, W. A., R. A. Howie and J. Zussman, Perovskite, In: *Rock-Forming Minerals*, vol 5, p. 48-55, 1962.

Demazeau,, G., B. Buffat, N. Chevreau, M. Pouchard, and P. Hagenmuller, Recent developments in the field of the high oxidation states in oxides: Stabilization of six-coordinated high spin-iron(IV) and iron(V), In: *Solid State Chemistry 1982*. R. Metselaar, H. J. M. Heijligers, and J. Schoonman, eds; Elsevier Sci. Publ. Co., *Studies in Inorganic Chemistry, 3*, 433-437, 1983.

Demazeau, G., M. Pouchard and P. Hagenmuller, Sur de nouveaux composes oxygenes du cobalt +III derives de la perovskite, *J. Solid State Chem., 9*, 202-209, 1974.

Eibschutz, M., S. Shtrikman, and D. Treves, Mossbauer studies of Fe^{57} in orthoferrites, *Phys. Rev., 156*, 562-577, 1967.

Gibb, T. C., Magnetic exchange interactions in perovskites solid solutions. Part 1. Iron-57 and ^{151}Eu Mossbauer spectra of $EuFe_{1-x}Co_xO_3$ ($0<x<1$), *J. Chem. Soc. Dalton Trans.*, 873-877, 1983a.

Gibb, T. C., Magnetic exchange interactions in perovskite solid solutions. Part 2. Iron-57 and ^{151}Eu Mossbauer spectra of $EuFe_{1-x}Cr_xO_3$ ($0<x<1$), *J. Chem. Soc. Dalton Trans.*, 2031-2038, 1983b.

Grenier, J.-G., L. Fournes, M. Pouchard, and P. Hagenmuller, Mossbauer resonance studies on the $Ca_2Fe_2O_5$-$LaFeO_3$, *Mat. Res. Bull., 17*, 55-61, 1982.

Grenier, J. C., M. Pouchard, and P. Hagenmuller, In: *Solid State Chemistry 1982*, R. Metzelaar, H. J. M. Heijligers, and J. Schoonman, eds; Elsevier Sci. Publ. Co., *Studies in Inorganic Chem, 3*, 347-351, 1983.

Jadhao, V. G., R. M. Singru, G. R. Rao, D. Bahadur, and C. N. R. Rao, Effect of the rare earth ion on the spin state equilibria in perovskite rare earth metal cobaltates, *J. Chem. Soc. Faraday Trans. II, 71*, 1885-1893, 1975.

Jadhao, V. G., R. M. Singru, G. N. Rao, D. Bahadur, and C. N. R. Rao,

^{151}Eu and ^{57}Fe Mossbauer studies of $EuCoO_3$, *J. Phys. Chem. Solids, 37*, 113-117, 1976.

Jeanloz, R., E. Knittle, Q. Williams, and X. Li, (Mg,Fe)SiO_3 perovskite: Geophysical and crystal chemical significance, This Volume, 1988.

Lam, D. J., B. W. Veal. and D. E. Ellis, Electronic structure of lanthanum perovskites with 3d transition elements, *Phys. Rev. B, 22*, 5730-5738, 1980.

Madhusudan, W. H., K. Jagannathan, P. Ganguly, and C. N. R. Rao, A magnetic susceptibility study of spin-state transitions in rare-earth trioxocobaltates(III), *J. Chem. Soc. Dalton Trans.*, 1397-1400, 1980.

Main, J. G., J. F. Marshall, G. Demazeau, G. A. Robbins, and C. E. Johnson, Spin state of the Co^{3+} ion in $RCoO_3$ compounds with the $GdFeO_3$ structure, *J. Phys. C: Solid State Phys., 12*, 2215-2224, 1979.

Marezio M., J. P. Remeika, and P. D. Dernier, The crystal chemistry of the rare earth orthoferrites, *Acta Cryst., B, 26*, 2008-2022, 1970.

Marx, R., Paramagnetic properties of the Mott isolator $LaCoO_3$, *phys. stat. sol. (b), 99*, 555-563, 1980.

Marx, R., and H. Happ, Optical absorption spectrum of Co^{III} in $LaCoO_3$, *phys. stat. sol. (b), 67*, 181-189, 1975.

Nishihara, Y., Effect of nearest neighbor ions on the hyperfine fields at ^{57}Fe nuclei in $TbFe_{1-x}Cr_xO_3$, *J. Phys. Soc. Japan, 38*, 710-717, 1975.

O'Neill, H. St. J., Thermodynamics of Co_3O_4: a possible electron spin unpairing transition in Co^{3+}, *Phys. Chem. Minerals, 12*, 149-154, 1985.

Ohnishi, S., A theory of the pressure-induced high-spin - low-spin transition of transition metal oxides, *Phys. Earth Planet. Interiors, 17*, 130-139, 1978.

Ohnishi, S., and S. Sugano, Strain interaction effects on the high-spin - low-spin transition of transition-metal compounds, *J. Phys. C: Solid State Phys., 14*, 39-55, 1981.

Olivella, R., J. Tejada, and T. Harami, Mossbauer emission spectroscopy of pure and doped MgO: ^{57}Co, *Radiation Effects, 73*, 179-183, 1983.

Raccah, P. M. and J. B. Goodenough, First-order localized-electron --- collective-electron transition in $LaCoO_3$, *Phys. Rev., 155*, 932-943, 1967.

Rajoria, D. S., V. G. Bhide, G. R. Rao, and C. N. R. Rao, Spin state eqiuilibria and localized versus collective d-electron behaviour in neodymium and gadolinium trioxocobaltate(III), *J. Chem. Soc. Faraday II, 70*, 512-523, 1974.

Rao, C. N. R., V. G. Bhide, and N. F. Mott, Hopping conduction in $La_{1-x}Sr_xCo_3$ and $Nd_{1-x}Sr_xCoO_3$, *Phil. Mag., 32*, 1277-1282, 1975.

Sherman, D. M., High-spin to low-spin transitions of iron(II) oxides at high pressures: possible effects on the physics and chemistry of the lower mantle, Ch. 6 in: S. Ghose and J. M .D. Coey, eds, *Adv. Phys. Geochem., 7*, 113-128, 1988.

Smith, P. A., C. D. Spencer, and R. P. Stillwell, Co^{57} and Fe^{57} Mossbauer studies of the spinels $FeCo_2O_4$ and $Fe_{0.5}Co_{2.5}O_4$, *J. Phys. Chem. Solids, 39*, 107-111, 1978.

Spencer, C.D., and D. Schroeer, Mossbauer study of several cobalt spinels using Co^{57} and Fe^{57}, *Phys. Rev. B., 9*, 3658-3665, 1974.

Takano, M., J. Kawachi, N. Nakanishi, and Y. Takeda, Valence state of the Fe ions in $Sr_{1-x}La_xFeO_3$, *J. Solid State Chem., 39*, 75-84, 1981.

Takano, M., and Y. Takeda, Electronic state of Fe^{4+} ions in perovskite-type oxides, *Bull. Inst. Chem. Res., Kyoto Univ., 61*, 406-425, 1983.

Tang, H., Z. Q. Qiu, Y. W. Du, G. Xiao, C. L. Chien, and J. C. Walker, Magnetic ordering in $GdBa_2(Cu_{0.94}Fe_{0.06})_3O_{9-\delta}$ below the superconducting transition temperature, *Phys. Rev. B, 36*, 4018-4020, 1987.

Thornton, G., B. C. Tofield, and A. W. Hewat, A neutron diffraction study of $LaCoO_3$ in the temperature range $4.2<T<1248$ K, *J. Solid State Chem., 61*, 301-307, 1986.

Thornton, G., B. C. Tofield, and D. E. Williams, Spin state equilibria and the semiconductor to metal transition in $LaCoO_3$, *Solid State Comm., 44*, 1213-1216, 1982.

STRUCTURE-PROPERTY RELATIONSHIPS IN PEROVSKITE ELECTROCERAMICS

R.E. Newnham

Materials Research Laboratory, The Pennsylvania State University, University Park, Pennsylvania 16802

Abstract. The multimillion dollar markets for multilayer capacitors, piezoelectric transducers, and PTC thermistors are based on ferroelectric ceramics made from oxide perovskites. Atomistic and electronic phenomena crucial to an understanding of these components are reviewed in this paper, together with a brief description of their electrical properties. All three devices make use of ferroelectric phase transformations and chemical dopants to optimize performance.

Introduction

A number of engineering applications of perovskite ceramics and crystals are listed in Table 1, but only the first three are profitable. Multilayer capacitors, piezoelectric transducers, and positive temperature coefficient (PTC) thermistors all make use of ferroelectric ceramics with the perovskite structure. The world market for ceramic capacitors is about two billion dollars per year, approximately twice that for transducers, and ten times the market for thermistors.

Origin of Ferroelectricity

The oxide perovskites used as capacitors, transducers, and thermistors have high dielectric constants and large piezoelectric coefficients. Corner-linked $(TiO_6)^{2-}$ octahedra are the highly polarizable active ions promoting ferroelectric phase transitions with large spontaneous polarization. Tetravalent titanium is a d^o ion with low-lying 3d, 4s, and 4p orbitals which combine with the σ- and π-orbitals of the six O^{2-} neighbors to form the molecular orbitals of the octahedral complex. The bond energy of the complex can be lowered by distorting the octahedron to a lower symmetry. This leads to dipole moments, ferroelectricity, and large electric permittivity.

It is interesting to note that such distortions occur not only in perovskite crystals, but in liquids as well. In aqueous solution, active ions such as Ti^{4+} and Nb^{5+} form asymmetric complexes with water molecules. In the $[TiO(H_2O)_5]^{2+}$ complex, titanium forms a short strong bond with a single oxygen ion and five longer weaker bonds to water molecules. The distorted octahedral complex possesses polar symmetry (point group 4mm) identical to that of tetragonal $BaTiO_3$.

Capacitor Dielectrics

Multilayer capacitors are made from tape-cast ceramic layers interleaved with silver-palladium electrodes. Modified barium titanate with its three ferroelectric phase transformations is generally used as the ceramic dielectric. Barium atoms are located at the corners of the cubic unit cell (Fig. 1) and oxygens at the face center positions. The Ba^{2+} and O^{2-} ions have radii of about 1.4Å, and together make up a face-centered cubic array with a lattice parameter near 4Å. Octahedrally coordinated Ti^{4+} ions at the center of the cubic unit cell are the active ions promoting ferroelectricity.

On cooling from high temperature, the crystal structure of $BaTiO_3$ undergoes three displacive phase transitions (Fig. 1) with atomic movements

TABLE 1. Electrical Applications of Perovskites.

Multilayer Capacitor	$BaTiO_3$
Piezoelectric Transducer	$Pb(Zr,Ti)O_3$
P.T.C. Thermistor	$BaTiO_3$
Electrooptic Modulator	$(Pb,La)(Zr,Ti)O_3$
Dielectric Resonator	$BaZrO_3$
Thick Film Resistor	$BaRuO_3$
Electrostrictive Actuator	$Pb(Mg,Nb)O_3$
Superconductor	$Ba(Pb,Bi)O_3$
Magnetic Bubble Memory	$GdFeO_3$
Laser Host	$YAlO_3$
Ferromagnet	$(Ca,La)MnO_3$
Refractory Electrode	$LaCoO_3$
Second Harmonic Generator	$KNbO_3$

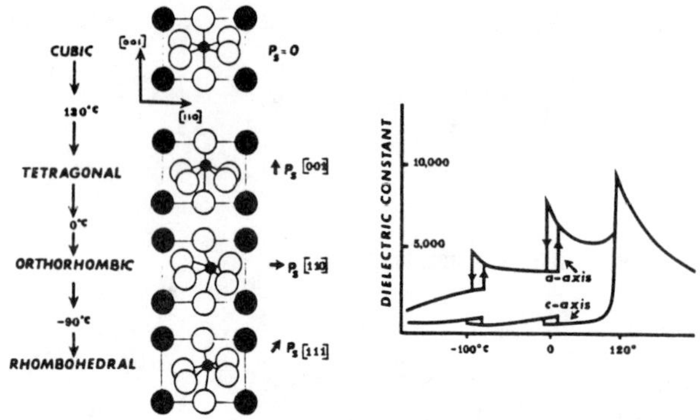

Fig. 1. Structural changes occurring at the three ferroelectric phase transformations in $BaTiO_3$ result in large values of the dielectric constant over a wide temperature range.

of 0.1Å or less. At the Curie temperature of 130°C, the point symmetry changes from cubic m3m to tetragonal 4mm. The tetragonal state with spontaneous polarization along [001] persists down to 0°C where it transforms to orthorhombic (mm2) symmetry as P_s shifts to a face diagonal direction. On further cooling, the orthorhombic structure transforms to rhombohedral (3m) near -90°C.

Changes in the perovskite structure are illustrated in Fig. 1, along with the associated peaks in the dielectric constant. In regard to capacitors, it is very important that the dielectric constant be high over a wide temperature range. The presence of two lower ferroelectric transformations ensures that the dielectric constant remains high below the Curie temperature. Note that the dielectric constant is highly anisotropic with much larger values along directions perpendicular to the polar axis. The instability of the structure makes it easy to tilt the spontaneous polarization vector with a transverse electric field.

Crystal Chemistry of $BaTiO_3$

Barium titanate ceramics have been used as capacitors for forty years, and there have been many studies of its solid solutions. There are basically three ways in which the structure of $BaTiO_3$ can be altered to modify its properties: (1) smaller divalent cations can be substituted for barium, (2) larger tetravalent ions can be substituted for titanium, and (3) smaller tetravalent cations be used to replace titanium. A fourth possibility in which barium is replaced by larger divalent ions is difficult because Ba^{2+} is the largest non-radioactive divalent cation.

Non-isovalent substitutions are also possible, of course, and these will be considered later as donor and acceptor dopants. Coupled substitutions such as $Ba_{1-x}Na_xTi_{1-x}Nb_xO_3$ are occasionally useful but will not be considered here.

Replacing barium with a smaller divalent cation modifies the transition temperatures. The three most commonly used "Curie Point Shifters" are Pb^{2+}, Sr^{2+}, and Ca^{2+}. Modest amounts of Pb^{2+} raise T_c, while Sr^{2+} drastically lowers T_c. Ca^{2+} has little effect on T_c, but decreases the two lower transitions by destabilizing the orthorhombic and rhombohedral phases. Divalent Pb is one of the very few additions which increases the Curie temperature; the tetragonal pyramidal coordination favored by Pb^{2+} stabilizes the tetragonal phase with respect to the adjacent cubic and orthorhombic phases.

A pinching together of the phase transitions occurs when titanium is replaced with larger tetravalent ions. Typical of this type of behavior is the $BaTi_{1-x}Zr_xO_3$ solid solution series used in high-permittivity capacitors. With increasing zirconium content, the Curie temperature drops while the two lower transitions are raised, causing the three transitions to converge near x = 0.1 and T_c = 50°C. As a consequence the three maxima in the dielectric constant merge together to give an immense peak of about K = 8000. Similar phase diagrams occur in the $BaTi_{1-x}Sn_xO_3$ and $BaTi_{1-x}Hf_xO_3$ solid solution series.

Substituting smaller ions for titanium causes the perovskite structure to destabilize in favor of one of its polytypes. The structures of $BaMnO_3$, $BaFeO_3$, $BaIrO_3$, and $BaCrO_3$ are hexagonal polytypes of the cubic perovskite structures. The dielectric constant of these structures is too small to be of interest in capacitor formulations.

Domain Walls and Dielectric Loss

Domain walls are the chief source of dielectric loss for temperatures below T_c. Under applied electric fields, the domain walls of ferroelectric $BaTiO_3$ change position, dissipating energy. A number of different types of domain

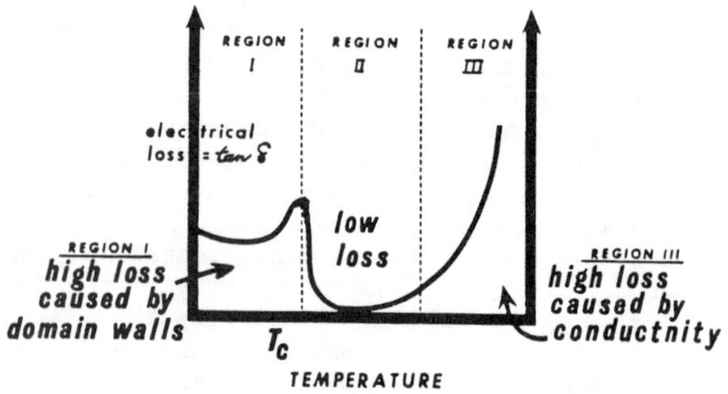

Fig. 2. Temperature dependence of the electrical loss tan δ of a ferroelectric crystal or ceramic. At high temperature the losses are caused by conduction while domain wall movements are responsible at temperatures below T_c.

walls are found in capacitor compositions, and each has a different mobility. Tetragonal $BaTiO_3$ has 180° walls, and both charged and uncharged 90° walls. Charged walls are important only in low-resistivity $BaTiO_3$ where currents can flow, neutralizing charge. The 180° walls are generally more mobile than 90° walls because of the mechanical strain associated with 90° walls.

Inhomogeneous capacitor compositions sometimes make use of the "Pinch Effect" in which the cubic, tetragonal, orthorhombic and rhombohedral phases coexist. This means that many different types of domain walls are present in the ceramic. Orthorhombic domain walls separate regions differing in the orientation of the spontaneous polarization vector by 60°, 90°, 120° and 180°. For the rhombohedral phase there are 70.5°, 109.5°, and 180° walls. Wall interactions are very complex when more than one type of phase is present.

In general, the electrical loss of ferroelectric ceramics display three identifiable temperature ranges (Fig. 2). Below T_c the losses are moderately high and are caused by domain wall motion. The magnitude of tan δ (the ratio of the imaginary part of the dielectric constant to the real part) increases rapidly with the applied field but does not depend strongly on frequency.

The second temperature--typically 100-200° above the Curie temperature--has very low dielectric loss. Above T_c there are no domains to cause dielectric loss, and the temperature is too low for appreciable conductivity loss.

At high temperature, the loss due to conductivity becomes important causing tan δ to increase rapidly with temperature. Conduction losses are inversely proportional to the measurement frequency.

Cation and Anion Vacancies

Vacancies are far more common than interstitials in oxide perovskites. Since the perovskite structure is based on close-packed layers of oxygen and barium ions, there is very little space for interstitial ions to enter the structure.

Vacancy concentration strongly influences electrical losses and can be controlled by doping. Donor dopants such as La^{3+} (for Ba^{2+}) or Nb^{5+} (for Ti^{4+}) create cation vacancies in $BaTiO_3$, although it is not always clear whether the vacancies occur in the barium or the titanium site. In earlier studies of semiconducting $BaTiO_3$, it was generally assumed that donor dopants were partly compensated by electronic defects (Ti^{3+} for Ti^{4+}), and partly by barium vacancies, but recent work suggests that titanium defects are also important at heavy doping levels.

Acceptor dopants such as K^+ (for Ba^{2+}) or Fe^{3+} (for Ti^{4+}) create oxygen vacancies in $BaTiO_3$: $(Ba_{1-x}K_x)Ti(O_{3-x/2}[]_{x/2})$ or $Ba(Ti_{1-x}Fe_x)(O_{3-x/2}[]_{x/2})$. Oxygen vacancies exert a much stronger influence on dielectric loss than do barium or titanium vacancies.

Domain Wall Pinning

The way in which oxygen vacancies affect tan δ depends on temperature and the dominant loss mechanism (Fig. 2). Below T_c where domain wall losses predominate, oxygen vacancies lower tan δ substantially, improving the capacitor. Lanthanum additions raise the dissipation factor while acceptor dopants such as Fe^{3+}, Cr^{3+}, or Mn^{3+} lower it.

The explanation of why acceptor and donor dopants affect tan δ differently involves the pinning of domain walls. Donor-doped perovskite ferroelectrics have lossy hysteresis loops and considerable domain wall motion whereas acceptor perovskites do not.

Several mechanisms for domain wall pinning by oxygen vacancies have been proposed: (1) Associated oxygen vacancies and dopant ions form electric dipole moments which align with the spontaneous polarization creating an internal bias field. (2) The oxygen vacancies diffuse to domain walls, locking them in position. (3) Defects in the grain boundary regions adjust in position to compensate local polarization charges, again stabilizing the domain configuration and pinning domain walls.

The reason why domain walls are pinned more effectively in acceptor perovskites can be seen from interatomic distances. Oxygen vacancies diffuse much faster than cation vacancies because of the proximity of oxygen sites. The distance between neighboring oxygens is only 2.8Å compared to 4Å for the shortest Ti-Ti or Ba-Ba separation.

Defect dipoles in acceptor-doped $BaTiO_3$ consisting of paired iron atoms and oxygen vacancies realign more easily than do the corresponding point dipoles in donor-doped ceramics. Thus the defect dipoles in acceptor-doped $BaTiO_3$ align with the spontaneous polarization of the domain structure to pin domain walls, thereby lowering the dissipation factor in the low temperature region below T_c.

Conduction Losses and Degradation

Oxygen vacancies are also important in the high temperature region where conduction losses dominate (Fig. 2). The rapid increase of dissipation factor with temperature is caused by free carrier conductivity, and the concentration of carriers depends on doping level.

Based on a number of experiments, the following picture has been developed for the DC degradation process in barium titanate ceramics. Polycrystalline titanates are appreciably reduced at the temperatures (~1300°C) required to sinter ceramic capacitors. On cooling, rapid reoxidation takes place above 1100°C but effectively

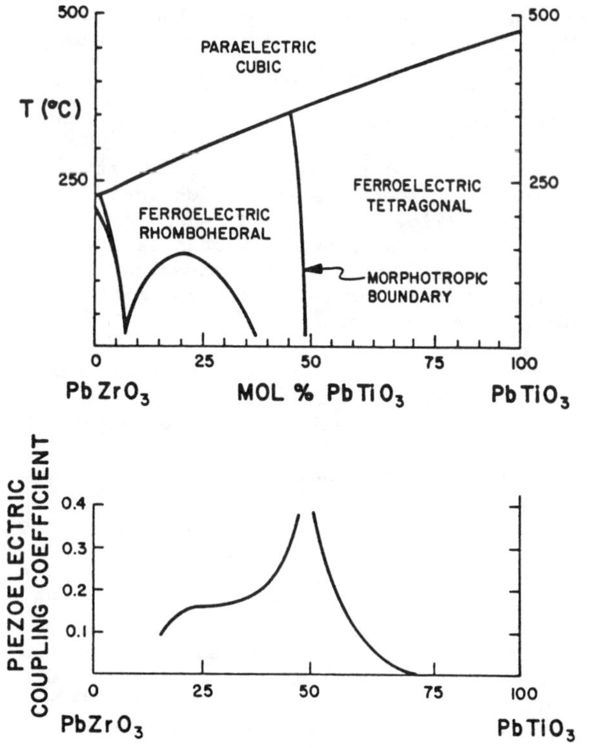

Fig. 3. (a) Binary phase diagram for the lead zirconate-lead titanate ceramics used as transducers. (b) Large electromechanical coupling coefficients are obtained for poled PZT ceramics near the morphotropic phase boundary.

stops at temperatures below 600°C. As a consequence the grain boundary regions are well oxidized, but the interior of each grain remains oxygen deficient. Oxygen vacancies carry an effective charge of +2e, which is compensated by 3d electrons belonging to the titanium atoms. Two Ti^{3+} ions form for every oxygen vacancy. At low tempera- tures, the oxygen vacancies and Ti^{3+} ions are bound by a weak attractive force with a coulomb energy of 0.1 to 0.2 eV, sufficiently large that only a few defect pairs are separated. Electrons associated with the unattached Ti^{3+} ions are responsible for conduction by means of a narrow 3d conduction band. Alternatively, the conduc- tion process can be described as electron hopping between titanium ions: $Ti^{3+} \longleftrightarrow Ti^{4+} + e^-$. Unattached oxygen vacancies also contribute to the conductivity but their mobility is much smaller than that of electrons. Over long periods of time, the vacancies migrate across the sample under DC fields to degrade the capacitor. To prevent degradation it is necessary to control vacancy migration. This is unfortunate because the oxygen vacancies required to lower domain wall losses enhance the high temperature conductivity losses.

Manganese, the Magic Additive

Trivalent manganese plays an important role in many capacitor compositions by controlling both domain and degradation losses. Mn^{3+} ions attach strongly to oxygen vacancies because of the Jahn-Teller effect. This prevents migration and degradation but allows the defect dipole reorientation so crucial to the prevention of domain wall losses.

The strong pairing of Mn^{3+} with oxygen vacancies can be seen in $CaMn(O_{2.5}[\,]_{0.5})$, a defect perovskite in which each Mn^{3+} ion is bonded to five oxygens in the form of a tetragonal pyramid. This unusual coordination is achieved by removing one oxygen from each octahedron of the ideal perovskite structure. Mn^{3+} has the $3d^4$ electron configuration and is widely known as a Jahn-Teller ion. By adopting a large tetragonal distortion its single e_g electron achieves a lower crystal field energy.

Piezoelectric Transducers

Piezoelectric transducers convert mechanical energy to electrical energy (the direct piezo-electric effect), or electrical energy to mechanical energy (the converse piezoelectric effect). Ferroelectric ceramics of lead zirconate titanate (PZT) become strongly piezo-electric when electrically poled. Poling is carried out under intense DC fields at tempera-tures just below the ferroelectric Curie temperature where the domains are most easily aligned.

Morphotropic Phase Boundary

The phase diagram of the $PbZrO_3$-$PbTiO_3$ system is shown in Fig. 3. A complete solid solution forms at high temperature with Zr and Ti randomly distributed over the octahedral sites of the cubic perovskite structure. On cooling, the structure undergoes a displacive phase transformation into a distorted polar phase. Titanium-rich compositions favor a tetragonal modification with a sizeable elongation along <001> and a large spontaneous polarization in the same direction. There are six equivalent domain states with P_s along the $[100]$, $[\bar{1}00]$, $[010]$, $[0\bar{1}0]$, $[001]$ and $[00\bar{1}]$ directions of the cubic paraelectric state. A rhombohedral ferroelectric state is favored for zirconium-rich compositions. Here the distortions and polarization are along <111> body diagonal directions, giving rise to eight domain states: $[111]$, $[\bar{1}11]$, $[1\bar{1}1]$, $[11\bar{1}]$, $[1\bar{1}\bar{1}]$, $[\bar{1}1\bar{1}]$, $[\bar{1}\bar{1}1]$, and $[\bar{1}\bar{1}\bar{1}]$.

The compositions which pole best lie near the morphotropic phase boundary between the rhombo-hedral and tetragonal ferroelectric phases. For these compositions near $Pb(Zr_{0.5}Ti_{0.5})O_3$ there are fourteen possible poling directions coexisting over a very wide temperature range. Extensive domain wall motion takes place during the poling operation at high temperature and a highly polar configuration is frozen into the PZT ceramic at room temperature. This explains why the piezoelectric coefficients are largest near the morphotropic phase boundary (Fig. 3b).

Morphotropic boundaries are relatively common in Pb-based perovskites, more so than in other perovskite phase diagrams. In solid solutions based on $BaTiO_3$, a different sequence of phase transformations appears. On cooling from high temperatures the cubic phase undergoes transformations to tetragonal, orthorhombic, and rhombohedral symmetries with polar vectors along <100>, <110>, and <111> respectively. The intervening orthorhombic phase makes it possible for the <100> polar vector of the tetragonal phase to swing to the <111> direction of the rhombohedral phase. Thus there is no vertical morphotropic phase boundary in $BaTiO_3$-based ceramics.

It appears that morphotropic boundaries occur in $PbTiO_3$-based systems because of the suppression of the orthorhombic phase. The Pb^{2+} ion plays a major role in the suppression because of its lone-pair $6s^2$ electron configuration which favors the type of pyramidal bonding found in the polymorphic PbO structures. In the tetragonal and rhombohedral perovskites, such bonding is readily accommodated in the form of tetragonal and trigonal pyramidal coordination, but not in the orthorhombic phase. Here the Pb^{2+} ion is forced to move in a [110] direction directly toward a neighboring oxygen, an extremely unfavorable distortion. As a result, the orthorhombic phase is destabilized, and it becomes very difficult for the tetragonal phase to transform to the rhombohedral form.

Piezoelectricity in PZT

Poled ceramic transducers have conical symmetry (point group ∞m), the symmetry of a polar vector. The physical properties are described by matrix coefficients referred to a set of property axes x_1, x_2, x_3. By convention, the x_3 axis is chosen along the polar axis with the orthogonal x_1 and x_2 axes perpendicular to x_3.

The piezoelectric coefficients relate electric polarization to mechanical stress. P_1, P_2, and P_3 are the components of the stress-induced polarization along x_1, x_2, and x_3. In matrix notation, stress components σ_1, σ_2, and σ_3 are applied tensile stresses along x_1, x_2, and x_3. Shear stresses about x_1, x_2, and x_3 are designated σ_4, σ_5, and σ_6, respectively.

For poled ceramics conical symmetry dictates that all piezoelectric coefficients are zero except $d_{31} = d_{32}$, d_{33}, and $d_{15} = d_{24}$. The direct piezoelectric effect can therefore be described by the following matrix expression:

$$\begin{pmatrix} P_1 \\ P_2 \\ P_3 \end{pmatrix} = \begin{pmatrix} 0 & 0 & 0 & 0 & d_{15} & 0 \\ 0 & 0 & 0 & d_{15} & 0 & 0 \\ d_{31} & d_{31} & d_{33} & 0 & 0 & 0 \end{pmatrix} \begin{pmatrix} \sigma_3 \\ \sigma_2 \\ \sigma_3 \\ \sigma_4 \\ \sigma_5 \\ \sigma_6 \end{pmatrix}$$

Multiplying out,

$$P_1 = d_{15}\sigma_5$$

$$P_2 = d_{15}\sigma_4$$

$$P_3 = d_{31}(\sigma_1+\sigma_2) + d_{33}\sigma_3.$$

Thus polarization charges along x_1 can only be generated by shear stresses about x_2. For a hydrostatic pressure p, $\sigma_1=\sigma_2=\sigma_3=-p$, and $\sigma_4=\sigma_5=\sigma_6=0$. The resulting piezoelectric polarization appears along x_3: $P_3 = -(2d_{31}+d_{33})p$.

Molecular mechanisms for piezoelectric coefficients d_{33}, d_{31} and d_{15} are pictured and explained in Fig. 4. For PZT compositions near the morphotropic phase boundary, $d_{33} \cong 400$ pC/N, $d_{31} \cong -170$, and $d_{15} \cong 500$.

Hard and Soft PZT Ceramics

The magnitudes of the piezoelectric coefficients depend markedly on dopants and defect structure because of their influence on domain wall motion. This in turn controls the nature of the hysteresis loop in the ferroelectric ceramic. Hard PZT ceramics are analogous to permanent magnets, while soft PZT has a high permittivity analogous to the high permeability of the soft magnetic materials used in transformers.

Donor ions create Pb vacancies in the PZT structure. As an example, when Nb^{5+} is substituted for Ti^{4+}, vacancies in the lead-site result:

$$(Pb_{1-x/2}\square_{x/2})(Ti_{1-x-y}Zr_yNb_x)O_3.$$

As in $BaTiO_3$, donor doping is not effective in pinning domain walls. Pinning is believed to result from the alignment of defect dipoles with the spontaneous polarization within a domain. The defect dipoles are formed from negatively charged Pb-vacancies paired with Nb^{5+} dopant ions. Since the defect dipoles are formed at high temperature, the dipoles are not aligned with P_s initially because the spontaneous polarization is zero in the cubic paraelectric state. Alignment can only take place below the Curie temperature (about 350°C for PZT) where the diffusion rates for Nb^{5+} ions and Pb-vacancies are very low. Such is the case for donor-doped

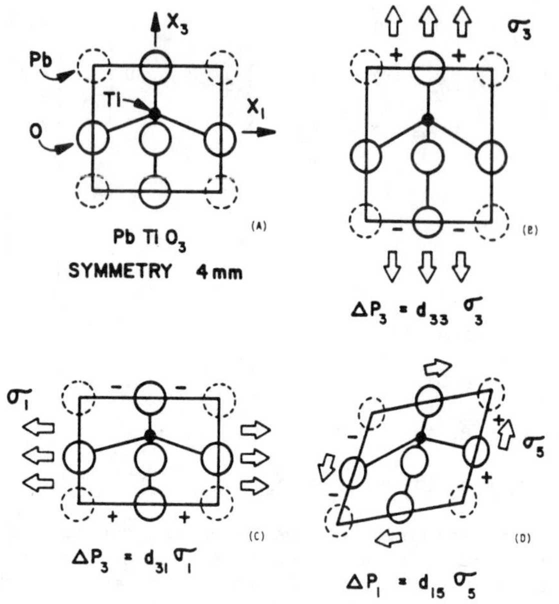

Fig. 4. (a) Electromechanical coupling in PZT ceramics can be illustrated with the tetragonal structure of $PbTiO_3$. The acentric perovskite unit cell belongs to point group 4mm with the titanium ion displaced from the center of the cell. (b) When a tensile stress σ_3 is applied parallel to x_3, the Ti^{4+} ion moves further off-center, creating a positive polarization in the same direction: $P_3 = d_{33}\sigma_3$. (c) If the stress is applied along x_1, the dipole moment of the unit cell is diminished, and a negative polarization appears. Hence d_{31} is negative while d_{33} is positive. (d) For a shear stress about x_2, the dipole moment is tipped, producing polarization in the x_1 direction: $P_1 = d_{15}\sigma_5$.

PZT, a soft PZT in which domain wall motion contributes to the dielectric and piezoelectric coefficients.

Soft PZT transducers are used as hydrophones and ultrasonic detectors where high sensitivity to weak pressure signals is needed. Adversely, however, soft PZT ceramics are easily depoled by large electric fields or mechanical stresses because the domain walls are not pinned. For this reason, hard PZT is used for sonar transmitters and spark generators.

Acceptor doping with lower valent ions such as K^+ for Pb^{2+}, or Fe^{3+} for Ti^{4+}, is employed to produce hard PZT. Oxygen vacancies are generated by acceptor doping. Domain walls are pinned in hard PZT because the defect dipoles are able to align in directions dictated by the domain structure. As pointed out earlier, oxygen sites are only 2.8Å apart in the perovskite structure, making it easy to move oxygen vacancies and realign defect dipoles. Because of the pinning, hard PZT transducers are difficult to pole, and even more difficult to depole.

PTC Thermistors

A thermistor is a temperature-dependent resistor, and a PTC thermistor has a positive temperature coefficient of resistance. Most commercial PTC thermistors are made from lightly doped $BaTiO_3$ ceramics. The resistance of PTC barium titanate shows a rapid increase of several orders of magnitude near the ferroelectric phase transition of 130°C.

When lightly doped with donor ions such as La^{3+} (for Ba^{2+}) or Nb^{5+} (for Ti^{4+}) the room-temperature resistivity of $BaTiO_3$ decreases sharply (Fig. 5a). If sintered and cooled in air (or oxygen) this low resistivity ceramic shows a pronounced PTC effect (Fig. 5b). Single crystals prepared in this way to not show a PTC anomaly, nor do ceramics prepared in a reducing atmosphere.

Conduction Mechanism

Based on these observations it is apparent that grain boundaries, the oxidizing atmosphere, donor dopants, and the ferroelectric phase transformation all play a role in the PTC effect, making it an extremely complex phenomenon. Explanation of the PTC effect rests upon understanding the electronic defect structure. When sintered at high temperatures (~1300°C), lanthanum-doped $BaTiO_3$ becomes an n-type semiconductor:

$$(Ba^{2+}_{1-x}La^{3+}_{x})(Ti^{4+}_{1-x}Ti^{3+}_{x})O_3^{2-} \quad \text{for } x \sim 10^{-3}.$$

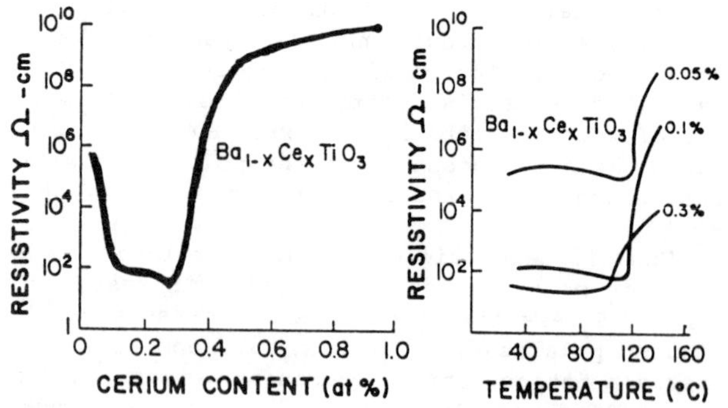

Fig. 5. (a) Room-temperature resistivity of cerium-doped barium titanate ceramics, $(Ba_{1-x}Ce_x)TiO_3$, plotted as a function of composition x. (b) Resistivity of a lanthanum-doped $BaTiO_3$ ceramic measured as a function of temperature. A large PTC anomaly is observed near the Curie temperature.

[Higher doping levels introduce some of the ionic defects described earlier, with an accompanying increase in resistivity (Fig. 5a)]. Conduction in lightly doped $BaTiO_3$ takes place via transfer of electrons between titanium ions: $Ti^{4+} + e^- \longleftrightarrow Ti^{3+}$. Thus the barium titanate grains in the n-type ceramic are conducting, and remain conducting on cooling to room temperature.

Grain Boundary Barriers

But the grain boundary regions change dramatically during cooling. Oxygen is adsorbed on the surface of the ceramic and diffuses rapidly to grain boundary sites, altering the electronic defect structure along the grain boundaries. The added oxygen ions attract electrons from nearby Ti^{3+} ions, thereby reducing the conductivity and creating an insulating barrier between grains. If y excess oxygens are added per formula unit the grain boundary region can be described as follows:

$$(\overset{2+}{Ba}_{1-x}\overset{3+}{La}_x)(\overset{4+}{Ti}_{1-x+2y}\overset{3+}{Ti}_{x-2y})\overset{2-}{O}_{3+y}.$$

As a result of this process the ceramic consists of semiconducting grains separated by thin insulating grain boundaries (Schottky barriers). The electrical resistance of the ceramic is inversely proportional to the grain size because there are more grain boundaries per unit thickness in fine-grained ceramics, and therefore higher resistance.

Influence of Ferroelectricity

To explain the PTC effect, it is necessary to consider the ferroelectric phase transition and its effect on the insulating barriers between grains. Barium titanate is cubic and paraelectric above 130°C, the Curie temperature of pure $BaTiO_3$. Below this temperature the perovskite structure distorts to a tetragonal ferroelectric state in which a large spontaneous polarization P_s develops on the (001) faces of each domain. The dielectric constant reaches a maximum at T_c and then falls off in the high-temperature paraelectric state following a Curie-Weiss law:

$$K \cong \frac{C}{T - T_c}$$

for T greater than T_c. The Curie constant C is about 10^5°C.

The PTC anomaly in doped $BaTiO_3$ occurs at temperatures near T_c and is strongly affected by the appearance of ferroelectricity. Both the spontaneous polarization and the Curie-Weiss law play important roles in the PTC effect.

PTC Anomaly in $BaTiO_3$

At room temperature the resistance of the PTC thermistor is low because the electron charge trapped in grain boundary regions is partly neutralized by spontaneous polarization charges. Wherever the domain structure is advantageously positioned, positive polarization charge will cancel the negatively charged barriers between conducting grains, thereby short-circuiting the Schottky barriers and establishing low resistance paths across the ceramic.

Above T_c the spontaneous polarization disappears and the resistivity increases giving rise to the PTC effect. At first the increase is very slow because of the high dielectric constant of the Curie point. The barrier height is inversely proportional to the dielectric constant of the surrounding medium; a highly polarizable medium shields the charges trapped at the grain boundary, reducing the height of the barrier and lowering its electric resistance. As the temperature increases further above T_c, the dielectric constant K decreases rapidly in accordance with the Curie-Weiss law. The reduction in K causes the Schottky barrier between grains to increase resulting in a large increase in resistance of several orders of magnitude. Eventually the decrease in K with temperature slows down and the resistance levels off until at high temperature the normal NTC effect of the semiconducting grains takes over.

Thermistor Applications

PTC thermistors are used as protection against current surges, temperature rises, and short circuits. When connected in series with the load, a PTC thermistor limits the current to safe levels. A surge in current causes the temperature of the thermistor to rise into the PTC range, thereby raising the resistance and reducing the current and protecting the load. A sudden loss of resistance or a sudden increase in ambient temperature has a similar effect on the thermistor. Two major markets are the electric choke and honeycomb heater used to reduce noxious emissions in automobiles.

Further Reading

Two recent books by John Herbert provide an in-depth study of the composition, physical properties, and manufacture of ceramic capacitors and transducers. Excellent literature surveys are included in the texts. Another new book, "Ceramic Materials for Electronics," edited by Relva Buchanan contains authoritative articles on PTC thermistors, piezoceramics, and multilayer capacitors. The most recent meeting on ferroelectrics is ISAF '86 and the Proceedings provide an up-to-date summary of recent advances in perovskite electroceramics.

References

Buchanan, R.C. (Editor). Ceramic Materials for Electronics. Marcel Dekker, Inc.: New York. 481 pp. (1986).

Herbert, J.M., Ferroelectric Transducers and Sensors. Gordon and Breach Science Publishers: New York. 437 pp. (1982).

Herbert, J.M., Ceramic Dielectrics and Capacitors. Gordon and Breach Science Publishers: New York. 264 pp. (1985).

ISAF '86. Proceedings of the Sixth IEEE International Symposium on Applications of Ferroelectrics, IEEE Publishing Services: New York. 741 pp. (1986).

DEFECT EQUILIBRIA IN PEROVSKITE OXIDES

D. M. Smyth

Materials Research Center, Lehigh University

Abstract. The basic features of the perovskite structure, for which the ideal composition is ABO_3, can persist over a wide range of compositional variation. Thus compositions over the entire range $A_2B_2O_5$ to $A_2B_2O_7$ may retain layers of various thickness having the perovskite structure, separated by layers of oxygen-deficiency or oxygen-excess. Similar structures have been observed in compositions that are AO-rich. The evolution of such structures from the ideal perovskite as a result of aliovalent doping or of oxidation-reduction reactions is reviewed.

Introduction

The general pattern of the perovskite structure is extremely tenacious and it persists over a wide range of cation substitutions, both isovalent and aliovalent, and over a wide range of nonmetal activities. Actually, the response to aliovalent doping and to oxidation-reduction are linked, since reduction, the loss of oxygen from an oxide, reduces the average formal oxidation state of the cations, and is thus equivalent to substitution of a lower-valent cation for one of the host cations. In addition, the extent of nonstoichiometry achieved by changes in the equilibrium oxygen activity depends very strongly on the chemical nature of the cations present. In order for an oxide to undergo extensive reduction, for example, it must contain a cation that is easily reduced. For modest deviations from the ideal perovskite composition, the oxides typically respond by the formation of lattice defects distributed randomly throughout the perovskite matrix. For large deviations, on the other hand, the defects may order into a superstructure that maintains major elements of the perovskite structure. This paper will review the ways in which perovskite oxides accommodate aliovalent substitutions, and how this affects their oxidation-reduction behavior.

The Compositional Line

It is customary to represent the perovskite oxides by the generic formula ABO_3, where A is usually considerably larger than B. It is then convenient to organize compositions along a compositional line that is described by the O/A ratio, as shown in Figure 1 [Smyth, 1985]. For convenience, we will use the specific case of $SrTiO_3$ as an example.

Acceptor-Doping or Reduction

Reduction of perovskite oxides results in the subtractive loss of oxygen and the creation of oxygen vacancies.

$$O_o \rightleftharpoons 1/2O_2 + V_o^{\cdot\cdot} + 2e' \qquad (1)$$

The defect notation is that of Kroger and Vink [1956] where the main symbol identifies the species, the subscript indicates the lattice site on which it is located, and the superscript represents the resulting difference in charge relative to the ideal lattice, with dots indicating extra positive charges (or missing negative charges) and slashes extra negative charges (or missing

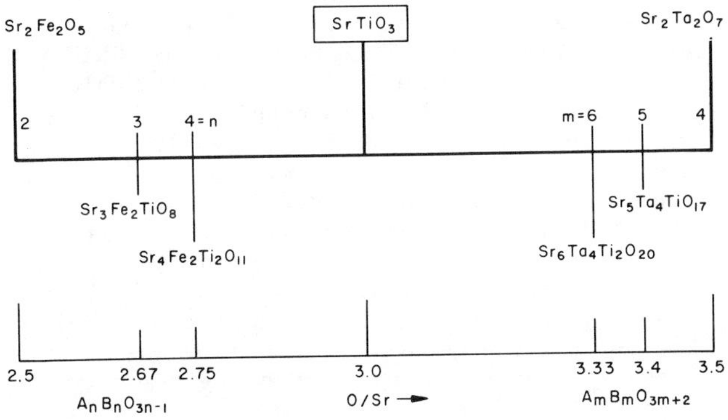

Fig. 1. The perovskite-related compositional line from $A_2B_2O_5$ to $A_2B_2O_7$, showing potentially ordered compositions.

positive charges). In this reaction neutral oxygen leaves the crystal, leaving the vacant oxygen site and the two electrons with which the atom had been combined in the form of an oxide ion. In the case of modestly reducible cations, such as Ti, Zr, or Ta, the oxygen loss can reach the percent level in equilibrium with CO/CO_2 mixtures, for example, and can be treated by standard dilute solution thermodynamics through the mass-action expression for the reduction reaction

$$[V_o^{\cdot\cdot}]n^2P(O_2)^{1/2} = K_n \qquad (2)$$

where K_n is the mass-action constant and contains the enthalpy of the reduction reaction, brackets denote the activity of the enclosed species, $n=[e']$, and $P(O_2)$ is the oxygen activity in the gas phase.

Substitutional lower-valent cations, frequently called acceptor impurities, also result in oxygen vacancies. Thus Fe^{+3} or Al^{+3} can be accommodated on Ti-sites according to the reaction

$$2BaO + Al_2O_3 \rightarrow 2Ba_{Ba} + 2Al'_{Ti} + 5O_o + V_o^{\cdot\cdot} \qquad (3)$$

For the addition of small amounts of acceptor impurities, these vacancies are randomly distributed and interact with the reduction reaction through Equation (2). Cation diffusion is very much restricted in the perovskite structure, so once the impurity cation is incorporated into the structure during the formation of the material, it remains fixed in place.

For large concentrations of acceptor impurities, the oxygen vacancies may become ordered, thus forming new structures. This has been observed in the Ca-Ti-Fe-O system, for example, where a series of structures corresponding to the generic formula $A_nB_nO_{3n-1}$ have been identified [Grenier, et al., 1976], e.g. $Ca_4Fe_2Ti_2O_{11}$, $Ca_3Fe_2TiO_8$, and $Ca_2Fe_2O_5$. In the corresponding Sr compositions, an ordered structure has been observed only in the endmember $Sr_2Fe_2O_5$. This member has the brownmillerite structure, which is shown in Figure 2. This consists of two-dimensional layers of the perovskite structure, one unit cell thick, separated by layers in which the oxygen deficiency relative to the ideal perovskite composition is ordered in lines so that the small cations in this layer are 4-coordinate in tetrahedral symmetry. In successive members of the series i.e. for n>2 the perovskite layers become stepwise thicker, such that they are n-1 unit cells thick. This is a striking example of the retention of as much of the perovskite structure as possible even when the composition is far from the ideal ABO_3.

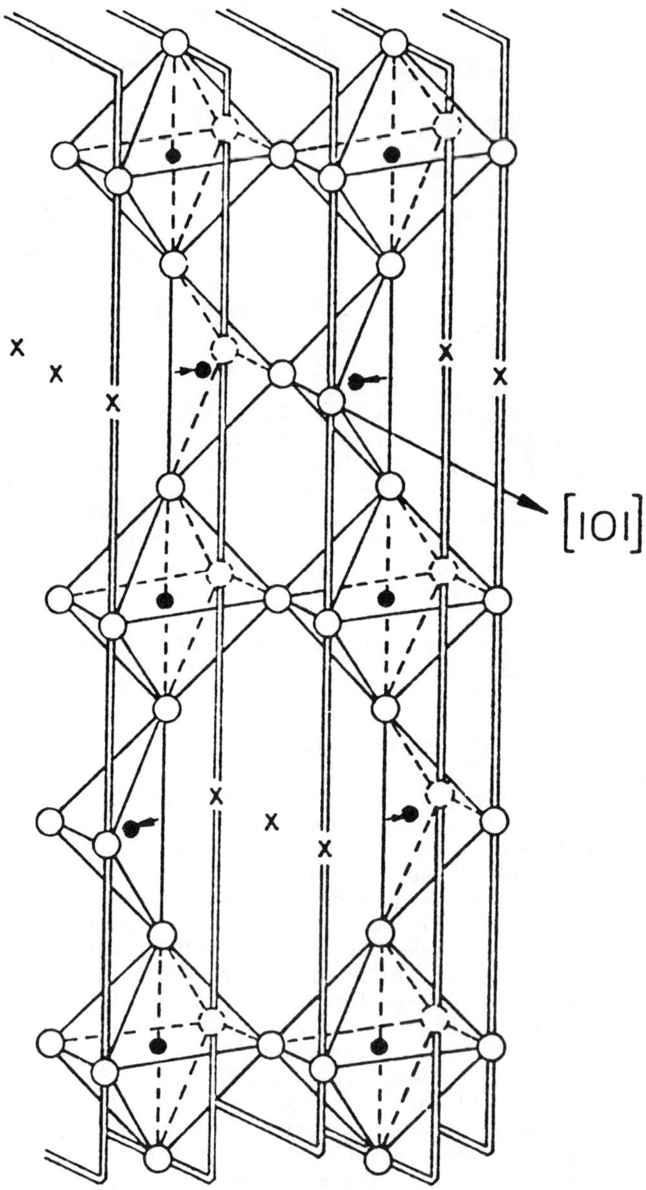

Fig. 2. The brownmillerite structure. Unoccupied oxygen sites relative to the perovskite structure shown as X in ordered rows.

The brownmillerite-perovskite structural sequence just described was originally studied in highly acceptor-doped materials, i.e. substitution of Fe^{+3} for Ti^{+4}. However, since the ratio of large cations to small cations remains fixed at unity throughout the series, in principle related compositions can also be obtained by oxidation-reduction reactions. In a study of $Sr_2Fe_2O_5$, it has been shown that the stoichiometric composition is in equilibrium with 10^{-10} atm of oxygen at 1000°C, and that at 950°C in 1 atm of oxygen, the composition, $Sr_2Fe_2O_{5.3}$, is very close to that of the next member of the homologous series: $Sr_3Fe_3O_8 = Sr_2Fe_2O_{5.33}$ [Chang, E. K., Lehigh University, unpublished]. In the range 700-800°C in pure oxygen, the composition will be close to $Sr_4Fe_4O_{11}$. These compositions can be viewed as mixtures of Fe^{+3} and Fe^{+4}, analogous to mixtures of Fe^{+3} and Ti^{+4}. In the reduction sense that we are equating to the substitution of acceptor impurities, these compositions are derived by reduction of $SrFeO_3$. Equilibrium conductivity data indicate that the compositions are best viewed as $Sr_2Fe_2O_{5+x}$, and that for x>0, they are oxygen-excess versions of the stoichiometric $Sr_2Fe_2O_5$. These are obtained by the oxidation reaction

$$V_I + 1/2O_2 \rightleftharpoons O_I'' + 2h^{\bullet} \qquad (4)$$

where V_I is a vacant oxygen site relative to the perovskite structure, but in the ordered $Sr_2Fe_2O_5$, they are interstitial sites that become partially filled by oxidation. The neutral oxygen atoms that enter these sites pick up electrons from the valence band to become oxide ions, and the resulting hole species contribute to a p-type conductivity that increases with increasing oxygen pressure.

Donor-Doping or Oxidation

Excess oxygen is not likely to occur in a pure perovskite phase because the energy required to form interstitial oxygen, or cation vacancies of both types, is prohibitive. The common observation that important electronic ceramics having the perovskite structure are oxygen-excess, p-type insulators is related to the naturally-occurring impurity content. The most abundant impurities have smaller charges than the cation they replace, e.g. Na^+ and K^+ for Ba^{++} and Fe^{+3} and Al^{+3} for Ti^{+4} in $BaTiO_3$ [Chan et al., 1981]. They are thus acceptor impurities and are compensated by oxygen vacancies as shown in Eq. (3). These oxygen vacancies can then be filled by a stoichiometric excess of oxygen until the oxygen sublattice is saturated

$$V_O^{\bullet\bullet} + 1/2O_2 \rightleftharpoons O_O + 2h^{\bullet} \qquad (5)$$

The holes are trapped out by the acceptor centers at ambient temperatures to give the familiar insulating behavior that is essential for most electronic applications.

The substitution of a donor impurity, such as Nb_2O_5 substituted for TiO_2 in $SrTiO_3$, represents an attempt to bring a stoichiometric excess of oxygen into the lattice. This may not be tolerated, in which case the extra oxygen is expelled, leaving behind electrons that contribute to n-type conductivity.

$$2BaO + Nb_2O_5 \rightarrow 2Ba_{Ba} + 2Nb_{Ti}^{\bullet} + 6O_O + 1/2O_2 + 2e' \qquad (6)$$

For large impurity concentrations, the extra oxygen is usually retained, and the impurity charge is compensated by cation vacancies. In the case of $BaTiO_3$, the compensating defect has been shown to be the Ti-vacancy, and may result in the separation of a Ti-rich phase [Chan et al., 1986].

$$5BaO + 2Nb_2O_5 + Ti_{Ti} + 2O_O \rightarrow 5Ba_{Ba} + 4Nb_{Ti}^{\bullet} + V_{Ti}'''' + 15O_O + TiO_2 \qquad (7)$$

In the case of $SrTiO_3$, it appears that Sr-vacancies are the compensating defects.

When large fractions of the B cations are replaced by donor impurities, the system may eliminate the defects by forming a superlattice that retains layers of the perovskite structure separated by oxygen-rich layers, as shown in Figure 3. The structures correspond to the homologous series of compositions $Sr_nNb_4Ti_{n-4}O_{3n+2}$, with n≥4, in the case of the Sr-Nb-Ti-O system. The two-dimensional perovskite slabs are n unit cells thick, and are separated by layers that contain the excess oxygen.

Since the A/B ratio is unity for all of these structures, it is again possible in principle to obtain them by oxidation-reduction reactions. Since it is not possible to oxidize a compound such as $SrTiO_3$ to oxygen-excess compositions, it is best to view the process as reduction of $Sr_2Nb_2O_7$. In a recent study of the oxidation-reduction equilibria in this material, it was found that reduction resulted in an intergrowth of the n = 4 and n = 5 members of the homologous series [Liu, D. H., Lehigh University, unpublished]. As oxygen is removed by reduction, there is less need for oxygen-rich layers and there is the appearance of perovskite slabs five unit cells thick dispersed in the matrix of slabs four unit cells thick. The response of the system is structural rather than by the formation of point defects. As in the case of the acceptor-doped materials, this demonstrates a strong tendency to order the structures to retain as much as possible of the perovskite pattern.

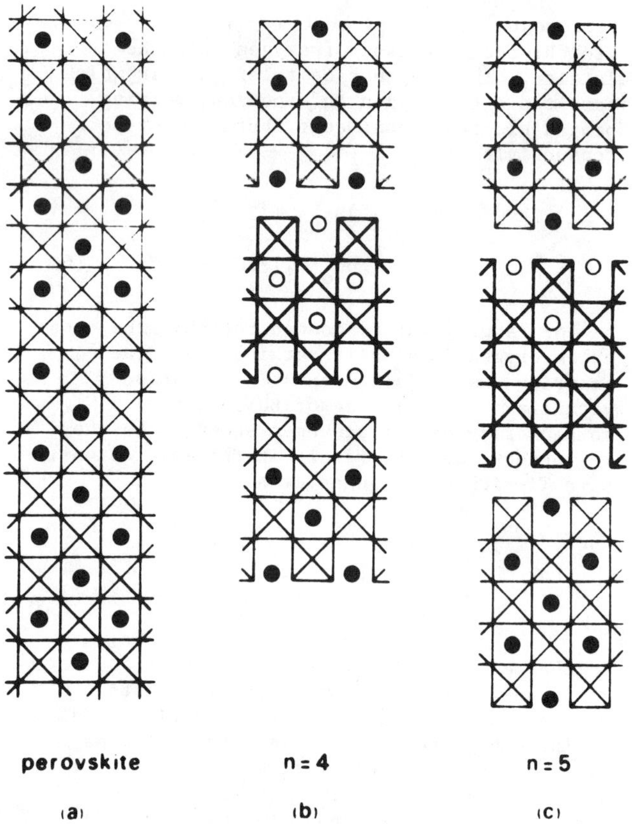

Fig. 3. The structure of members of the $A_nB_nO_{3n+2}$ homologous series with n = 4, 5, and ∞. The structures can be derived by a splitting of the perovskite structure into two-dimensional blocks, with extra oxygen terminating the separated corners.

Excess AO

Thirty years ago, Ruddlesden and Popper described a series of structures in the Sr-Ti-O system that corresponded to compositions having the generic formula $SrO\text{-}nSrTiO_3$ [Ruddlesden and Popper, 1958]. The structures consist of two-dimensional perovskite slabs, n unit cells thick, separated by single layers of excess AO in a NaCl structural sequence. The structure of the member with n = 2, $Sr_3Ti_2O_7$, is shown in Figure 4. Interest in these structures has picked up recently, because they are very closely related to the structures of the newly discovered high temperature superconductors. It has been shown that all of the high temperature superconducting oxides discovered to date can be described in terms of an expanded Ruddlesden and Popper series with the generic formula $mAO\cdot nABO_3$, [Smyth, 1988]. The structures have two dimensional slabs of the perovskite structure, n unit cells thick, separated by layers of excess AO in the NaCl structure, m layers thick.

$MgSiO_3$

Of particular interest to geophysicists, $MgSiO_3$, a major component of the earth's mantle, is squeezed into a slightly distorted perovskite structure under the temperature-pressure conditions of the mantle. Under more normal conditions, Mg^{++} is much too small to occupy the 12-coordinate A-site in the perovskite structure, and in such normal

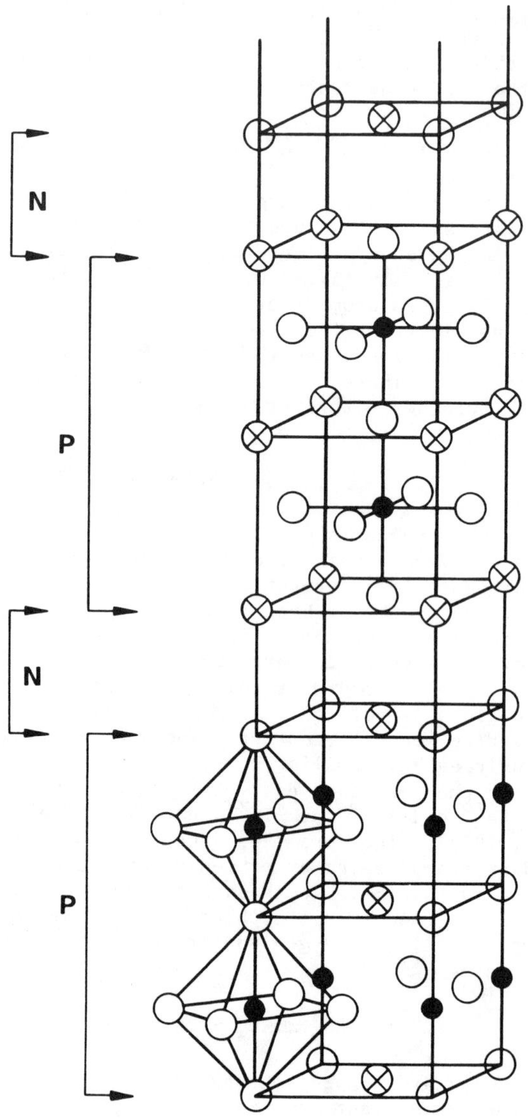

Fig. 4. The structure of $Sr_3Ti_2O_7$, the member of the Ruddlesden-Popper series, $A_{n+1}B_nO_{3n+1}$, with n = 2.

perovskites as $BaTiO_3$ and $SrTiO_3$, Mg^{++} preferentially substitutes for Ti on the B-sites. $MgTiO_3$ itself has the ilmenite structure where both cations are octahedrally coordinated in a hexagonally-close-packed oxygen array.

The defect chemistry of $MgSiO_3$ in the perovskite structure can be expected to follow the usual perovskite pattern. Insofar as Si^{+4} can be considered to be a reducible cation, $MgSiO_3$ should lose oxygen at low oxygen activities with the formation of oxygen vacancies and n-type conductivity [see Eq.(1)]. It should achieve a stoichiometric excess of oxygen and p-type conductivity at high oxygen activities only when acceptor-doped so that compensating oxygen vacancies are available for the accommodation of excess oxygen [see Eqs. (3) and (5)]. Excess MgO may be easily accommodated by the structure because of the ease with which Mg^{++} can fit into the octahedral B-sites

$$2MgO \rightarrow Mg_{Mg} + Mg_{Si}'' + 2O_o + V_o^{\cdot\cdot} \qquad (8)$$

No comparable reaction for excess SiO_2 is to be expected because Si is never found in 12-coordinated sites; however, excess SiO_2 may be possible with the formation of Mg vacancies

$$SiO_2 \rightarrow V_{Mg}'' + Si_{Si} + 2O_o + V_o^{\cdot\cdot} \qquad (9)$$

In the complex composition of the earth's mantle, the $MgSiO_3$ phase is more accurately described as $(Mg_{0.9}Fe_{0.1})SiO_3$, where the Fe is the major component of the group of impurities that include Ni, Co, etc. It has been suggested that in the mantle, the Fe may largely reside on the B-sites, and this would not be surprising given the propensity of Fe^{++} for octahedral coordination [Jeanloz, et al., 1988]. The movement of the Fe to the B-sites can be compensated by an uptake of Mg^{++} from the coexisting magnesio-wustite phase in the mantle

$$Fe_{Mg} + 2MgO[(Mg\text{-}Fe)O] \rightarrow 2Mg_{Mg} + Fe_{Si}'' + 2O_o + V_o^{\cdot\cdot} \qquad (10)$$

If this is the case, the mineral should show an excess of Mg+Fe over Si, i.e. (Mg+Fe)/Si>1. The material also becomes acceptor-doped by this transfer of the lower valent cation to the B-site.

Summary

For large deviations from stoichiometry in perovskite-based systems, there is a strong tendency to confine the nonstoichiometry to single, ordered layers that separate layers that retain the perovskite structure. As discussed, this may involve oxygen-deficiency, oxygen-excess, or an excess of AO. The responses are linked for all of these cases, since the common feature of the ordered structures is the two-dimensional slab of perovskite structure of variable thickness. This behavior is a demonstration of the high degree of energetic stability characteristic of the perovskite pattern. The systems try to retain as much of this pattern as possible, even when the composition deviates grossly from the ideal.

Acknowledgement. The opportunity to concentrate on the basic solid state chemistry of perovskite-related systems has been made possible by continuous support from the Division of Materials Research of the National Science Foundation. The author is grateful.

References

Chan, H. M., M. P. Harmer, and D. M. Smyth, Compensating defects in highly donor-doped $BaTiO_3$, J. Am. Ceram. Soc., 69, 507-510, 1986.

Chan, N.-H., R. K. Sharma, and D. M. Smyth, Nonstoichiometry in undoped $BaTiO_3$, J. Am. Ceram. Soc., 64, 556-562, 1981.

Grenier, J.-C., J. Darriet, M. Pouchard, and P. Hagenmuller, Mise en evidence d'une nouvelle famille de phases de type perovskite lacunaire ordonnee de formule $A_3M_3O_8$ $(AMO_{2.67})$, Mat. Res. Bull., 11, 1219-1226, 1976.

Jeanloz, R., E. Knittle, Q. Williams, and X. Li, $(Mg, Fe)SiO_3$ perovskite: geophysical, and crystal chemical significance, this volume.

Kroger, F. A., and H. J. Vink, in Solid State Physics, Vol. 3, edited by F. Seitz and D. Turnbull, p. 307, Academic Press, New York, 1956.

Ruddlesden, S. N., and P. Popper, The compound $Sr_3Ti_2O_7$ and its structure, Acta Cryst., 11, 54-55, 1958.

Smyth, D. M., Defects and order in perovskite-related oxides, in Ann. Rev. of Mat. Sci., edited by R. A. Huggins et al., pp. 329-357, Annual Reviews Inc., Palo Alto, 1985.

Smyth, D. M., Structural patterns in high T_c superconductors, Advances in Ceramics, Proceedings of the Superconductor Symposium, 90th Annual Meeting of the American Ceramic Society, in press, 1988.

LOW TEMPERATURE SYNTHESIS OF OXYGEN DEFICIENT PEROVSKITES

John B. Wiley and Kenneth R. Poeppelmeier

Department of Chemistry and Ipatieff Catalytic Laboratory

Northwestern University

Abstract. Isopiestic and thermal gradient techniques have been used to study oxygen nonstoichiometry in polycrystalline $CaMnO_3$, Ca_2MnO_4 and single crystals of $CaMnO_3$. The topotactic nature of oxygen depletion and reoxidation that occurs at temperatures below 700°C in these manganese perovskites has been a particular focus.

Introduction

Oxygen deficient forms of perovskite (ABO_{3-x}) and perovskite-related ($AO \cdot n(ABO_{3-x})$; n=1,2,3,...∞) compounds can show a variety of structural and electronic properties. The importance of such compounds has most recently been made evident by the 90K superconductor, $YBa_2Cu_3O_{7-x}$. As the oxygen content of this compound varies from x=0 to x=1 structural changes are accompanied by a decrease in the transition temperature, T_c, until superconductivity is no longer observed [Gallagher et al., 1987]. Since the variation in such properties is dependent on the material's oxygen content, the study of this phenomenon could be expedited by a method that can conveniently synthesize phases of intermediate stoichiometry.

Oxygen deficient perovskites are generally prepared by reduction at high temperature in air or with a buffered gas mixture. The dependence of these methods on the oxygen partial pressure makes it difficult to prepare compounds of a selected stoichiometry without prior knowledge of the equilibrium pO_2. An isopiestic method, however, can be used for the synthesis of intermediate phases without exact knowledge of the equilibrium partial pressure.

With the technique used in this work, zirconium metal is used as an oxygen scavenger. In an evacuated system at elevated temperatures, the oxygen released from mixed metal oxides irreversibly reacts with zirconium metal to form ZrO_2. If the amount of zirconium is strictly controlled then an oxygen deficient phase of desired stoichiometry is readily prepared.

In general the structures of oxygen deficient perovskites have been determined from polycrystalline samples. The availability of single crystals would make structure determination, and other physical measurements easier. Attempts to topotactically reduce single crystal with this isopiestic method are described in this paper.

Experimental

Sample Preparation

Polycrystalline oxides. Polycrystalline $CaMnO_3$ and Ca_2MnO_4 were prepared by the decomposition of carbonate precursors [Horowitz and Longo, 1978]. Solid solution precursors were precipitated from aqueous solutions of calcium and manganese nitrate. The solutions of calcium and manganese nitrate were prepared by dissolving $CaCO_3$ in dilute nitric acid and combining it with $Mn(NO_3)_2$ solution (Mallinkrodt, 50% W/W). The $CaCO_3$ was reagent grade and thermogravimetric analysis showed the sample to be > 99.9% $CaCO_3$. Reagent grade $MnCO_3$ was unacceptable because it always was found to contain a significant fraction of oxidized manganese and other noncarbonate phases. The Mallinkrodt $Mn(NO_3)_2$ solution was stable with respect to oxidation and so proved superior to other commercial sources of manganese (II) compounds. Compleximetric titrations with EDTA were used to verify the concentration of the $Mn(NO_3)_2$ solution [West, 1969].

Solid solutions $Ca_{1-x}Mn_xCO_3$ ($0.0 \le x \le 1.0$) were precipitated by the slow addition of the appropriate premixed solution of calcium and manganese nitrate to a large excess of approximately 2M ammonium carbonate solution with good mixing. The off-white precipitates were washed with water, dried at 100°C in a vacuum oven, and fired in air in an open crucible in a preheated furnace at 1000°C for two days. It should be noted that the manganese(II) in the solid solution carbonates is sensitive to oxidation in air. Because premature oxidation can lead to a nonequilibrium mixture of manganese oxide and calcium-rich manganates [Longo et al., 1980], the precursors were not exposed to air for prolonged periods.

Single Crystals. Crystals of $CaMnO_3$ were grown in scrupulously dry $CaCl_2$. Water, by reacting with molten $CaCl_2$ to form CaO [Ohsato et al., 1980], leads to the growth of undesired calcium-rich phases. To prevent this, efforts were made to exclude any water from the flux. Reagent grade $CaCl_2 \cdot 2H_2O$ was heated past its melting point (782°C) under vacuum to form the anhydrous salt. This $CaCl_2$, along with $CaMnO_3$, was stored in a dry nitrogen environment.

The charge consisted of 10 mole% $CaMnO_3$ in dry $CaCl_2$ (Aldrich, reagent). Platinum boats, sealed into quartz tubes under vacuum ($< 10^{-4}$ torr), were used to contain the mixture. The temperature program for crystal growth consisted of a soak time of three days at 825°C, a slow cool of 1°C per hour to 700°C, and a final, more rapid cooling to room temperature. A water wash was used to isolate the crystals from the melt. The shiny black cube-shaped crystals averaged 0.15 mm. on a side (See Figure 1(a)).

Isopiestic equilibration. Oxygen deficient perovskites were synthesized by reduction with zirconium metal. Stoichiometric amounts of mixed metal oxide (ca. 0.5 g) and zirconium metal (-20 +60 mesh, Alfa 99.9%) were weighed out separately into small quartz vials (7 mm. o.d.) on an analytical balance. The vials were placed into a 12 mm. o.d. quartz tube (See Figure 2) that was then sealed under vacuum ($<10^{-4}$ torr). Samples were slowly heated over two days to 600°C. A chromel-alumel thermocouple was placed within a few millimeters of the reaction end of the sealed tube to monitor the temperature. Typical reaction times were seven days, with the completion of the reaction judged by the conversion of the grey, coarse grained zirconium metal to a white fine powder of ZrO_2. Oxidation of zirconium was observed to occur near 580°C. Samples were reweighed after the quartz tubes were opened to verify that complete reaction had occurred.

Single crystals were reduced in a similar manner, except, to conserve the number of crystals used, a small amount (ca. 0.001g) was placed into the sealed tube along with polycrystalline mixed metal oxide and the zirconium metal.

Initially all the syntheses were done in an isothermal manner at 600°C. This was necessary because the zirconium powder used did not oxidize until at least 580°C. With some compounds an isothermal approach such as this is not ideal, i.e. when the thermal stability of the product is exceeded. An alternative would be to position the thermally unstable oxide at a lower temperature while maintaining the zirconium at or above 580°C. This would allow the zirconium charge to be at its required temperature, while the mixed metal oxide, at a lower temperature would avoid thermal decomposition. To employ this approach, two zone furnaces were used. Typically the zirconium metal was heated to 600°C while the end of the sealed tube containing the mixed metal oxide was kept significantly cooler. To determine the minimum temperature required for reduction, the temperature of the oxide was raised until oxidation of the zirconium became discernible.

Characterization

X-ray Diffraction. Polycrystalline oxide samples were characterized before and after reduction by X-ray powder diffraction (XRD). A Rigaku Geigerflex diffractometer, equipped with Cu radiation and a Ni filter, was used to obtain room temperature data in the range $5° \leq 2\theta \leq 80°$ in increments of 0.02°.

Oscillation photographs of the crystals before and after reduction, obtained on a Supper Weissenberg camera, were used to observe the changes in the single crystals. $CaMnO_3$ was found to be orthorhombic (Pnma) with a = 5.279(1), b = 7.448(1) and c = 5.264(1) Å.

Thermogravimetric Analysis. Thermogravimetric measurements were performed on a Du Pont 9900 thermal analysis system. Oxygen contents were determined by reduction of the mixed metal oxides in dilute hydrogen (8.5% in helium). Generally flow rates of 25 mL./min. were used. Weight loss was attributed to manganese with an oxidation state greater than two. Oxygen content of the oxidized (Mn^{4+}) compounds were within 1 at.% of the theoretical values $CaMnO_{3.0}$ and $Ca_2MnO_{4.0}$.

Scanning Electron Microscopy. An Hitachi S-570 Scanning Electron Microscope (SEM) was used to study the $CaMnO_3$ crystals before and after reduction.

Results

Polycrystalline $CaMnO_{3-x}$. To determine the utility of this method, systems known to contain oxygen deficient phases were initially examined. The first studied was $CaMnO_{3-x}$, which exhibits the unusual feature of reversible oxygenation over the range $0 \leq x \leq 0.5$ [Poeppelmeier et al. 1982a]. The end-member (Mn^{3+}) with x = 0.5 and intermediate (Mn^{4+}/Mn^{3+}) compositions x = 0.2, 0.25, and 0.33 have been discussed in the literature [Kuroda et al., 1981; Reller et al., 1984]. Synthetic efforts at 600°C successfully yielded $CaMnO_{2.5}$ and $CaMnO_{2.8}$ as determined by TGA and X-ray diffraction (Table I). TGA indicated that the two intermediate compositions

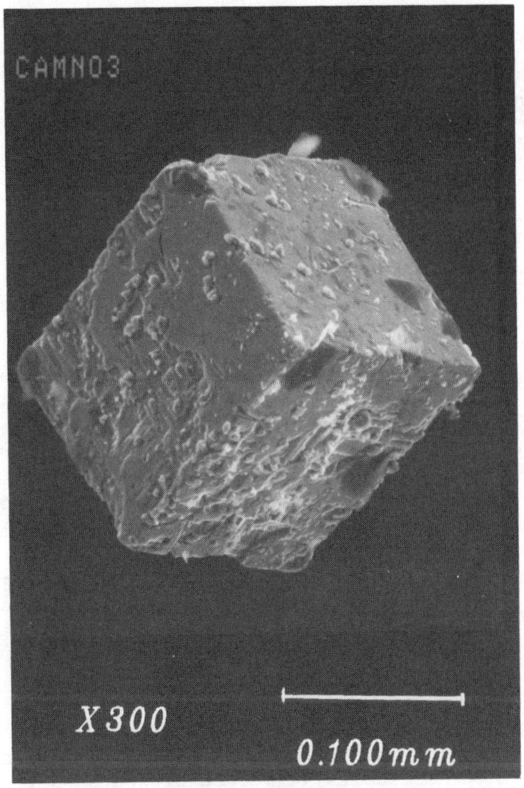

Fig. 1a

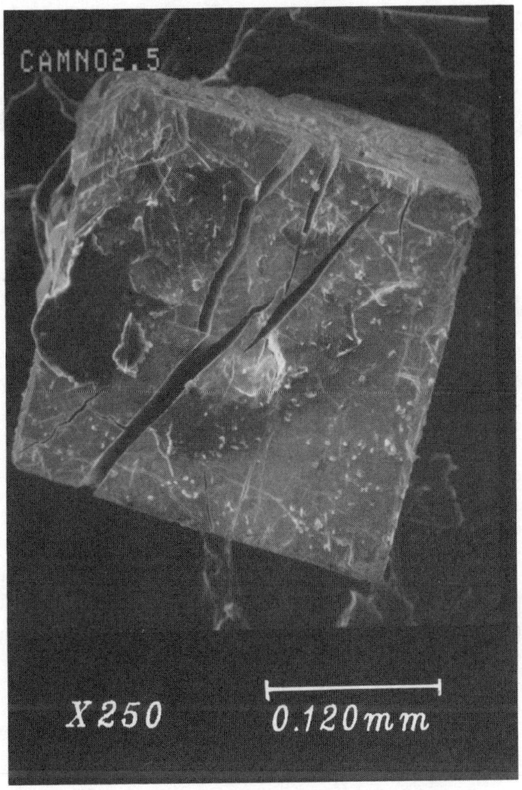

Fig. 1b

Figure 1. SEM photograph of a) $CaMnO_3$ single crystal b) reduced single crystal.

x = 0.25 and 0.33 contained the correct overall oxygen content but X-ray diffraction showed in fact the samples were mixtures of $CaMnO_{2.5}$ and $CaMnO_{2.8}$.

Polycrystalline Ca_2MnO_{4-x}. $Ca_2MnO_{3.5}$ is the only known oxygen-deficient compound in the Ca_2MnO_{4-x} system [Poeppelmeier et al., 1982a, Leonowicz et al., 1985]. It could not be synthesized at 600°C. XRD revealed that unreacted Ca_2MnO_4 was accompanied by several reduction products; $CaMn_2O_4$, $CaMn_2O_3$, $Ca_{0.9}Mn_{0.1}O$ and $Ca_{0.15}Mn_{0.85}O$. It is known that some oxygen-deficient phases are thermally unstable [Poeppelmeier, unpublished work] and there are multiple reduction paths (eqns. 1-3).

$$4Ca_2MnO_4 + (Zr) \longrightarrow 4Ca_2MnO_{3.5} + (ZrO_2) \quad (1)$$

$$2Ca_2MnO_{3.5} \longrightarrow 3CaO + CaMn_2O_4 \quad (2)$$

$$2CaMn_2O_4 + (Zr) \longrightarrow 2CaMn_2O_3 + (ZrO_2) \quad (3)$$

The formation of $Ca_{0.9}Mn_{0.1}O$ and $Ca_{0.15}Mn_{0.85}O$ is probably due to a miscibility gap in the CaO-MnO solid solution [Poeppelmeier et al., 1986]. This collection of compounds is clearly a non-equilibrium mixture.

In an attempt to alleviate any problems that may have been associated with thermal decomposition, a two-zone heating method was employed. The zirconium was placed at the hot end of the tube while the mixed metal oxide was placed at the cooler end. In the case of Ca_2MnO_4, it was found that a temperature of 400°C successfully yielded $Ca_2MnO_{3.5}$ (Table I).

Single Crystals. Attempts to reduce single crystals of $CaMnO_3$ resulted in a loss of crystallinity. Instead of the usual set of intense spots, oscillation photographs showed circles of uneven intensity characteristic of partially oriented crystallites. Even though the crystals did not seem to change appearance, scanning electron micrographs showed that cracks had formed during the reduction process (see Figure 1(b)).

Discussion

Figure 3(a) is an ORTEP drawing of the room temperature structure of $CaMnO_3$ based on a

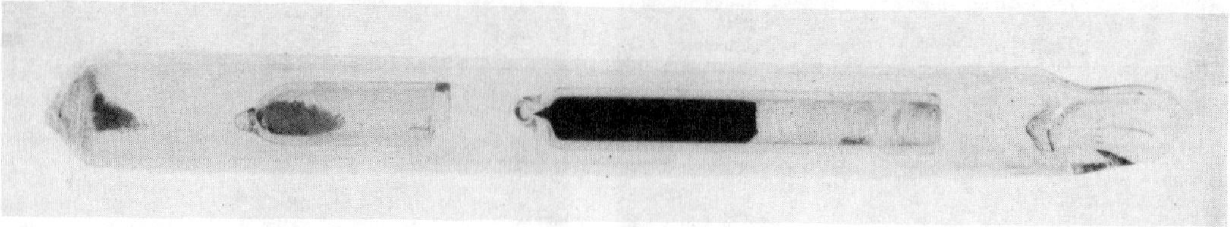

Figure 2. Sealed quartz tube. Reduced mixed oxide (right), oxidized zirconium (left).

single crystal X-ray determination. The structure of $CaMnO_{2.5}$ was determined [Poeppelmeier, et al., 1982b] by neutron diffraction on a polycrystalline sample. Figure 3(b) illustrates the respective polyhedra in the oxygen-defect phase $CaMnO_{2.5}$. Notice that in $CaMnO_{2.5}$ all the $MnO_{5/2}$ polyhedra remain connected through Mn-O-Mn linkages similar to those in $CaMnO_3$, and that the oxygen vacancies form parallel channels throughout the structure. The oxygen vacancies are generated by the reduction of Mn^{4+} in $CaMnO_3$ to Mn^{3+} in the phase $CaMnO_{2.5}$. In Figure 4 the ordering pattern of the oxygen vacancies is illustrated in schematic form. The vacancies order in every (001) BX_2 plane of cubic perovskite with alternate $[110]_c$ strings of oxygen atoms (based on the simple 3.7-Å cubic cell) that are missing every other oxygen. The vacancies are displaced by $a_o/2$ in alternate strings. Thus the manganese ions are coordinated by five oxygen anions in $CaMnO_{2.5}$ (see Figure 3(b)) as opposed to six in $CaMnO_3$ (see Figure 3(a)).

Table I. Thermogravimetric Analysis (TGA) and X-ray powder diffraction (XRD) data for $CaMnO_{3-x}$ and Ca_2MnO_{4-x}.

$CaMnO_{3-x}$ (600°C)

x	TGA[a]	XRD
0.00	0.00	$CaMnO_3$
0.20	0.19	$CaMnO_{2.8}$
0.25	0.26	$CaMnO_{2.8}$ $(CaMnO_{2.5})$[b]
0.33	0.34	$CaMnO_{2.8}$, $CaMnO_{2.5}$
0.50	0.49	$CaMnO_{2.5}$

Ca_2MnO_{4-x} (400°C)

x	TGA	XRD
0.00	0.01	Ca_2MnO_4
0.50	(0.49)[c]	$Ca_2MnO_{3.5}$

[a]Propagation of error calculation estimated the error to be ± 0.02.

[b]Minor component.

[c]Sample air sensitive w.r.t. oxidation. Value based on bulk sample weight immediately after reduction and after exposure to ambient conditions.

The intermediate phases in the $CaMnO_{3-x}$ systems can be viewed as a combination of end members, $CaMnO_3$ and $CaMnO_{2.5}$. These phases consist of the appropriate ratios of $MnO_{6/2}$ and $MnO_{5/2}$ polyhedra while oxygen vacancies also order in parallel channels. Reller et al. [1984] have proposed several structural models based on electron diffraction and HREM studies.

The isopiestic method presented here allows for the simple control of oxygen partial pressure above perovskite and perovskite related phases. At elevated temperatures oxygen dissociates from the oxide (eqn. 4) to combine with zirconium to form the more refractory phase ZrO_2 (eqn. 5).

$$ABO_3 \rightleftharpoons ABO_{3-x} + x/2\ O_2 \quad (4)$$

$$Zr + O_2 \longrightarrow ZrO_2 \quad (5)$$

$$ABO_3 + x/2\ (Zr) \longrightarrow ABO_{3-x} + x/2\ (ZrO_2) \quad (6)$$

The overall reaction leads to the conditions needed to make an oxygen deficient phase of selected stoichiometry (eqn. 6).

This premise, however, makes the assumption that the oxygen partial pressure and temperature are the only factors that determine the formation of these defect structures. This may be true in some cases, but in others, factors such as structural reorganization and temperature effects and, for unknown systems, whether the oxide has the ability to support an oxygen deficient structure are equally important.

The point of structural reorganization relates to the ability of one oxygen deficient phase to reorganize to form another with a small additional change in composition. In this instance a possibly significant barrier inhibits

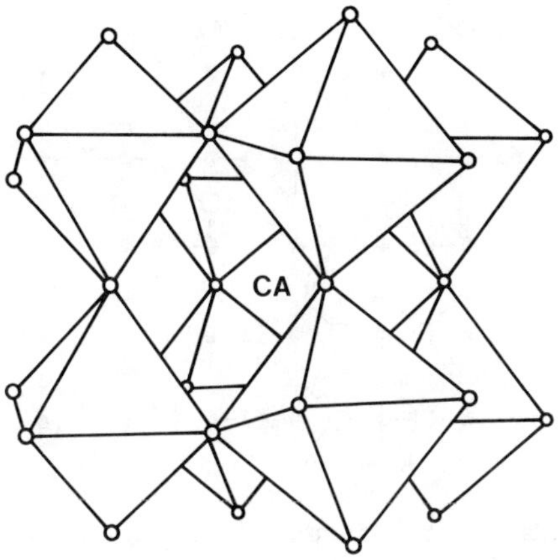

Fig. 3a Fig. 3b

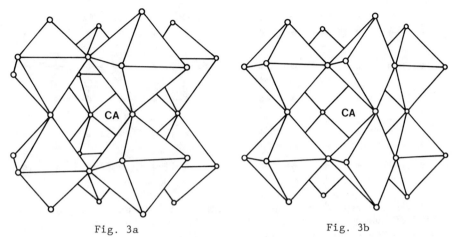

Figure 3. The $MnO_{6/2}$ - connected polyhedra of $CaMnO_3$ (a) and the $MnO_{5/2}$ polyhedra found in $CaMnO_{2.5}$ (b).

the transformation, even for compounds where the stoichiometries are very close. Consider the two phases $CaMnO_{2.8}$ and $CaMnO_{2.75}$. The suggested structures of $CaMnO_{2.8}$ and $CaMnO_{2.75}$ [Reller et al., 1984] are shown in figure 5. Both materials consist of a combination of octahedra and square pyramids but the ordering of oxygen vacancies is different. To transform one structure to the other is not just a question of the simple addition (to $CaMnO_{2.75}$) or removal (from $CaMnO_{2.8}$) of selected oxygen anions. A more significant rearrangement in the relative positions of the established oxygen vacancies must occur.

Our synthetic results appear to give some support to this argument. The two phases $CaMnO_{2.8}$ and $CaMnO_{2.5}$ predominate the $CaMnO_{3-x}$ ($0.2 \leq x \leq 0.5$) system (Table I). These results indicate that the decrease in pO_2 at 600°C necessary to cause the reduction of $CaMnO_{2.8}$ to $CaMnO_{2.75}$ or $CaMnO_{2.67}$ is sufficient to reduce a fraction of the oxide to $CaMnO_{2.5}$ instead of to the intermediates $x = 0.25$ or 0.33.

One might ask at this point, if a barrier to structural reorganization exists then how is it that these phases have been prepared at 600°C [Reller et al., 1982] in buffered gas systems such as H_2/H_2O. The answer may lie in the fact that the hydrogen and water come in physical contact with the bulk material and this combination can move oxygen from one vacancy to another. Hydrogen forming water to create a vacancy at one surface site, while water dissociates to fill a vacancy at another may assist in the lattice reorganization.

The question of thermal stability must also be addressed when considering these systems because some oxygen deficient phases are thermally sensitive. We have shown that in the Ca_2MnO_{4-x} system, $Ca_2MnO_{3.5}$ is not stable at 600°C (eqns. 1,2). At these higher temperatures

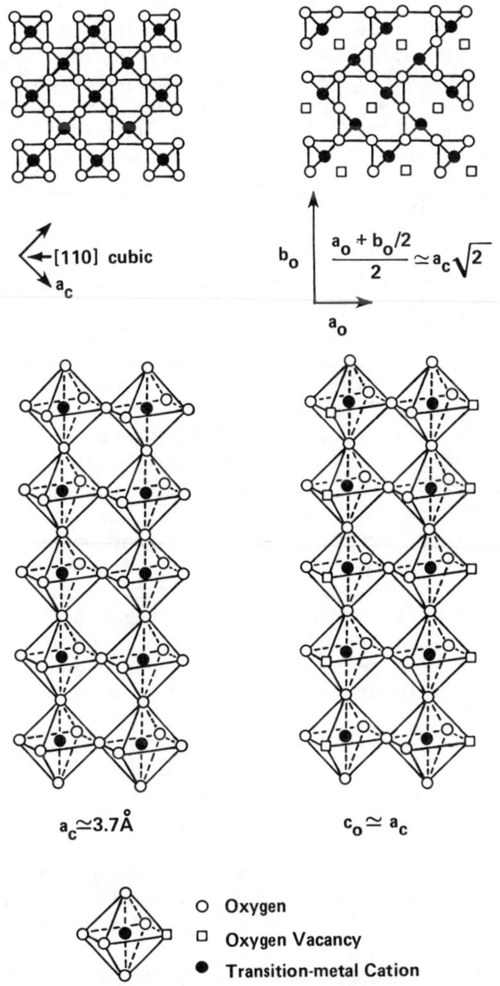

Figure 4. Schematic structures of stoichiometric $CaMnO_3$ (left) and $CaMnO_{2.5}$ (right).

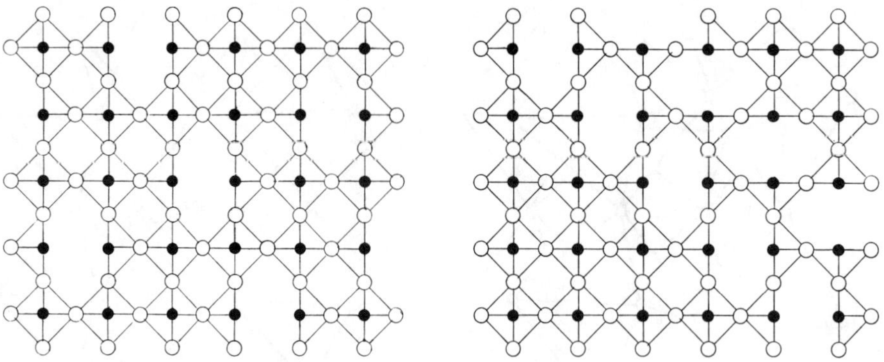

Figure 5. Schematic representations of suggested structures ($MnO_{4/2-x}$ planes) [Reller et al., 1984] of $CaMnO_{2.8}$ (left) and $CaMnO_{2.75}$ (right). Open circles represent oxygen and dark circles represent manganese.

a decomposition reaction (see eqn. 2) to form the compounds CaO and $CaMn_2O_4$ occurs. Lowering the temperature of the oxide to 400°C, however, made it possible to synthesize $Ca_2MnO_{3.5}$.

This example suggests that other metastable oxygen deficient materials can be prepared by this technique. In other systems it may be found that only small accessible temperature ranges exist where the oxygen deficient compound can be obtained, and low temperatures and long reaction times will be necessary to prepare them.

Reduction of a single crystal of $CaMnO_{3.0}$ to $CaMnO_{2.5}$ occurs without change in its external form (pseudomorphic) even though the reaction has proceeded throughout the crystal (see Figure 1). However, the structure has been disrupted and there are small and irregularly oriented crystallites after completion of the reaction [Dent Glasser et al., 1962]. Although the three dimensional nature of the original structure is in all likelihood retained in the individual crystallites, the large degree of misorientation most likely results from the discontinuous relationship between pO_2, temperature and composition leading to epitactic nucleation on external surfaces in the early stages of reaction [Günter and Oswald, 1977]. Work is in progress to circumvent these problems and to study this interesting reaction in more detail.

Conclusions

The method of isopiestic equilibration is a useful technique for the synthesis of oxygen-deficient mixed oxides such as those found with the perovskite ABO_{3-x} framework. Accessibility of various intermediates will depend on their relative stabilities, kinetic, and nucleation phenomena. The latter appear to be especially important at temperatures below 600°C. Particularly noteworthy is the use of temperature gradients which allow syntheses otherwise not possible because of thermal decomposition. These techniques are also adaptable to gram amounts. The application to single crystals should provide a method for studying the fundamental nature of oxygen release and reoxidation of perovskite-like mixed oxides.

Acknowledgments. We acknowledge support for this research from NSF (DMR) - Solid State Chemistry Program 8610659. We would like to thank D. M. Smyth for his critical review of this paper. Also, we are grateful to A. J. Jacobson and J. P. Thiel for helpful discussions.

References

Dent Glasser, L. S., F. P. Glasser, and H. F. W. Taylor, Quart. Rev., 16, 343, 1962.

Gallagher, P. K., H. M. O'Bryan, S. A. Sunshine, and D. W. Murphy, Mat. Res. Bull., 22, 995 1987.

Günter, J. R. and H. R. Oswald, J. Solid State Chem., 21, 211, 1977.

Horowitz, H. S. and J. M. Longo, Mat. Res. Bull., 13, 1359, 1978.

Kuroda, K., N. Fujie, N. Mizutani, and M. Kato, J. Chem. Soc. Japan, 12, 1855, 1981.

Leonowicz, M. E., K. R. Poeppelmeier, and J. M. Longo, J. Solid State Chem., 59, 71, 1985.

Longo, J. M., H. S. Horowitz, and L. R. Clavenna, in S. L. Holt, J. B. Milstein, and M. Robbins (Eds.) Solid State Chemistry: A Contemporary Overview Adv. Chem. Ser., 186, American Chemical Society, Washington D. C., 1980, p.139.

Ohsato, H. T. Sugimura, and K. Kageyama, J. Crystal Growth, 48, 459, 1980.

Poeppelmeier, K. R., M. E. Leonowicz, and J. M. Longo, J. Solid State Chem., 44, 89, 1982.

Poeppelmeier, K. R. M. E. Leonwicz, J. C. Scanlon, J. M. Longo, and W. B. Yelon, J. Solid State Chem., 45, 71, 1982b.

Poeppelmeier, K. R., H. S. Horowitz, and J. M. Longo, J. Less-Common Met., 116, 219, 1986.
Reller, A., D. A. Jefferson, J. M. Thomas, R. A. Beyerlein, and K. R. Poeppelmeier, J. Chem. Soc. Chem. Commun., 1378, 1982.
Reller, A., J. M. Thomas, D. A. Jefferson, and M. K. Uppal, Proc. R. Soc. Lond., A 394, 223, 1984.
West, T. J., Complexometry, 3rd Ed. British Drug Houses Chemicals Limited, Poole, England, 1969.

K. Poeppelmeier and J. Wiley, Department of Chemistry and Ipatieff Catalytic Laboratory, Northwestern University, Evanston, Illinois 60208.

INTERMEDIATE COMPOSITIONS IN THE SERIES Bi_2O_3-M_2O_5 STRUCTURES INCORPORATING ELEMENTS OF PEROVSKITE AND FLUORITE

Wuzong Zhou and David A. Jefferson

Department of Physical Chemistry, University of Cambridge, UK.

John M. Thomas

The Royal Institution of Great Britain, London, UK.

Abstract. $Bi_5Nb_3O_{15}$ has been identified by high resolution electron microscopy (HREM) as a stepped intergrowth of NbO_4^{3-} and $Nb_2BiO_7^{1-}$ perovskite units and $Bi_2O_2^{2+}$ fluorite units. A similar structure has also been observed in $Bi_{31}Ta_{17}O_{89}$. The unit cells of both were determined by selected area electron diffraction (SAED) and the details of the structures were elucidated by computer simulation matching of HREM images. The structural principles of these stepped arrangements are discussed.

Introduction

In 1962, using X-ray powder diffraction (XPD), Roth and Waring deduced the structure of the compound $Bi_5Nb_3O_{15}$ to be a tetragonally distorted pyrochlore phase with a=1.0912 and c=1.0496 nm. They also reported some extra diffraction lines in the XPD spectrum which they could not however index on this unit cell [Roth and Waring, 1962]. Gopalakrishnan et al., in 1984, restudied this compound using HREM and reported it as a regular intergrowth of n=1 and n=2 members of the Aurivillius oxide family [Aurivillius, 1950] at the unit cell level, with alternate structural components having stoichiometries of $Bi_2NbO_6^-$ and $Bi_3Nb_2O_9^+$ respectively. A schematic drawing of this structure is shown in Fig. 1.

In our previous work in the system Bi_2O_3-Nb_2O_5, we have investigated the structures of the bismuth-rich phases using HREM. Roth and Waring reported a solid solubility of up to 23 mole per cent Nb_2O_5 in Bi_2O_3, but our studies indicated that ordered arrangements of Nb^{5+} cations existed, forming characteristic structural units inside a host lattice of δ-Bi_2O_3. With a Bi:Nb ratio of 60:1, a β-Bi_2O_3 structure was found, this being a distorted 2x2x1 superstructure of the fluorite-like δ-Bi_2O_3. On increasing the niobium content, this altered to a body-centred cubic 2x2x2 superstructure in which Nb^{5+} cations were isolated from each other at compositions up to 15:1, this structure being designated type I. With further addition of Nb^{5+}, this structure in turn altered into a phase with a large cubic cell containing pyrochlore-like structural units with Nb_4O_{18} stoichiometry, designated type II [Zhou et al., 1986] and this transformed, at Bi:Nb ratios of less than 3:1, into a type III phase which contained these Nb_4O_{18} units linked by perovskite-like units of four corner-sharing NbO_6 octahedra [Zhou et al., 1987]. At a Bi:Nb ratio of 5:3, all fluorite-like character appeared to disappear in the structure, and XPD indicated a perovskite-like structure, designated type IV. This would be in agreement with the intergrowth Aurivillius phase of Gopalakrishnan et al., but further investigations revealed that the layers in these structures were not such intergrowths, but contained regular steps. The cation positions in two such superstructures were determined by HREM, as outlined below. In addition, a type IV-like structure was observed in the Bi_2O_3-Ta_2O_5 system.

Experimental

The compound $Bi_5Nb_3O_{15}$ was prepared by solid-solid reaction of intimately mixed Bi_2O_3 and Nb_2O_5, both of 99.9 % purity, at 1000 °C/O_2 for 48 hours followed by quenching directly to room temperature. Initial characterization of the product was by XPD which was performed on a Philips diffractometer using Cu Kα radiation with operating conditions of 40 KV and 40 mA. High temperature XPD studies, using the same X-ray tube, were performed on a Philips ANTON PAAR HTK 10 high-temperature attachment, the temperature range being variable from room temperature to 1200 °C at atmospheric pressure. The powder sample was placed on a heating strip made of platinum which itself was heated directly by the current. A Pt/10%Rh-Pt thermocouple which was welded on the heating strip, enabled temperature measurement of up to 1600 °C. Pure oxygen was allowed to pass through the system to protect the sample from reduction. The normal mode of scanning was programmed with interval times 1 second and step size 0.05°(2θ). Compositional homogeneity was confirmed by energy dispersive X-ray microanalysis (EDS) within the electron microscope. The observed ratio of the Bi Lα emission line to the Nb Kα line from 20 chosen particles was 2.89 $\pm$ 0.10, corresponding to the cation ratio in the sample of Bi:Nb as 5:3, as calculated from the standard material of $BiNbO_4$ which was freshly prepared. SAED patterns were taken on a Jeol-200CX electron microscope operating at 200 KV with a $\pm$45° tilt goniometer stage. High resolution images were recorded by using a new type of side-entry specimen stage (C_s=0.52 mm, C_c=1.05 mm, information limit ca. 0.18 nm) and computer simulated model images were

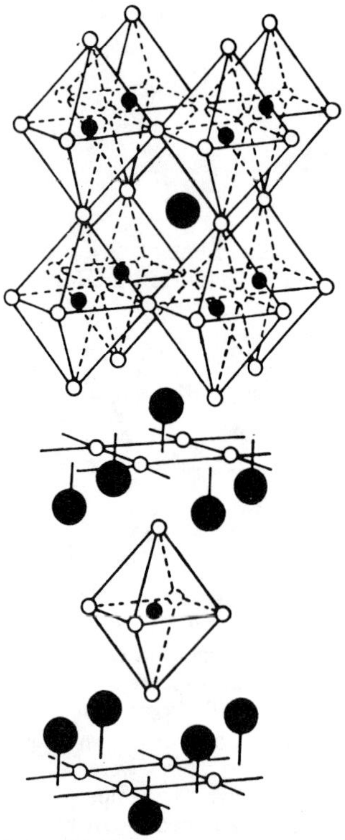

Fig. 1. The model for the regular intergrowth of n=1 and n=2 members of Aurivillius oxide family for $Bi_5Nb_3O_{15}$.

calculated according to the multislice method [Cowley and Moodie, 1957; Goodman and Moodie, 1974], using programs developed for use with very large unit cells [Jefferson et al., 1976].

Results and Discussion

The room temperature XPD spectrum of $Bi_5Nb_3O_{15}$ obtained in the present work was very similar to the previous work (Fig. 2a). The basic structure could be determined by the main peaks as a tetragonal unit cell with **a**=1.0936 and **c**=1.048 nm. However, if the high resolution XPD spectrum was examined, there were many extra lines which could not be indexed (Fig. 2b), this suggesting that a more complicated structure existed in the sample. All these main lines and the extra lines were stable at any temperature below the melting point.

In the electron microscope, many SAED patterns were recorded. Three of them down the principal projections of the perovskite sublattice are shown in Fig. 3a, 3b and 3c. Almost all the diffraction spots could be indexed on an orthorhombic lattice with **a**=3.179, **b**=0.545 and **c**=4.102 nm, although an incommensurate nature in the **a*** direction was noted and the **a** parameter could be 6.358 nm or larger. Compared with other superstructures in the system [Zhou et al., 1986, 1987], this type IV structure shows strong perovskite character rather than any relationship with fluorite or pyrochlore. Fig. 3d shows the SAED pattern and the corresponding HREM image in the [110] projection of the perovskite-like sublattice or [610] projection of the superstructure determined above. In this projection, the structure looks like a regular Aurivillius phase (Fig. 1), but when viewing down the [010] axis, a stepped nature was revealed (Fig. 4a and 4b). In the crystal of Fig. 4a, the structure is partially stepped and partially layered, forming an orthorhombic lattice which was designated as type IV. In Fig. 4b, however, the structure is completely stepped with no continuous layers, forming a monoclinic lattice, designated type IV*. In many cases, these two phases intergrew together, with usually the type IV structure being found in the central region of the particle and the type IV* at the edge.

Models for the type IV and type IV* structures are shown in Fig. 5. The type IV model has a unit cell identical to that determined from the SAED patterns, and has a structure where continuous layers of NbO_6 octahedra are flanked on one side by strips of further octahedra, such that the strips on adjacent layers point towards each other. Linkage between layers and adjacent strips is provided by $Bi_2O_2^{2+}$-like components, as in the Aurivillius phases, although that linking the adjacent strips is of necessity "stepped". The type IV* model, on the other hand, is monoclinic with **a**=7.010, **b**=0.545, **c**=4.114 nm and

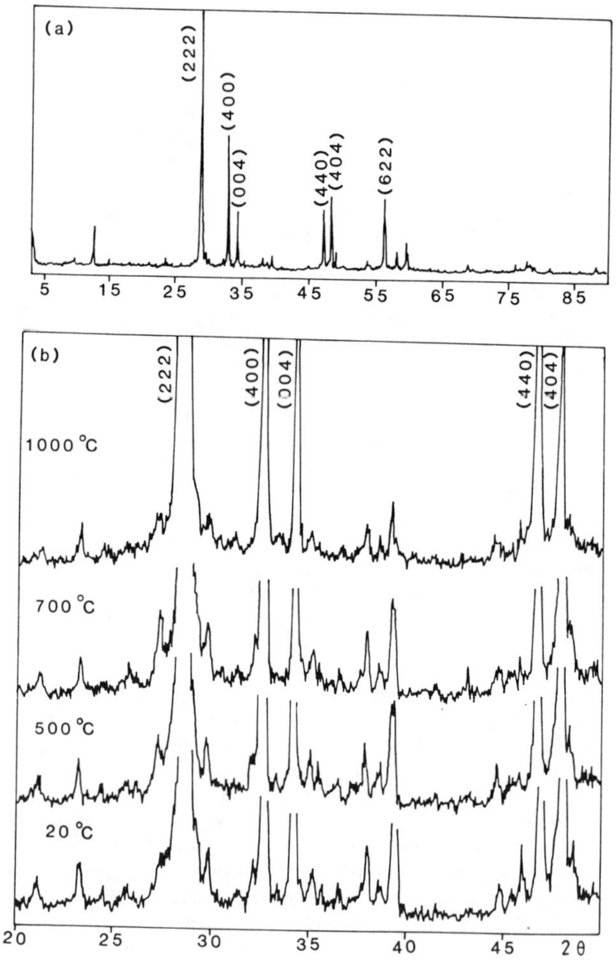

Fig. 2 (a) The room temperature XPD spectrum of $Bi_5Nb_3O_{15}$. (b) High temperature XPD spectra of $Bi_5Nb_3O_{15}$. The main peaks are indexed into the tetragonal subunit cell with **a**=1.0936 and **c**=1.048 nm.

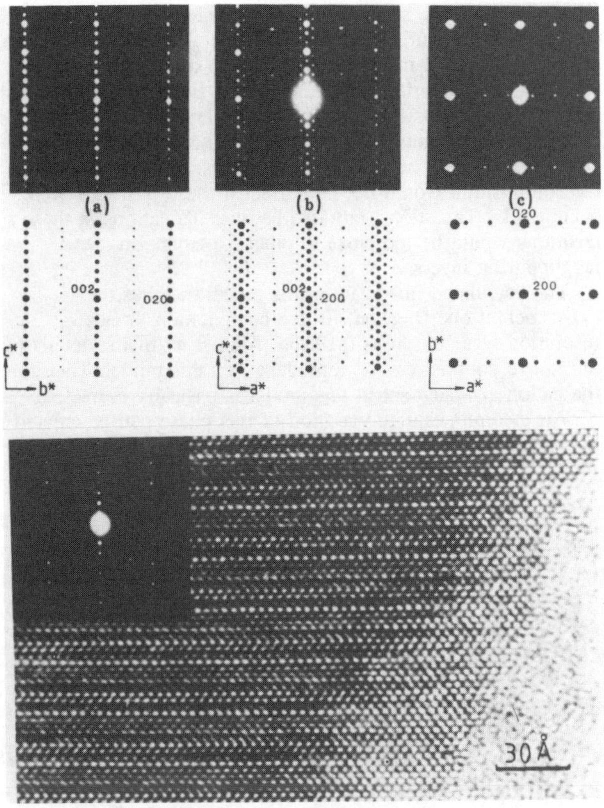

Fig. 3 Three SAED patterns from $Bi_5Nb_3O_{15}$ in the (a) [100], (b) [010] and (c) [001] projections of perovskite sublattice. (d) SAED pattern and HREM image of $Bi_5Nb_3O_{15}$ in the [610] projection of the type IV orthorhombic superunit cell.

β=81.59°. In this model, there are no continous layers of octahedra, but wider strips which overlap to form "stepped" layers. Once again, linkages are provided by $Bi_2O_2^{2+}$ components. When compared to the idealised perovskite arrangement, the structures of both models are compressed in a direction perpendicular to the layers or strips, such that the $Bi_2O_2^{2+}$ components and the NbO_6 octahedra link together with Bi^{3+} and Nb^{5+} cations on approximately the same plane perpendicular to [001], necessitating some distortion of the NbO_6 octahedra.

In general, for ABO_3 materials, when a factor $\mathbf{t} = (R_A+R_O)/\sqrt{2}(R_B+R_O)$, R_A, R_B and R_O being the radii of the relevant cations, lies between 0.8 and 1.0, a perovskite type structure is formed, but for **t** between 0.8 and 0.89, some deformation is invariably present. For example, **t** =0.89 in $CaTiO_3$, and although the cations are in idealised positions, the TiO_6 octahedra, although still regular, are tilted relative to one another and form a puckered network [Deer et al., 1962]. In $Bi_5Nb_3O_{15}$, the value of t is 0.80, and a similar, although possibly greater, effect can be expected. Such a distortion is in keeping with the apparent compression of the lattice perpendicular to the layers in type IV, but in type IV*, where there is no complete extension of the NbO_6 octahedra layers, a further condition imposed is that the number of NbO_6 octahedra in the positions where strips overlap must be even.

The model shown in Fig. 5b shows a 6x5 arrangement of overlapped and non-overlapped octahedra and does not fit the correct stoichiometry exactly. Consequently, the true arrangement is probably a mixture of 6x5 and 6x7. Local variations in this arrangement might well produce a slight compositional uncertainty in this phase.

Good image matches calculated from the type IV and the 6x5x6x7 type IV* models have been obtained (see the insets of Fig. 4a and 4b). The thickness dependence of image contrast from the type IV structure was reproduced. However, both of these are still only the most probable models. In fact the unit cell of both type IV and type IV* structures are variable at least along [100] as indicated in SAED patterns.

For preparing the type IV structure in the Bi_2O_3-Ta_2O_5 system, a $5Bi_2O_3$-$3Ta_2O_5$ composition was chosen. As indicated by EDS studies, the product always contained two phases, one being $BiTaO_4$ and another a phase of composition $Bi_{31}Ta_{17}O_{89}$. Although the real type IV structure is possibly incommensurate, as implied by the SAED patterns, it is still possible to propose a simplified unit cell that is monoclinic with **a**=2.272, **b**=0.385, **c**=1.926 nm and β=101.9°.

One HREM image in the [010] projection is shown in Fig. 6a, taken on the 400 kV electron microscope in Arizona State

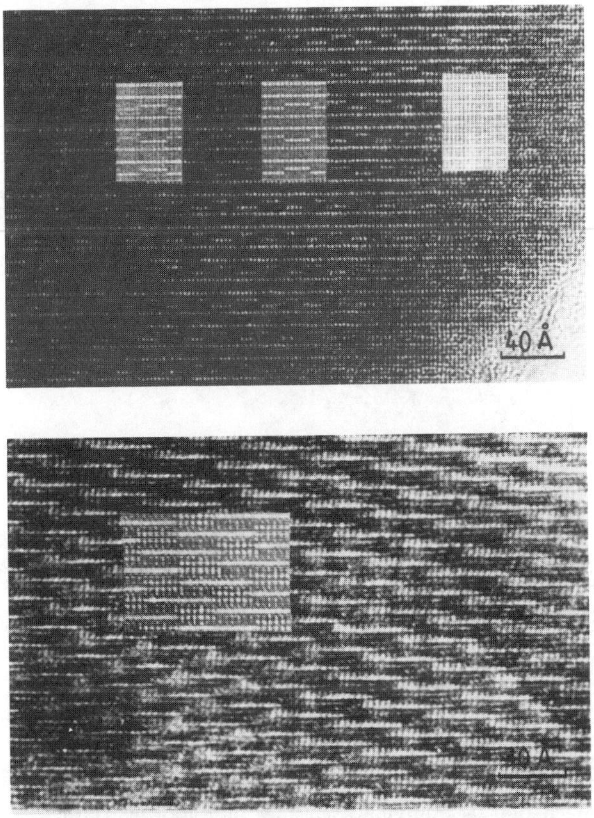

Fig. 4 (a) and (b) HREM images of the type IV and type IV* structures in the [010] projection. Computer simulations for (a) correspond to 8, 6, 2 nm thick (from left to right) and defocus 80, 80, 120 nm. Computer simulation for (b) corresponds to a specimen thickness of 5 nm and defocus of 120 nm.

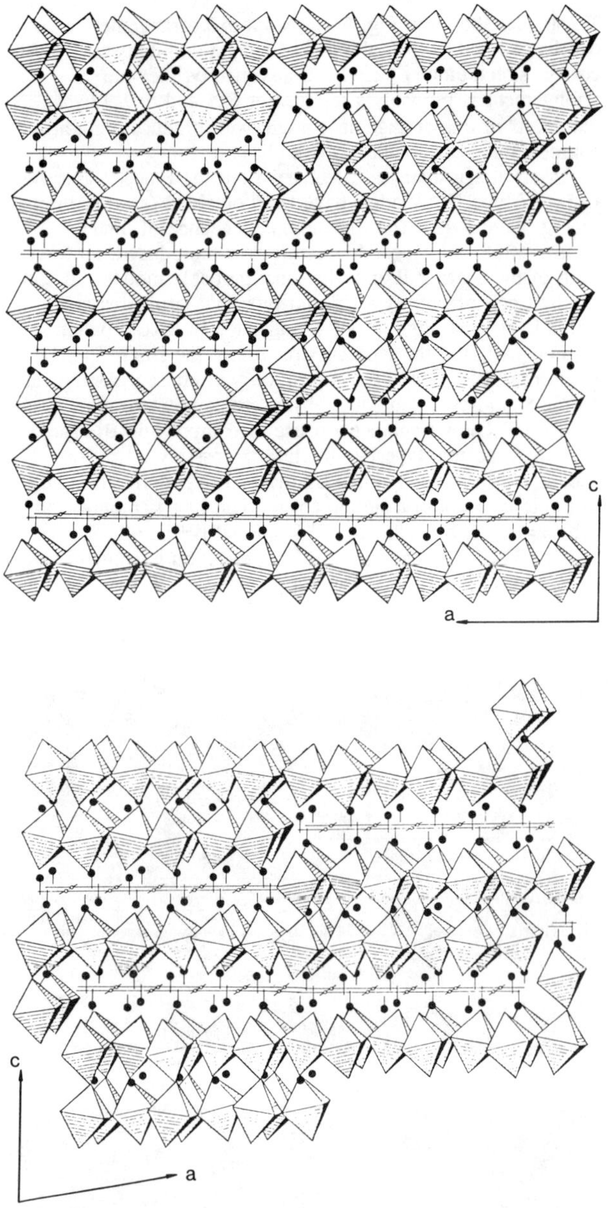

Fig. 5. The models for the type IV structure (a) and the type IV* structure (b). Only about one quarter of the unit cell of type IV* is shown.

University. It is very obvious that the structure is close to the type IV* of $Bi_5Nb_3O_{15}$, which has a stepped structure built by n=1 and n=2 members of an Aurivillius phase as indicated by distinct discontinuities perpendicular to the **c** axis on the image. However, the shortest axis **b** of this phase is along the [100] direction of the Aurivillius sublattice instead of the [110] direction as in $Bi_5Nb_3O_{15}$ as discussed above.

Using the same method as for the type IV* $Bi_5Nb_3O_{15}$, a model has been suggested, as shown in Fig. 6b. The composition of the whole unit cell of this model is $Bi_{31}Ta_{17}O_{89}$ which is in agreement with the EDS result. In the [001] direction, the structure is almost exactly n=1 and n=2 members of the Aurivillius phases, while along [100] direction, these n=1 and n=2 members intergrow in an arrangement of 3x3 octahedral overlap for the first layer and 2x4 for the second instead of extending through the crystal in an uninterrupted fashion. It was noted that the positions "A" and "B" in Fig. 6b are occupied by Bi cations, not only because the composition would not be correct if these positions were occupied by Ta cations, but also because Ta cations in these positions would be too close to other Ta atoms on neighbouring layers.

The calculated image from this model matches the experimental one (Fig. 6a). In the computation, image resolution was chosen as 0.18 nm. Almost all of the details of the image contrast can be reproduced by this model. Therefore, the cation arrangement in this model is basically correct.

An incommensurate variation of this phase can be created by alternation of the arrangement of the n=1 and n=2 members

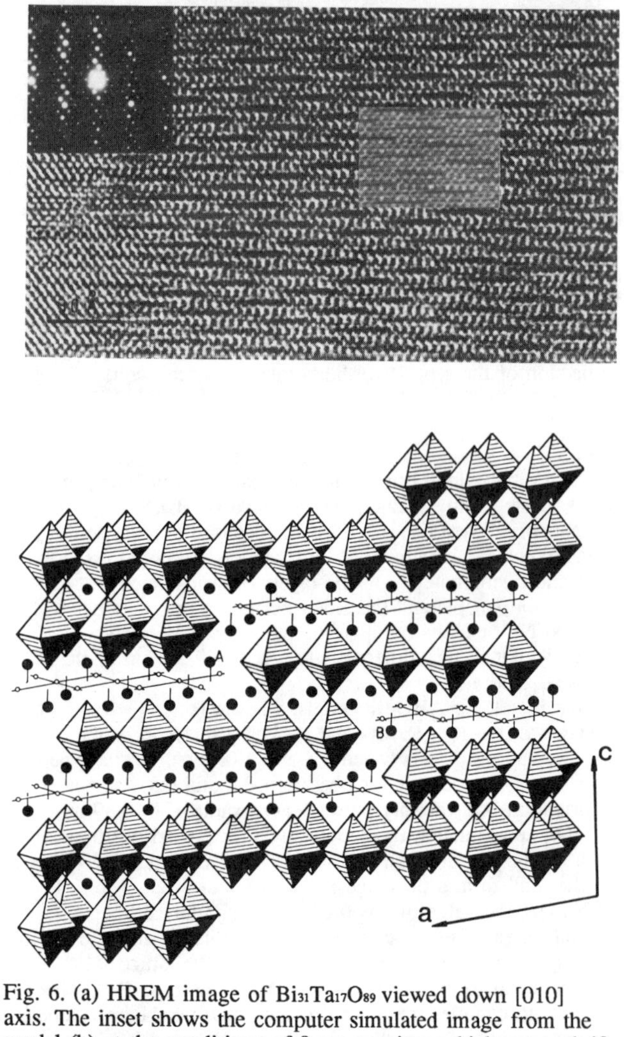

Fig. 6. (a) HREM image of $Bi_{31}Ta_{17}O_{89}$ viewed down [010] axis. The inset shows the computer simulated image from the model (b) at the conditions of 8 nm specimen thickness and 40 nm lens defocus.

as that in $Bi_5Nb_3O_{15}$ structure. The factor $t=(R_A+R_O)/\sqrt{2}(R_B+R_O)=0.81$, where R_A, R_B and R_O are ionic radii of Bi^{3+}, Ta^{5+} and O^{2-} respectively. In this case, a distortion of TaO_6 octahedra similar to that of $Bi_5Nb_3O_{15}$ structure is also possible. Such a distortion is also satisfactory in the fitting of the steps, but because the octahedral distortion is usually along the [110] direction of the Aurivillius sublattice as that shown on the model of type IV and type IV* $Bi_5Nb_3O_{15}$, in this model, it is difficult to show such a distortion in this projection. Therefore a regular arrangement of TaO_6 octahedra is used in the drawing.

In summary, the $Bi_5Nb_3O_{15}$ composition has a basic structure comprising a regular intergrowth of n=1 and n=2 members of the Aurivillius phases, but exists in a stepped configuration instead of in a regular phase. This phenomenon appeared to be caused by the presence of charges on the $Bi_2NbO_6^-$ layers and $Bi_3Nb_2O_9^+$ layers. Similar behaviour was also observed in the Bi_2O_3-Ta_2O_5 system. The orientations of the steps of the type IV structures in the Bi_2O_3-Nb_2O_5 and Bi_2O_3-Ta_2O_5 systems were different, although as yet no explanation for this difference is known.

It was noted that normal Aurivillius phases, even those of the regular intergrowth variety, did not have stepped structures. For example, Bi_2WO_6 was a regular n=1 Aurivillius phase, in which a WO_4^{2-} layer alternated with $Bi_2O_2^{2+}$ layers. The complete double layer, however, was electrically neutral. However, when some Nb cations were introduced into the lattice, the composition being $Bi_2(W_{0.5}Nb_{0.5})O_{5.75}$, a very similar stepped structure to the type IV structure discussed above was observed [Lieven et al., 1984]. $Bi_9Ti_6CrO_{27}$ is an Aurivillius phase having alternating $Bi_4Ti_3O_{12}$ and $Bi_5Ti_3CrO_{15}$ layers, but both of these two component layers are neutral. Similarly, $BaBi_8Ti_7O_{27}$ has alternating $Bi_4Ti_3O_{12}$ and $BaBi_4Ti_4O_{15}$ layers, and $Bi_7Ti_{4.5}W_{0.5}O_{21}$ contains $Bi_4Ti_3O_{12}$ and $Bi_3Ti_{1.5}W_{0.5}O_9$ layers. All of these layers are electrically neutral, and do not need to adopt a stepped structure for neutralization of the charges between layers [Gopalakrishnan et al., 1984]. However in the Bi_2O_3-Nb_2O_5 and Bi_2O_3-Ta_2O_5 systems, there is no way to construct such neutral layers. It was then concluded that Aurivillius phases having a single n member, e.g. Bi_2WO_6, will not form stepped structures. Those with different n numbers but with neutral layers inside the unit cells, e.g. $Bi_9Ti_6CrO_{27}$, will also not form a stepped structure, but those with different n numbers and without neutral layers inside the unit cell, e.g. $Bi_5Nb_3O_{15}$, will. To prove this structural principle, further work on other related meterials are being carried out.

Acknowledgments. One of the authors (W.Z.) thanks Fudan University in Shanghai for the financial support. Part of this work was carried under N.S.F. program P 15999. The authors are grateful to D.J. Smith for assistance with HREM work and to L. Eyring and R.S. Roth for critical reviews of this paper.

References

Aurivillius, B., Mixed oxides with layer lattices. III. Structure of $BaBi_4Ti_4O_{15}$, Ark. kemi., 2, 519-527, 1950.

Cowley, J.M. and Moodie, A.F., The scattering of electrons by atoms and crystals. I. A new theoretical approach, Acta Crystallogr., 10, 609-619, 1957.

Deer, W.A., Howie, R.A. and Zussman, J., Perovskite, in Rock-Forming Minerals Vol 5: Non-Silicates, pp 49-55, Longmans, Green & Co., London, 1962.

Goodman, P. and Moodie, A.F., Numerical evaluation of N-beam wave functions in electron scattering by the multislice method, Acta Crystallogr., A30, 280-290, 1974.

Gopalakrishnan, J., Ramanan, A., Rao, C.N.R., Jefferson, D.A. and Smith, D.J., A homologous series of recurrent intergrowth structures of the type $Bi_4A_{m+n-2}B_{m+n}O_{3(m+n)+6}$ formed by oxides of the Aurivillius family, J. Solid State Chem., 55, 101-105, 1984.

Jefferson, D.A., Millward, G.R. and Thomas, J.M., The role of multiple scattering in the study of lattice images of graphitic carbons, Acta Crystallogr., A32(5), 823-828, 1976.

Lieven, J.L., Reller, A. and Jefferson, D.A., Structural consequences of Nb-substitution into Bi_2WO_6 and Bi_2MoO_6: high-resolution electron microscopic observations, Mat. Res. Bull., 19, 571-575, 1984.

Roth, R.S. and Waring, J.L., Phase-equilibium relations in the binary system bismuth sesquioxide-niobium pentoxide, J. Res. Nat. Bur. Stand. Sect., A 66, 451-463, 1962.

Zhou, W., Jefferson, D.A. and Thomas, J.M., Defect fluorite structures containing Bi_2O_3: the system Bi_2O_3-Nb_2O_5, Proc. R. Soc. Lond. A, 406, 173-182, 1986.

Zhou, W., Jefferson, D.A. and Thomas, J.M., A new structure type in the Bi_2O_3-Nb_2O_5 system, J. Solid State Chem., 70, 129-136, 1987.

D.A. Jefferson and W. Zhou, Department of Physical Chemistry, Lensfield Road, Cambridge CB2 1EP, UK.

J.M. Thomas, The Royal Institution, 21 Albemarle Street, London W1X 4BS, UK.

DEFORMATION MECHANISMS OF CRYSTALS WITH PEROVSKITE STRUCTURE

Jean-Paul Poirier, Solange Beauchesne and François Guyot

Laboratoire des Géomatériaux, Institut de Physique du Globe de Paris

Abstract. The slip systems and dislocations compatible with the geometrical requirements of the perovskite structure are listed and the published transmission electron microscopy observations of dislocations in various perovskites are reviewed.

The results of recent high-temperature creep experiments on single crystals of fluoride and oxide perovskites are reported and discussed in terms of a systematics.It is tentatively suggested that perovskites that are ferroelectric at room temperature and those that are ideal cubic have a different high temperature creep behavior.

Introduction

Mineralogical models of the Earth's lower mantle fall into two classes:

i) The pyrolytic models (Ringwood,1975; Weidner & Ito,1987), for which the lower mantle has the same composition as the upper mantle and contains about 80 vol % $(Mg,Fe)SiO_3$ perovskite and 20 % (Mg,Fe)O magnesiowüstite.

ii) The chondritic or pyroxenitic models (Liu, 1982; Anderson,1984) for which the lower mantle is more silicic than the upper mantle and contains from 90 to 100 vol% $(Mg,Fe)SiO_3$ perovskite.

In any case, the dominant mineral phase of the lower mantle is $(Mg,Fe)SiO_3$ perovskite: even in the more perovskite-poor models, there is at least 80 vol % perovskite phase. As the percolation threshold for connectedness of a phase in a two-phase mixture, assuming no surface tension effect, is about 20 vol % (Cahn, 1966), perovskite is, of necessity, a connected phase that must control the deformation of the lower mantle.In the more magnesiowüstite-rich models, the magnesiowüstite content is just about or below the percolation threshold and it is therefore probably unconnected. If it is more deformation-resistant than perovskite, it will be passively entrained in the deformation of the more ductile perovskite matrix; if it is less resistant, the connected matrix will anyway have to be deformed. So, in all cases,for the current mantle models,it is reasonable to consider that the viscosity of the lower mantle is controlled by perovskite. Some knowledge of the deformation mechanisms and creep laws of perovskites is then a prerequisite to informed speculation on the viscosity of the lower mantle.It is therefore unfortunate that almost nothing is known on the deformation (and especially high-temperature deformation) of crystals with perovskite structure, despite a wealth of information on their physical properties at low temperatures.

In what follows, we will, first, set up a framework for the investigation of the deformation mechanisms of crystals with perovskite structure, by listing, a - priori the potential slip systems and dislocations, together with their possible splitting schemes, compatible with the crystallographic requirements of the perovskite structure. We will, then, review the scarce observations of dislocations made by transmission electron microscopy on various perovskites. Finally, we will report on recent high-temperature creep experiments that constitute a starting basis for a systematics.

Potential Slip Systems and Dislocations

Although the crystals with perovskite structure are usually distorted at low temperature and/or high pressure, we will first consider here the ideal cubic structure ABO_3 consisting in a 3-dimensional framework of vertex-sharing BO_6 octahedra and A cations. The role of the distortions will be assessed later and anyway, all perovskites are cubic at ambient pressure and high enough temperature.

An interesting characteristic feature of the ideal perovskite structure is that the oxygen ions and the A cations are of comparable size and, taken together, form a face-centered cubic lattice.Thus,the perovskite structure can conveniently be seen as a ReO_3 structure stuffed with small B cations at the center of the octahedra (Wells,1984). Conversely,it can be seen as an intermetallic CuZn (B2) structure stuffed with oxygen ions at the center of the cube faces (O'Keefe & Hyde, 1985)(Fig. 1).

Consideration of the face-centered cubic or intermetallic B2 lattice,whose slip systems and dislocations are known, can help us to make educated guesses on the possible slip systems and Burgers vectors of perovskites.

The ReO_3 stuffed lattice

The lattice of mixed A cations and oxygen ions could, in principle, slip on the close-packed slip planes available to the fcc lattice: {111}, {110} and {100}. The shortest Burgers vectors are <100> and <110> (not 1/2 <110> as for unstuffed fcc, due to the presence of the B ion at the center of the cube).

The potential slip systems are therefore: (111) $[1\bar{1}0]$, (110) $[1\bar{1}0]$ and (100) [001].

Speculation as to which systems are more probable can be made only on the basis of geometrical assessment of the splitting schemes of dislocations since nothing is known on the

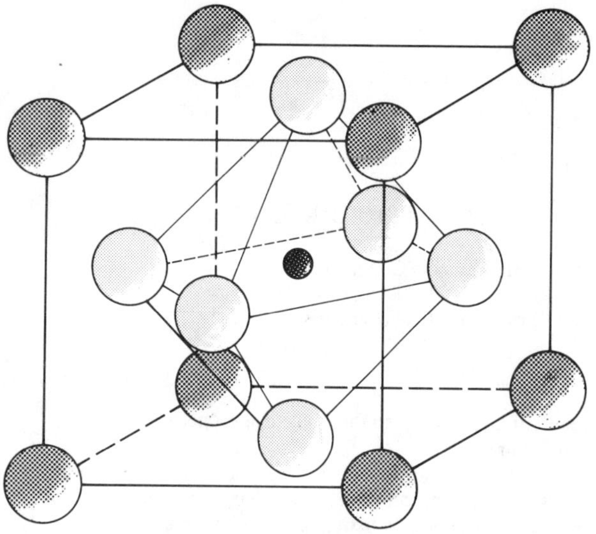

Fig. 1. Ideal cubic ABO_3 perovskite unit cell: The A cations at the corners of the cube and the O anions at the vertices of the octahedron form a FCC lattice.The A and B (at the center of the cube) cation form a CuZn (B2) lattice.

deformation of the very few crystals which have the ReO_3 structure.

Dislocations with [110] Burgers vector, responsible of slip on {111} can be expected to dissociate into Shockley partials as in the fcc lattice; however,due to the fact that the Burgers vector is twice as long,there are 4 partials and the screw dislocations can dissociate on 2 intersecting {111} planes,with a sessile stair-rod dislocation at the intersection,blocking their glide:

$$[110] \rightarrow 1/6\,[121] + 1/3\,[210] + 1/6\,[12\bar{1}] \quad (1)$$

Slip on {100} can be envisaged with a dissociation of the <100> Burgers vector into two collinear partial vectors(Fig.2):

$$[100] \rightarrow 1/2\,[100] + 1/2\,[100] \quad (2)$$

Finally, a favorable slip system would appear to be (110) $[1\bar{1}0]$, also with a collinear splitting:

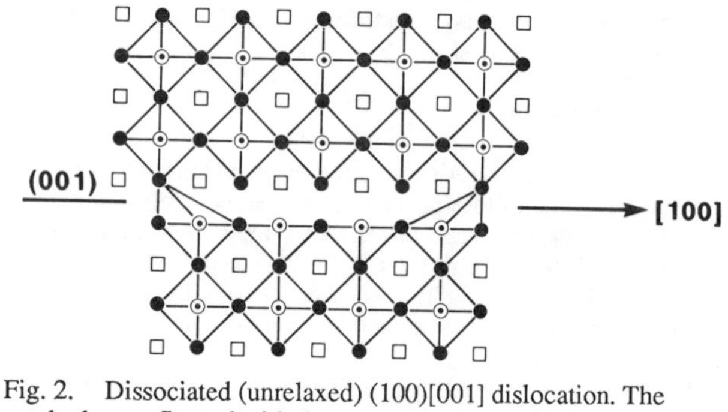

Fig. 2. Dissociated (unrelaxed) (100)[001] dislocation. The octahedra are figured with the oxygen ions represented as circles, the A cations are represented as squares.

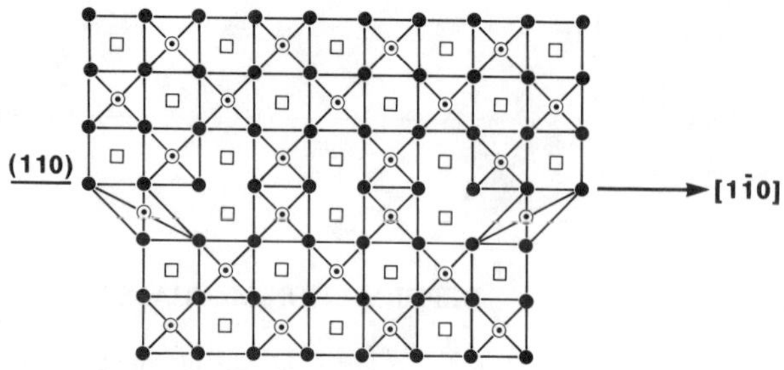

Fig. 3. Dissociated (unrelaxed) $(110)[1\bar{1}0]$ dislocation.The framework of octahedra is preserved across the stacking fault.

$$[1\bar{1}0] \rightarrow 1/2\,[1\bar{1}0] + 1/2\,[1\bar{1}0] \quad (3)$$

The corresponding stacking fault is probably of rather low energy, since it conserves the original oxygen sublattice (Fig.3).It must be noted however that the propagation of the partial dislocations nevertheless necessarily involves the breaking of Si-O bonds.

The CuZn stuffed lattice

There is no need here to speculate on the slip systems of this structure, since the deformation of many intermetallic compounds (CuZn, NiAl, CoAl, FeAl) is quite well known.At low temperatures,dislocations with <111> Burgers vector glide on {110} planes, whereas at high temperatures both (001)[100] and (110) $[1\bar{1}0]$ slip systems may be operative (e.g. Saka et al.,1985 ; Rudy & Sauthoff, 1986).

The perovskite structure

From the considerations developed above, we can surmise that crystals with the perovskite structure may deform by slip on (110) $[1\bar{1}0]$ and (001)[100] systems. Splitting schemes other than the ones presented above are also theoretically possible:It is in particular interesting to investigate the climb dissociations, which are now found to exist in many structures and may be responsible for a paradoxical strengthening with temperature increase, noticed in some compounds (Poirier, 1985). In the perovskite structure, it is possible to envisage that edge dislocations with [100] Burgers vector can dissociate in their climb plane (Fig. 4), according to the reaction:

$$[100] \rightarrow 1/2\,[101] + 1/2\,[10\bar{1}] \quad (4)$$

Finally, let us consider the distortions of the non-ideal lattice: $MgSiO_3$ perovskite, like $CaTiO_3$, is distorted according to the $GdFeO_3$ distortion, i.e. the octahedra are tilted so that the lattice becomes orthorhombic, with a unit cell containing 8 octahedra. In principle,the Burgers vectors of the dislocations are twice as large as in the cubic perovskites and could be dissociated into two partials with collinear Burgers vectors; there is, however, no clear evidence for such a dissociation in $CaTiO_3$ (Doukhan & Doukhan, 1984). Another obvious consequence of the orthorhombic distortion is the existence of domains (or twins) that persist up to high temperatures at high pressures; their boundaries evidently act as obstacles to dislocations and are likely to play the same role as ordinary grain boundaries.

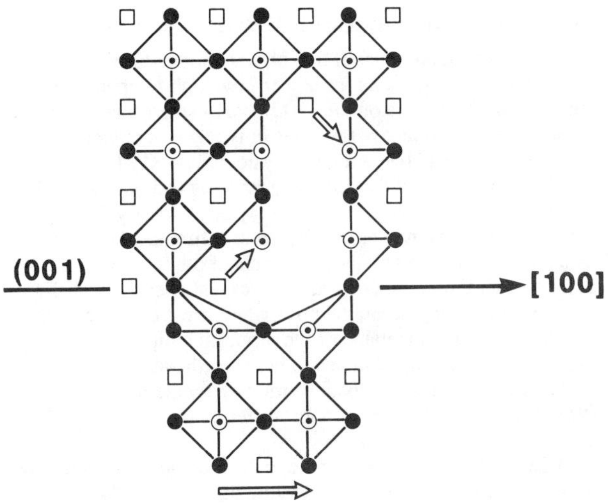

Fig. 4. Climb dissociation of a (100)[001] dislocation.

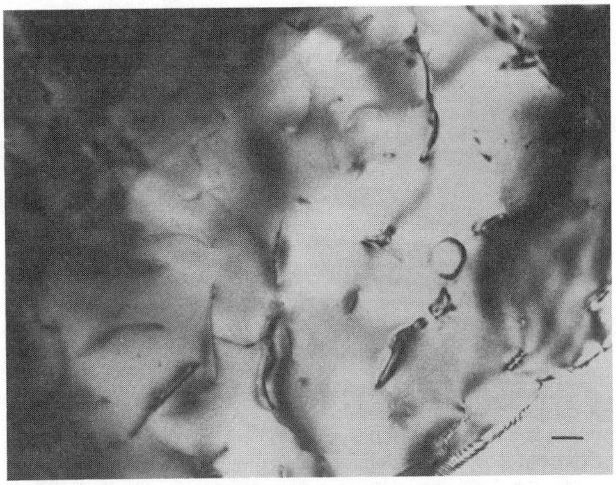

Fig. 6. Prismatic loops and dislocations with <100> Burgers vector (pseudo-cubic) in $MnGeO_3$ perovskite prepared in the diamond anvil cell at 40 GPa. Scale bar: 100 nm.

TEM observation of dislocations

Few crystals with perovskite structure have been deformed and observed in transmission electron microscopy.The observations generally agree with the speculations made on a crystallographic basis: The dislocations have <110> or <100> Burgers vectors, but it appears that, while <110> dislocations can be found both at room temperature and at high temperatures, <100> dislocations are generally found only at high temperatures, where they tend to be predominant.

Slip bands (110)[1$\bar{1}$0] have been observed in $KNbO_3$ at room temperature (Tanaka & Himiyama , 1975; Li et al.,1983), while Ammann et al. (1984) found <110> screw and <100> edge dislocations in $KTa_{1-x}Nb_xO_3$.

$KTaO_3$ has been deformed in creep at high temperature (Beauchesne & Poirier, in preparation) and both <110> and <100> dislocations have been observed (Fig. 5). Similar observations were made in $BaTiO_3$ crystals after creep: The <110> dislocations are present at all temperatures, along with the <100> dislocations responsible for the observed slip on the (001)[100] system. All dislocations rearrange in low energy polygonization walls (Beauchesne & Poirier, in preparation).The climb dissociation of [100] dislocations (eq.4) is observed, but it may not be characteristic of the high-temperature creep and the observed configurations have been interpreted as resulting from vacancy precipitation during cooling (Doukhan & Doukhan, 1984).

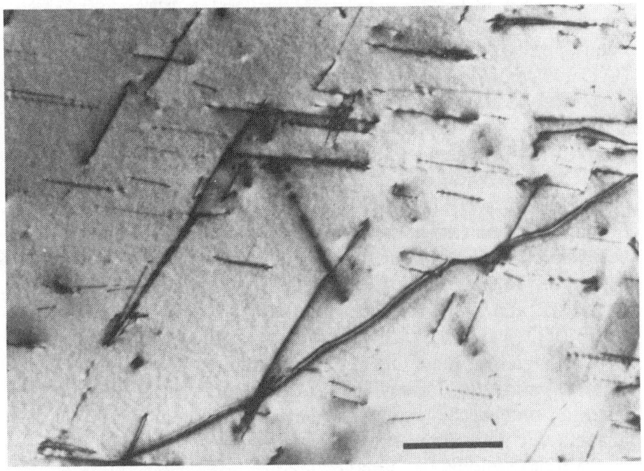

Fig. 5. Slip dislocations in $KTaO_3$ (TEM). Dislocations project along <110> directions in a {111} plane. Scale bar: 500 nm.

In $CaTiO_3$, <110> dislocations were observed in samples where (110) [1$\bar{1}$0] slip (in the pseudo-cubic lattice) had been induced by indentation at room temperature (Doukhan & Doukhan, 1984).

The quenched high-pressure perovskite phase of $MnGeO_3$, an analogue of $MgSiO_3$, prepared in the diamond-anvil cell at 400 kbar was observed in TEM and showed both <100> and <110> dislocations (in the pseudo-cubic lattice)(Fig.6).

The only discrepant observation in the literature concerns $PbTiO_3$ where [0$\bar{2}\bar{1}$] dislocations were reportedly seen (Dobrikov, 1980) but the evidence seems dubious.

High-Temperature creep experiments

Creep experiments have so far only been performed on single crystals or very fine-grained polycrystals.

Single crystals

The first creep experiments on single crystals with perovskite structure were performed on the cubic fluoperovskite $KZnF_3$ at temperatures between 0.8 and 0.95 T_m (melting temperature) (Poirier et al.,1983). Although slip occurred on {110} planes at all investigated temperatures, it was found that at T > 0.95 T_m, slip on {100} planes became extraordinarily easy and the electrical conductivity increased by 2 to 3 orders of magnitude. It was suggested that the ease of slip on {100} planes was caused by the transition to the solid electrolyte state, in which the oxygen sublattice would , so to speak, melt (O'Keefe & Bovin,1979) and the remaining lattice would have the CuZn structure.

TABLE 1. Activation energy Q, stress exponent n and scaling parameter g for the creep of various perovskites.

Perovskite	T/T_m	Q (kJ/mol)	n	$g = Q/RT_m$
$KZnF_3$	0.81-0.95	213	1.8	22
	0.95-0.98	122	1.1	
$KTaO_3$	0.89-0.994	339	1.5	25
$KNbO_3$	0.87-0.993	407	3.4	37
$BaTiO_3$	0.75-0.92	469	3.6	30

The creep data have been reanalyzed, using a global inversion technique (Sotin & Poirier, 1984; Poirier et al., in press) and the values of the activation energy Q and stress exponent n were calculated for a rheological power law of the form: $\dot{\varepsilon} = A\,\sigma^n \exp(-Q/RT)$, where $\dot{\varepsilon}$ is the creep rate, T is the temperature and A is a constant, (Table 1).

Experiments recently performed on oxide perovskites $BaTiO_3$ and $KTaO_3$ (Beauchesne & Poirier, in preparation),in an orientation favoring slip on {100} planes, failed to exhibit the same catastrophic increase in creep rate at high temperature. In the case of $BaTiO_3$, the same high relative temperature T/T_m could not be obtained due to the phase transition to hexagonal perovskite that takes place at T= 0.95 T_m . For $KTaO_3$ however there is no phase transition and the crystal stays cubic up to the melting point but normal creep behavior persists up to the highest temperature (0.994 T_m). $KNbO_3$ shows the same behavior, although the creep parameters are quite different (Table 1).

In all cases, the creep data are best fitted by a power law (Table 1)(see Poirier,1985) but the stress exponents differ by a large amount, from values smaller than 2 for $KZnF_3$ and $KTaO_3$ to the more typical values of about 4 for $BaTiO_3$ and $KNbO_3$.

Due to the lack of diffusion data on these perovskites,it is impossible to compare the activation energies of creep and diffusion as it is traditionally done.

Polycrystals

A few creep experiments have been performed on very fine-grained polycrystalline perovskites, at temperatures generally lower than 0.7 or 0.8 T_m.

$BaTiO_3$,with a grain size of 0.5 µm, tested in compression between 1150 and 1250°C , exhibits a superplastic behavior with a stress exponent n=2 and an activation energy of 800 kJ/mol (Carry & Mocellin,1986). The superplastic behavior is indeed to be expected in very fine-grained material of any structure.

$CaTiO_3$ and $SrZrO_3$, with grain sizes less than 3 µm, were tested in 4-point bending; activation energies of 837 kJ/mol and 706 kJ/mol were found for $CaTiO_3$ (Yamada,1984) and $SrZrO_3$ (Nemeth et al.,1972) respectively. These data are, in our opinion, of dubious value, due to the unreliability of the bending creep method for obtaining rheological equations.

Discussion and perspectives

There is now ample evidence showing that all crystals with perovskite structure, slip on the {110}<1$\bar{1}$0> system at room temperature,irrespective of whether they are ideal, or distorted with the $BaTiO_3$ or $GdFeO_3$ distortion. At high temperature, they slip both on {110}<1$\bar{1}$0> and {100}<001> systems with a predominance of the latter at the higher temperatures.

Rheological equations for power-law, high-temperature creep of single crystals on the {100}<001> slip system have been obtained for 4 different perovskites only and it is obviously not yet possible to establish a reliable systematics: $KZnF_3$, $BaTiO_3$, $KTaO_3$ and $KNbO_3$ differ in many respects and not two of them share the same characteristics. We may,however, try and discern trends that may be confirmed or infirmed when more data come in on creep of single crystals of other perovskites. We will consider the presence or absence of enhanced {100} slip close to the melting point, the power-law stress exponent n ($n<2$ or $3.5<n<4.5$) and the value of the proportionality constant g that expresses the scaling relationship between the activation energy for creep and the absolute melting temperature (Weertman,1970; Liebermann & Poirier,1984): $g=Q/RT_m$ (see Table 1).

$KZnF_3$ is a 1-2 fluoperovskite,ideal cubic at all temperatures, it exhibits a value of $n<2$ and a value of g that is also rather low for a non metal. Its spectacular ease of slip close to the melting point has been related to the appearance of a solid electrolyte transition.

$BaTiO_3$ is a 2-4 oxide perovskite, with ferroelectric distortion at room temperature, that becomes cubic at high temperature. It behaves in creep as a typical ceramic material with a stress exponent close to 4 and a value of g close to 30. It is impossible to know whether it would become a solid electrolyte close to the melting point since it undergoes a phase change.

$KTaO_3$ is a 1-5 oxide perovskite, ideal cubic at all temperatures.It exhibits a value of $n<2$ and a rather low value of g. It definitely shows no increase in the ease of {100} slip close to the melting point.

$KNbO_3$ is also a 1-5 oxide perovskite but it is ferroelectic at room temperature and becomes cubic at high temperature.It behaves, like $BaTiO_3$ as a typical ceramic with n close to 4 and g close to 30. It shows no increase in the ease of {100} slip close to the melting point.

Even though the data set is much too sparse for comfort,we may now formulate questions and make conjectures compatible with our limited data.

- Solid electrolyte behavior and enhanced ease of slip close to the melting point exist in a 1-2 ideal cubic fluoperovskite and it does not exist in 1-5 oxide perovskites, ideal or with ferroelectric distortion at room temperature. Solid electrolyte behavior also exists in $NaMgF_3$,with the $GdFeO_3$ distortion (O'Keefe & Bovin,1979). Does the anomalous ease of slip exist also in oxides? If so,it must be related to the cationic valences and it should be looked for in 2-4 perovskites. What about 3-3 perovskites (aluminates) ?
- Creep with low n and Q (possibly the little known Harper-Dorn creep) is found for $KZnF_3$ and $KTaO_3$, which have little in common except the fact that both are ideal cubic perovskites at all temperatures. Would all ideal cubic perovskites,fluorides or oxides, creep in the same way?
- Typical power-law (Weertman) creep is found for $KNbO_3$ and $BaTiO_3$, which are both ferroelectric at room temperature. Even though they are cubic at the high temperatures of the creep experiments, do they keep some memory of their former distorted state at the level of the dislocation cores? Would all ferroelectric perovskites creep in the same way?
- What would be the creep behavior of oxide and fluoride perovskites with $GdFeO_3$ distortion ($CaTiO_3$,$NaNbO_3$,$NaMgF_3$)?

The systematics for creep of perovskites turns out to be more complicated and more interesting than expected. If the trends suggested above are confirmed, the bond characters reflected in the various room temperature distortions,would be important parameters that could determine the isomechanical series and the selection of proper analogues to $MgSiO_3$ perovskite, but it would be rash at this point to offer a physical explanation of a behavior that is still largely conjectural.Besides,the role of possible non-stoichiometry has not been investigated yet. Many more experiments on single crystals are needed.

Acknowledgments. We warmly thank Jean-Claude and Nicole Doukhan for many discussions on the electron microscopy of perovskites. This study was partly supported by CNRS (UA 734 and RCP 782 "Perovskites à hautes températures". This is IPG contribution n° 1015.

References

Amman,J.J., P.Buffat, D.Rytz and P.Stadelmann. Nature des dislocations dans KTN, Frühjarstagung der Schweiz.Phys. Gesell.,57, 474-477, 1984.

Anderson,D.L., The Earth as a planet:Paradigms and Paradoxes,Science,223,347-355,1984.

Cahn,J.W., A model for connectivity in multiphase structures, Acta Metall.,14, 477-480,1966.

Carry,C. and A.Mocellin,Superplastic creep of fine-grained $BaTiO_3$ in a reducing environment,J.Am.Ceram.Soc., 69, C215-216.

Dobrikov,A.A., O.V.Presnyakova, V.I.Zaitsev, V.V.Prisedskii and G.F.Pan'ko,Investigation of $PbTiO_3$ crystal latice defects by transmission electron microscopy, Kristall & Technik, 15, 207-212,1980.

Doukhan,N. and J.C.Doukhan, Dislocations in perovskites $BaTiO_3$ and $CaTiO_3$,Phys.Chem.Minerals,13,403-410,1986.

Li,Q.,J.Chen,B.L.Liao and D.Feng, The interaction between dislocations and ferroelectric domains in $KNbO_3$ crystals,Rad. Effects,74,307-313,1983.

Nemeth,J., W.V.Youdelis and J.Gordon Parr, High-Temperature creep behavior of polycrystalline $SrZrO_3$, J.Am.Ceram. Soc.,55,125-129,1972.

O'Keefe,M. and J.O.Bovin,Solid electrolyte behavior of $NaMgF_3$: Geophysical implications, Science,206,599-600.

O'Keefe,M. and B.G.Hyde,An alternative approach to non-molecular crystal structures, Structure and Bonding, 61,78-144,1985.

Poirier,J.P., Creep of Crystals, Cambrige U.Press,Cambridge, 260 pp.,1985.

Poirier,J.P.,J.Peyronneau,J.Y.Gesland and G.Brebec, Viscosity and conductivity of the lower mantle;an experimental study on a $MgSiO_3$ perovskite analogue, $KZnF_3$. Phys.Earth Planet.Int.,32,273-287,1983.

Ringwood,A.E., Composition & Petrology of the Earth's mantle, McGraw Hill,New York,1975.

Rudy,M. and G.Sauthoff, Dislocation creep in the ordered intermetallic (Fe,Ni)Al phase, Mater. Sci. Eng.,81, 525-530, 1986.

Saka,M.,Y.M. Zhu, M.Kawase, A.Nohara and T.Imura, The anomalous strength peak and the transition in slip direction in β-CuZn, Philos.Mag.,A 51,365-371,1985.

Tanaka,M. and Y.Himiyama, The observation of interaction between domains and dislocations in $KNbO_3$, Acta Cryst. A31,S264, 1975.

Weertman,J.,The creep strength of the Earth's mantle, Rev. Geophys.Space Phys.8,145-168,1970.

Weidner,D.J. and I.Ito, Mineral Physics constraints on a uniform mantle composition, in High Pressure Research in Geophysics,M.H.Manghnani,S.Syono eds,AGU,1987.

Wells,A.F., Structural Inorganic Chemistry (5th ed), Clarendon Press, Oxford,1984.

Yamada,H.,Viscous creep deformation of polycrystalline $CaTiO_3$ at elevated temperatures,J.Mater.Sci,19,2639-2642,1984.

ELASTICITY OF $SrTiO_3$ PEROVSKITE UNDER HIGH PRESSURE

Myriam Fischer[1], Bernard Bonello[1], Alain Polian[2], and Jean-Michel Leger[3]

Département de Recherches Physiques (CNRS UA 71)
Université Pierre et Marie Curie T.22
4, place Jussieu F 75252 Paris Cedex 05

Alain Polian

Physique des milieux condensés (CNRS UA 782)
Université Pierre et Marie Curie T. 13
4, place Jussieu F 75252 Paris Cedex 05

Jean-Michel Léger

Laboratoire de Physico Chimie des Matériaux CNRS
1, place A. Briand F 92190 Meudon

Abstract. The equation of state and the elastic behaviour of $SrTiO_3$ under high pressure are studied by X-ray diffraction and Brillouin scattering in the [111] direction in a diamond anvil cell. The cubic-tetragonal phase transition occurs at about 6 GPa and has been observed by Brillouin scattering only. The pressure dependence of a transverse acoustical phonon is measured up to 25 GPa and a longitudinal acoustical one between 13 and 36 GPa. At the transition, there is a negative jump of the Brillouin shift of the transverse acoustical mode

Introduction

The evolution of the elastic properties under high pressure of compounds having the ideal perovskite structure is of great interest for the determination of the properties of the Earth's lower mantle materials.

The development of Brillouin spectroscopy afforded the measurements of elastic properties at room conditions of the high pressure phase of the magnesium silicate compounds. Particularly, new interesting results have been just obtained by A. Yeganeh Haeri and D. Weidner on the elastic constants of the perovskite structure of $MgSiO_3$ [A. Yeganeh Haeri and D. Weidner 1987].

It is still difficult to measure the elastic properties of this compound under high pressure. So it's reasonable to examine the elasticity under high pressure of analogous perovskites.

$SrTiO_3$ crystal is a particularly important example of the ideal perovskite structure AMX_3. It is considered as a model for the study of ferroelastic phase transition. This phase transition occurs at T_c of about 105 K.
The high temperature phase is cubic and the low temperature one tetragonal.

This phase transition produces no change in the cell volume, thermal expansion coefficient or dielectric constant but is accompanied by anomalies in the elastic constants [K.A. Muller and H. Thomas,1981 ; F.W. Lyttle, 1964 ; G.A. Samara, 1966].

In terms of the soft mode theory, the ferroelastic phase transition of $SrTiO_3$ involves the instability of a Brillouin zone boundary mode [K.A. Muller and H. Thomas, 1981].

It is well known that the vanishing (or near vanishing for a first order transition) of the soft mode frequency ω_s at T_c results from the cancellation of two terms. One short range interac-

[1]Departement de Recherches Physiques (CNRS UA 71), Universite Pierre et Marie Curie T.22, 4, place Jussieu F 75252, Paris Cedex 05.
[2]Physique des milieux condenses (CNRS UA 782), Universite Pierre et Marie Curie T.13, 4, place Jussieu F 75252, Paris Cedex 05.
[3]Laboratoire de Physico Chimie des Materiaux CNRS, 1, place A. Briand F 92190, Meudon

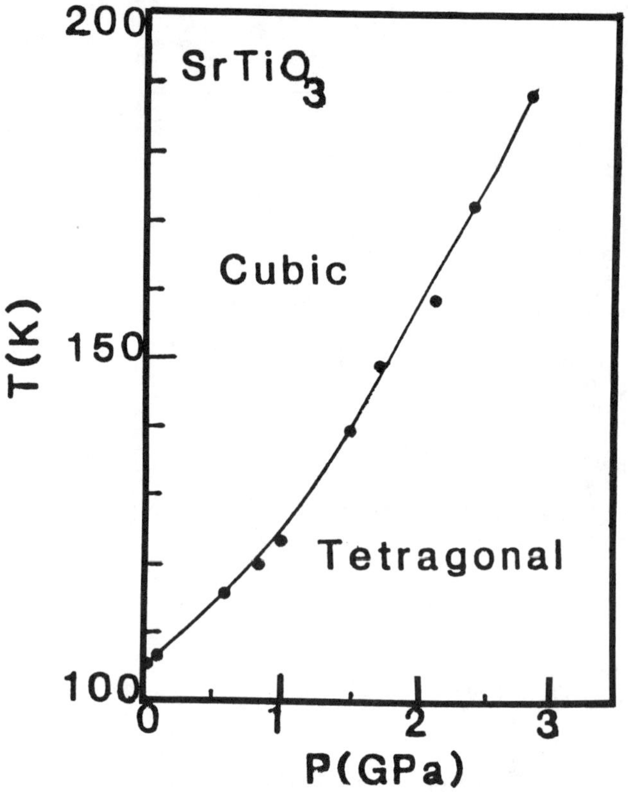

Fig. 1. Phase diagram of $SrTiO_3$. (from Okai, B., and J. Yoshimoto, 1975).

tion term (SR) and one long range interaction term (LR).

$$\omega_s^2 = (SR) + (LR) \qquad (1)$$

with (SR) <0 and (LR) >0 for $SrTiO_3$.

What happens to T_c when a hydrostatic pressure is applied ? The answer was given by G.A. Samara et al, 1975. High pressure strongly affects the balance of (SR) and (LR) forces. The pressure decreasing of the interionic distance increases the absolute value of the (SR) interaction term much more rapidly than the (LR) interaction term. In other words, there is decrease of ω_s for a ferroelastic transition and an increase of T_c under pressure.

Indeed, an increase of T_c with pressure was observed [B. Okai and J. Yoshimoto 1975]. The variation of T_c is linear up to 0.85 GPa and increases non linearly up to 3 GPa (figure 1). Consequently this phase transition must be observable at room temperature under high pressure.

The purpose of the present study is to show the elastic behaviour of $SrTiO_3$ under high pressure on both sides of the pressure phase transition.

In the first section we describe our experimental technique which uses Brillouin scattering and X-ray diffraction in a diamond anvil cell (D.A.C.). The second part is devoted to the analysis of the experimental results and their discussion.

Experimental technique

Brillouin Scattering under High Pressure

The high pressure is generated in a D.A.C. of Block Piermarini type [G.J. Piermarini and S. Block, 1975] which allows experiments in the back scattering geometry only. Typically the experimental volume is a cylinder of 200 µm in diameter and about 30 µm thick. So the typical dimensions of the sample is 20 µm in thickness and 100 µm in the largest dimension. The pressure transmitting medium is the 4:1 methanol-ethanol mixture and the pressure is measured using the non linear ruby scale[H.K. Mao and P.M. Bell, 1978]. Brillouin scattering experiments were performed using a piezoelectrically scanned five pass Fabry-Pérot interferometer. Spectra were stored in a multichannel analyser and data were fed into a microcomputer. The exciting light was the 514.5 nm line of a single moded Ar^+ laser with 100 mW output power. Due to the small size of the sample and the decrease of the apparent Brillouin efficiency a large number of accumulations was necessary.

The free spectral range (FSR) of the Fabry-Pérot was chosen as the very intense Brillouin lines coming from the diamond anvils ($\Delta\sigma_D \simeq 5.4$ cm^{-1}) were hidden in the elastically scattered Rayleigh lines, i.e. $\Delta\sigma_D$ was a multiple of the FSR.We used in these experiments FSR = 2.7273 cm^{-1} and FSR = 5.3868 cm^{-1}.

High pressure X-ray diffraction

The pressure volume relationship of $SrTiO_3$ was determined by angular dispersive X-ray diffraction up to ~ 21 GPa using a modified Bell and Mao type cell [J.M. Léger 1984].

The powdered sample was mixed with a pressure marker. Silver metal was chosen because its diffraction lines do not overlap the sample ones. The pressure transmitting medium was either the 4:1 methanol-ethanol mixture or a silicone grease. The first one produces hydrostatic pressure up to 10.5 GPa, but the shear stresses become rapidly large at higher pressure. The silicone grease crystallises at low pressure (P < 1 GPa) but does not produce large uniaxial stresses. The X-ray beam was produced by a conventional X-ray tube. The zirconium filtered Mo K α ray (λ = 0.7093 Å) was used and the beam was collimated to 0.1 mm. The diffraction pattern was recorded on a film situated at ~ 25 mm from the sample. One record requires approximatively 24 hours.

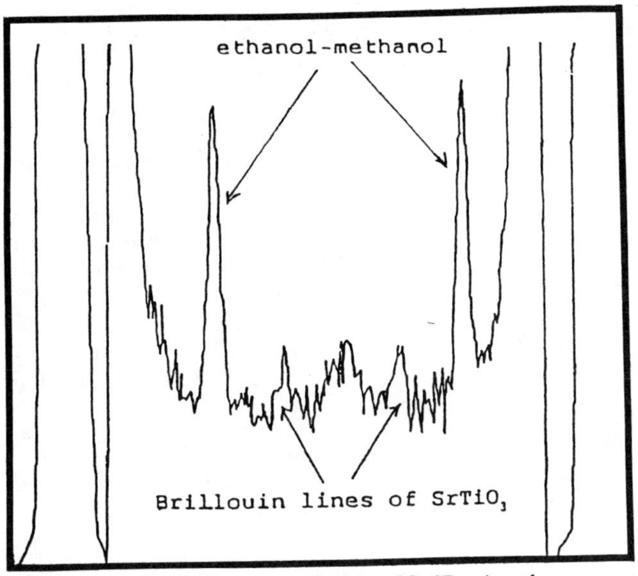

Fig. 2. Brillouin spectrum at 22 GPa in the back scattering geometry.
F.S.R. = 2.7273 cm^{-1}. The Brillouin lines correspond to a T.A. mode.

Experimental method

In back scattering geometry, the Brillouin wawe number shift $\Delta\sigma$ (in cm^{-1}) is related to the sound velocity v by

$$\Delta\sigma = \frac{2nv}{\lambda c} \quad (2)$$

where n is the refractive index, λ the wavelenght of the exciting laser light, c the light velocity in vacuum and v the sound velocity. In turn, the sound velocity is related to a combination of elastic constants C by

$$C = \rho v^2 \quad (3)$$

where ρ is the density. The combination of elastic constants involved in C depends on the propagation direction of the phonon relative to the crystallographic axes, and on the polarization of the phonon (longidudinal or transverse).

Under high pressure, the refractive index and density must be known. $\rho(P)$ is determined by X-ray diffraction measurements. Generally the pressure dependence of n is not known and is difficult to measure, so it must be evaluated using theoretical calculation of ionic polarizabilities.

For a cubic crystal the Lorentz Lorenz relation is

$$\frac{n^2-1}{n^2+2} = \frac{4}{3}\frac{\pi}{V}\alpha_0 \quad (4)$$

where α_0 is the optical contribution to the macroscopic polarizability of the crystal and V the volume.
And hence

$$\frac{dn^2}{dP} = \frac{36\pi V\alpha_0\left(\chi_T + \left(\frac{\partial \operatorname{Log} \alpha_0}{\partial P}\right)_T\right)}{(3V - 4\pi\alpha_0)^2} \quad (5)$$

χ_T is the isothermal compressibility coefficient.

The sign of $\frac{dn^2}{dP}$ depends on wether χ_T is larger or smaller than $-\left(\frac{\delta \operatorname{Log} \alpha_0}{\delta P}\right)_T$.

Results

Brillouin scattering measurements

In these experiments the phonon direction of propagation was parallel to the [111] direction of the cube.

We observed simultaneously the Brillouin lines of the crystal and those of the pressure transmitting medium. Around 10 GPa these two kinds of lines were superposed and the Brillouin shift measurement in the crystal was impossible.

At high pressure the Brillouin scattering is weak and the Brillouin peaks are obtained after 200 scans at least (figure 2). The quantive results which we obtained for the [111] direction are the following :

- The transverse acoustical mode has been measured up to 25 GPa. The variation of the Brillouin shift $\Delta\sigma$ with pressure is not continuous (figure 3). Around 6 GPa, its slope becomes smaller. A negative jump of $\Delta\sigma$ occurs between 5.5 and 7.5 GPa. The jump value is about 0.05 cm^{-1} . These two observations are consistent with a pressure phase transition. The accurate value P_c of the pressure of transition has not been determined because the points are too much scattered in the pressure range 5.5-7.5 GPa.

The observed transition is the cubic-tetragonal one. Indeed, the extrapolation of the pressure dependence of T_c towards high pressure (figure 1) is consistent with a transition around 5.5 GPa. Moreover, the cubic-tetragonal phase transition observed by Brillouin scattering at room pressure and low temperature produces a discontinuity in $\Delta\sigma$ of the same order of magnitude with the same sign [W. Kaiser and R. Zurek, 1966] (figure 4).

- Above 9 GPa when the pressure is increased new lines appear gradually (figure 3). The high frequency Brillouin shift corresponds to a longitudinal or quasi-longitudinal mode of the high pressure phase.

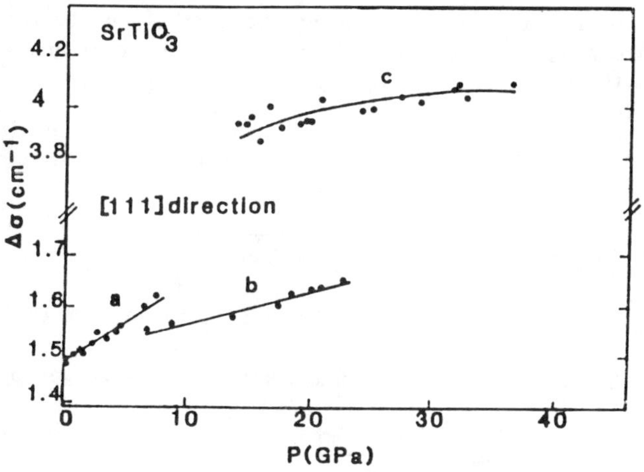

Fig. 3. Brillouin shift versus pressure. The phonon direction of propagation is parallel to the [111] direction of the cube. The pressure dependence of a transverse (cubic phase) and quasi transverse (tetragonal phase) mode is shown on the (a), (b) curves respectively. The pressure dependence of a quasi longitudinal mode is shown on (c) curve.

X-ray diffraction measurements

The diffraction patterns at high pressure in the DAC are consistent with those obtained under normal conditions. 4 or 6 lines were observed up to 5 GPa. So, it is possible to deduce the evolution of the cell parameter with a good accuracy in this pressure range. Above 5 GPa, the contrast of the lines decreases when the pressure is increased. But no new lines appear corresponding to a phase transition. The experimental points are compatible with an indexing on a cubic cell up to 22 GPa. The relative density $\frac{\rho_0}{\rho}$ versus pressure is plotted on figure 5. No discontinuity in the cell volume is observed between 5.5 and 7.5 GPa. This is compatible with a second order phase transition such as the cubic-tetragonal one.

The present results can be compared with those of L.R. Edwards and R.W. Lynch obtained in a previous work [L.R. Edwards and R.W. Lynch, 1970]. Their experimental techniques were rather different. Indeed, these authors used neither gasket between the anvils to confine the sample, nor pressure transmitting medium. Consequently, this experimental method induced anisotropic stresses throughout the sample and there is a strong non-hydrostaticity.

Discussion

The X-ray diffraction measurements were fitted by the Birch Murnagham equation of state:

$$P = \frac{3}{2} B_0 \left[\left(\frac{\rho_0}{\rho}\right)^{\frac{-7}{3}} - \left(\frac{\rho_0}{\rho}\right)^{\frac{-5}{3}} \right] \left[1 + \frac{3(B_0'-4)}{4} \left[\left(\frac{\rho_0}{\rho}\right)^{\frac{-2}{3}} - 1 \right] \right] \quad (6)$$

where B_0 and B_0' are respectively the bulk modulus and its pressure deravitive at room pressure (figure 5). The value B_0 = 174.2 GPa is calculated from [R.O. Bell and G. Rupprecht, 1963]. Using this B_0 value, the best fit of our data obtained by a least square method give us B_0' = 5.3. This value is more consistent with B_0' = 5.74 deduced from ultrasonic measurements [A.G. Beattie and G.A. Samara,1970] than B_0' = 4.4, obtained by Edwards and Lynch. This last result can be explained by the conditions of their experiments.

The low pressure dependence of the refractive index [G.A. Samara,1966] was extrapolated towards high pressure following the linear equation

$$n = 2.4612 - 0.0012\,P \quad (7)$$

with P in GPa.

This extrapolation is valid up to 5 GPa at least because the dielectric constant does not produce any anomaly in this pressure range [G.A. Samara and A.A. Giardini, 1965]. The negative value of $\frac{dn}{dP}$ is in agreement with sign calculated using the relation (5) and the volume dependence

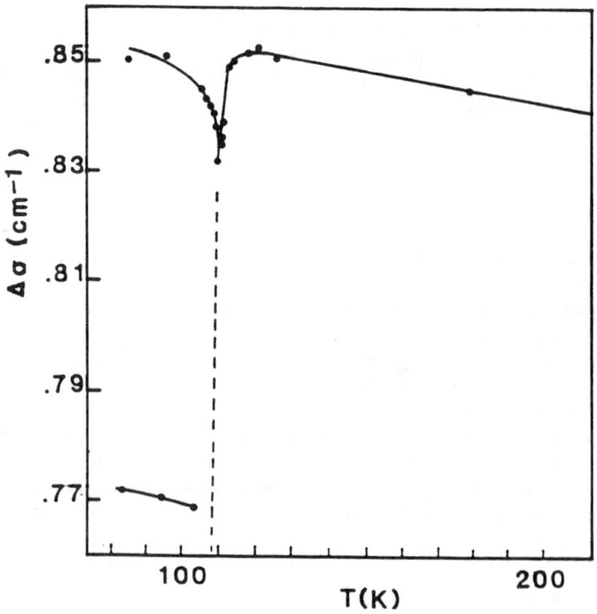

Fig. 4. Brillouin shift versus temperature. The phonon direction of propagation is parallel to the [100] direction âf the cube. (from Kaiser W. and R.Zurek, 1966).

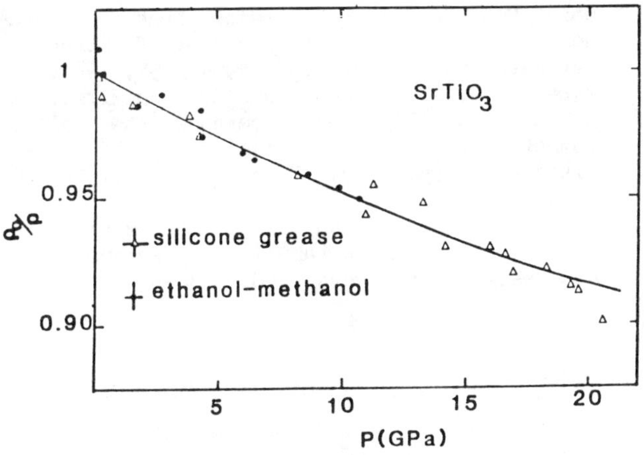

Fig. 5. Equation of state of $SrTiO_3$ in relative units. ρ_0 is the density under normal conditions. Various symbols correspond to different pressure transmitting media.

of optical polarizabilities [W.N. Lawless and H. Granicher, 1967]. Equation (7) was extrapolated in the whole studied pressure range.

In the experimental geometry, the elastic moduli combination was

$$\rho V_T^2 = \frac{C_{11} + C_{44} - C_{12}}{3} \tag{8}$$

Between 0 and 7.5 GPa ρV_T^2 increases linearly with a slope

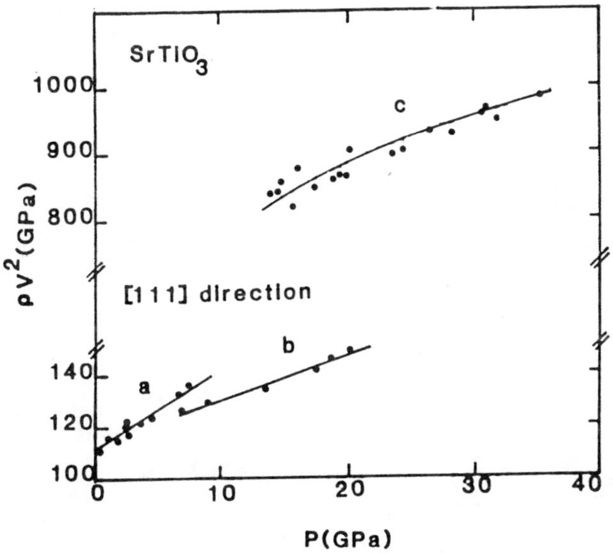

Fig. 6. Elastic moduli versus pressure. (a), (b) and (c) have the same meaning as on fig. 3.

$$\frac{d\rho V_T^2}{dP} \simeq 3.2 \tag{9}$$

In the high pressure phase, the slope of the quasi-transverse modulus $(\rho V_T^2)'$ becomes smaller and is equal to 1.7.

The longitudinal or quasi-longitudinal mode observed in the high pressure phase is shown in figure 6. Because the exact orientation of this phase is not known, the combination of elastic moduli cannot be deduced.

Conclusion

We measured the equation of state and the pressure dependence of the Brillouin shift in the [111] direction of $SrTiO_3$ at room temperature. This study shows that the cubic ↦ tetragonal phase transition occurs at about 6 GPa with a non measurable variation of the volume cell. The pressure dependence of a transverse acoustical phonon was measured in both phases up to 25 GPa and a longitudinal acoustical one in the high pressure phase between 13 and 36 GPa. The Brillouin efficiency in this compound is quite small and 200 scans or more are often necessary to obtain acceptable spectra. Nevertheless, the Brillouin scattering technique in a diamond anvil cell is a key for the study of elastic properties of the perovskite structre under high pressure.

References

Beattie, A.G., and G.A. Samara, Pressure dependence of the elastic constants of $SrTiO_3$, J. Appl. Phys., 42, 2376-2381, 1971.

Bell, R.O., and G. Rupprecht, Elastic constants of Strontium titanate, Phys. Rev., 129, 90-98, 1963.

Kaiser, W., and R. Zurek, Brillouin and critical light Scattering in $SrTiO_3$ crystals, Phys. Letters, 23, 668-670, 1966.

Lawless, W.N., and H. Gränicher, Temperature dependent polarizabilities in paraelectric $BaTiO_3$ and $SrTiO_3$, Phys. Rev., 157, 440-447, 1967.

Léger, J.M., High pressure X-ray diffraction of cerium compounds in an improved diamond anvil cell, Revue Phys. Appl., 19, 815-818, 1984.

Lytle, F.W., X-ray diffractometry of low temperature phase transformations in strontium titanate, J. Appl. Phys., 35, 2212-2215, 1964.

Mao, H.K., P.M. Bell, J.W. Shaner, and D.J. Steinberg, Specific volume measurements of Cu, Mo, Pd and Ag and calibration of the ruby R_1 fluorescence pressure gauge from 0.06 to 1 Mbar, J. Appl. Phys., 49, 3276-3283, 1978.

Muller, K.A., and H. Thomas, Topics in current physico-structural phase transition, 184 pp., 1981.

Okai, B., and J. Yoshimoto, Pressure dependence of the structural phase transition temperature in $SrTiO_3$ and $KMnF_3$, J. Phys. Soc.Japan, 39, 162-165, 1975.

Piermarini, G.J., and S. Block, Ultra-high pressure diamond anvil cell and several semiconductor phase transition pressures in relation to the fixed point pressure scale, Rev. Sci. Inst.,46, 973-979, 1975.

Samara, G.A., Pressure and temperature dependences of the dielectric properties of the perovskites $BaTiO_3$ and $SrTiO_3$, Phys. Rev., 151, 378-386, 1966.

Samara, G.A., and A.A. Giardini, Pressure dependence of the dielectric constant of Strontium Titanate, Phys. Rev., 140, A954-A957, 1965.

Samara, G.A., T. Sakudo, and K. Yoshimitsu, Important generalization concerning the role of competing forces in the displacive phase transitions, Phys. Rev. Letters, 35, 1767-1769, 1975.

Singh, A.K., and C. Balasingh, Uniaxial stress component in diamond anvil high-pressure X-ray cameras, J. Appl. Phys., 48, 5338-5340, 1977.

EQUATION OF STATE AND RHEOLOGY IN DEEP MANTLE CONVECTION

David A.Yuen and Shuxia Zhang

Minnesota Supercomputer Institute and Dept. of Geology and Geophysics
Univ. of Minnesota, Minneapolis, MN 55415

Abstract. We present here an overview of the nonlinear aspects of compressible convection with variable viscosity and variable thermodynamical parameters in the equation of state. We have employed the single-mode, mean-field convection equations for a spherical-shell geometry as a simple tool in understanding the basic physical mechanisms of deep mantle convection. We have extended our approach to include depth-dependences of the thermal conductivity, thermal expansivity and the Grüneisen parameter. We have also considered the temperature-dependence of thermal conductivity. Results of the interior temperature and the convective distortion of the core-mantle boundary show that for the Newtonian rheology the activation energy in the bulk mantle cannot be greater than about 40 kcal/mole and the activation volume must be smaller than 3 cm^3/mole. The temperatures of the lower mantle increase with depth-dependent and temperature-dependent thermal conductivity, decrease with depth-dependent expansivity, and increase again when both depth-dependent expansivity and conductivity are taken into account. The dynamical effects from a density-varying Grüneisen parameter are very small. The temperature-dependence of the mantle thermal conductivity can not be strongly nonlinear, as even a quadratic dependence of the conductivity with temperature would produce melting in the lower mantle. The thermal structure of the seismically resolved anomalous zone at the base of the mantle (D") is strongly influenced by these variable properties both in the equation of state and in the rheology.

Introduction

The last few years have seen progress made in the understanding of three-dimensional seismic structure in the mantle (Dziewonski, 1984; Woodhouse and Dziewonski, 1984; Tanimoto, 1987), mantle flow processes (Jarvis and Peltier, 1986; Machetel and Yuen, 1987a) and flow and magnetic field generation in the core (Bloxham and Gubbins, 1987). At the same time there have been significant strides made in high-pressure experimental techniques, which are shedding new light into the thermal regimes of the lower mantle (Heinz and Jeanloz, 1987, Williams and Jeanloz, 1988) and the core (Williams et al., 1987). Experiments with laser-heated diamond cell now show the possibilities for reactions between the liquid iron and solid oxides and silicates at conditions simulating the CMB (Knittle and Jeanloz, 1986), which is known from seismology to be extremely heterogeneous (e.g. Young and Lay, 1987). All of these developments have helped to put dynamical studies of the lower mantle and of the CMB into the limelight of solid-earth geophysics.

One important aspect of mantle dynamics which has been ignored in most past studies is the role played by equation of state. Yet it is well known from seismic studies (e.g. Dziewonski and Anderson, 1981) and thermodynamical arguments (e.g. Birch, 1952; Anderson, 1987) that the density of the mantle is stratified with a scale height of about 5000 km and that there exist significant variations of thermal parameters, such as the coefficient of thermal expansion α, across the mantle because of pressure effects. With the exception of the work by Jarvis and McKenzie (1980) there has been a virtual standstill in work on compressible fluids pertaining to mantle dynamics. Part of this may have stemmed from the small effects arising from compressibility obtained in the calculations of Jarvis and McKenzie (1980). The shortcomings in this model include (a) constant viscosity (b) specified heat-flux boundary conditions at the bottom of the mantle (c) aspect-ratio one configuration. From mean-field theory (Yuen, Quareni and Hong, 1987) it is shown that variable viscosity can promote strong nonlinear interactions in compressible convection. The constant temperature boundary conditions at the CMB may be more appropriate from considerations of the energetics of the geodynamo (Stevenson, Spohn and Schubert, 1983) and the nature of the compositional discontinuity of the CMB as being the chemical reaction zone (Knittle and Jeanloz, 1986). Recently it has been demonstrated that expanding the computational domain to incorporate long wavelength can alter the style of convection dramatically (Christensen, 1987a; Machetel and Yuen, 1987a).

The compressible models (Jarvis and McKenzie, 1980; Quareni and Yuen, 1988; Yuen et al., 1987; Zhang and Yuen, 1987) are based on constant thermodynamical parameters. This assumption of constant properties may not be reasonable from a physical standpoint. There is an awakening interest on the variation of thermodynamical parameters throughout the mantle (Anderson,1987) and the dynamical implications of these properties on mantle plumes (Zhao and Yuen, 1987). The effects of decreasing thermal expansivity and increasing thermal conductivity with pressure (Anderson, 1987) on mantle convection must be properly assessed by going beyond

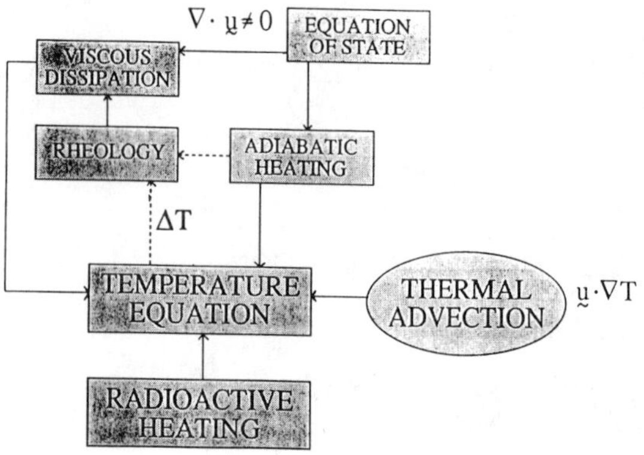

SOURCES OF NONLINEARITIES AND THEIR INTERACTIONS IN THE ENERGETICS OF MANTLE DYNAMICS

Fig. 1. Schematic diagram depicting the various nonlinear terms which contribute to the energy equation of compressible convection. $\nabla \cdot \underset{\sim}{u}$ refers to divergence of the velocity field, ΔT represents a change in the temperature and $\underset{\sim}{u} \cdot \nabla T$ is the thermal advection term.

the common incompressible equations for convection. For variable thermodynamical properties one cannot simply employ scaling relationships drawn from Boussinesq boundary-layer theory (e.g. Turcotte and Oxburgh, 1967; Olson and Corcos, 1980) to make predictions of the boundary-layer characteristics, since the variable thermal parameters, such as thermal expansivity, appear not only in the Rayleigh number, but also in other dimensionless parameters in both the energy and momentum equations. The interaction of rheology with equation of state in mantle convection is strongly nonlinear, as indicated from the mean-field results of Yuen et al. (1987) and Zhang and Yuen (1987). Radiogenic heating also serves to fuel this nonlinear process. Fig. 1 portrays schematically the nature of these nonlinear interactions in the energetics of mantle flow.

The main purposes of this paper are to acquaint the reader with the importance of rheology and equation of state in mantle convection and to communicate the need for the acquisition of certain parameters that are of particular importance to dynamicists. Up to now, the nonlinear roles played by rheology, mantle compressibility and internal heating in mantle convection have not been widely appreciated. Given all the parameters associated with this problem, including time-dependence and three dimensions into a rigorous formulation leads to an intractable problem, since only very few cases can be studied even with the present generation of supercomputers. For now it seems more important to map out the important relevant parameters associated with the nonlinear elements shown in Fig. 1 and to understand the basic physics produced by variations of the rheological and equation of state parameter values. We will employ the mean-field method (Quareni and Yuen, 1988) because it has the immediate advantage of being able to reach the range of parameters relevant to the mantle. It is an approximate method but is fast computationally. This method will enable us to conduct a parameter space sweep wide enough to address several important issues in lower mantle dynamics. The next section of this paper gives a brief look into the nonlinear nature of compressible mantle convection. This is followed by discussions of the equation of state and rheology along with an overview of the mathematical formulation of the compressible convection problem. We then present the results of these compressible mean-field models and we close with a discussion of the geophysical implications.

Nonlinear Nature of the Energetics in Mantle Convection

Mantle convection by its very nature is a highly nonlinear phenomenon. For highly viscous fluids, such as the Earth's mantle, thermal advection (see Fig. 1), constitutes the sole source of nonlinearity in the case of constant viscosity, incompressible convection. This nonlinearity involving the coupling between the temperature and velocity fields represents a source term in the energy equation and has been identified as being responsible for precipitating chaotic thermal convection in the mantle (Vincent and Yuen, 1988). With the introduction of compressibility, one must include both viscous dissipation and adiabatic heating in the energy equation. Viscous dissipation Φ is another source of nonlinearity and involves the product of the viscosity η and quadratic terms of the velocity gradients. For temperature-dependent viscosity $\eta(T) \sim \exp(A/T)$ where A is related to the activation energy Q^*, a rise in the temperature locally from viscous dissipation would decrease the viscosity, leading to higher velocities locally. Adiabatic heating also contributes to the overall energetics but occurs over broader spatial scales than viscous heating (Jarvis and McKenzie, 1980; Quareni and Yuen, 1988). As adiabatic heating is dependent linearly on the thermal expansivity α, its strength locally would then be controlled by the depth variations of α, which are governed by the equation of state of the mantle. With compressibility taken into account, the velocity field u has no longer zero divergence (see Fig. 1). A non-zero divergent term in the continuity equation would support an additional amount of viscous dissipation (e.g. Batchelor, 1967). The magnitude of this new contribution is governed by the logarithmic derivative of the density (see eqn. (9) below) in the vertical direction (Quareni and Yuen, 1988). Thus effects from equation of state come prominently into the picture of mantle energetics.

Internal heating represents another source term which is responsible for producing time-dependent convection in Boussinesq fluids (McKenzie, Weiss and Roberts, 1974). It has recently been found (Machetel and Yuen, 1988) that effects of compressibility, when acted in concert with radiogenic heating, give rise to much more nonlinear time-dependent situations. Radioactive heating also interacts nonlinearly with viscous dissipation because of the basic asymmetrical character of the streamlines in internally heated convective flows (e.g. McKenzie et al., 1974; Jarvis and Peltier, 1982) which produce narrow descending limbs with strong shear, thus providing a strong source of viscous dissipation. This nonlinear interaction between radioactive heating and viscous dissipation is further enhanced by the presence of variable viscosity. Fig. 1 gives a summary of all of these nonlinearities involved in the overall energetics of mantle convection.

Equation of State and Rheology

Knowing the appropriate equations of state to use depends upon the particular geophysical application in mind. For extrapolating elastic properties to zero pressure (e.g. Knittle, Jeanloz and Smith, 1986) in order to compare with laboratory data, one would need a very accurate equation of state, given to fourth order in the strain (e.g. Birch, 1978). But these finite-strain equations are considerably more complicated to apply in convection calculations than the Adams-Williamson E.O.S., which has been employed in previous compressible convection studies (Jarvis and McKenzie, 1980; Yuen et al., 1987; Quareni and Yuen, 1988).

Higher order E.O.S. has been employed in a limited convection study (Baumgardner, 1985) in which no significant dynamical effects from its usage were found. In what follows we will employ the Adams-Williamson E.O.S. in which an expansion about an adiabatic state is assumed (Birch, 1952). We will also make use of the seismic equation of state for the lower mantle derived by Anderson (1987) from lateral variations of seismic velocities in the mantle. From lattice dynamical theory, the thermal property variations have been cast in terms of the density dependences. In non-dimensional form the depth-dependence of the density ρ for variable thermodynamical properties is given by

$$\frac{1}{\rho}\frac{d\rho}{dr} = \frac{-\alpha(\rho)gd}{\gamma(\rho)C_p(\rho)} \tag{1}$$

where r is the radius, which has been non-dimensionalized with respect to d, the depth of the mantle, g is the gravitational acceleration, γ is the Grüneisen parameter, which depends weakly on density (e.g. McQueen et al., 1970; Stacey et al., 1981; Anderson, 1987) and C_P is the specific heat, which should be reasonably constant in the mantle (e.g. Kieffer, 1979). From the seismic anomalies Anderson (1987) deduced that thermal expansivity, lattice thermal conductivity k and Grüneisen parameter γ should vary with ρ according to

$$\alpha = \alpha_o\left(\frac{\rho_o}{\rho}\right)^3 \tag{2}$$

$$k = k_o\left(\frac{\rho}{\rho_o}\right)^3 \tag{3}$$

$$\gamma = \gamma_o\left(\frac{\rho_o}{\rho}\right) \tag{4}$$

where the subscripts "o" denote the values at the surface.There are experimental indications that thermal expansivity decreases with pressure faster than eqn. (2) (Böhler et al., 1988).

In the main we will take γ to be a constant, except in one instance (see Fig. 11). For constant thermodynamic properties, the density profile is

$$\rho(r) = \rho_o \exp\left(\frac{D}{\gamma}(r_o - r)\right) \tag{5}$$

where r_o is the nondimensionalized Earth's radius and $D = \alpha gd/C_P$ is the dissipation number (e.g. Jarvis and McKenzie, 1980), which is related to the inverse of the scale-height of the adiabatic temperature.

For the type of dependences of the thermodynamic properties on density, the density profile assumes the canonical form

$$\frac{1}{\rho}\frac{d\rho}{dr} = f(\rho) \tag{6}$$

For simple power law dependences of γ and α on ρ, simple analytical expressions of ρ(r) can easily be derived. In the case of γ constant and α(ρ), given by eqn (2), one finds

$$\rho(r) = \left[1 + \frac{3D_o}{\gamma}(r_o - r)\right]^{\frac{1}{3}} \tag{7}$$

where D_o is the dissipation number based on the thermodynamical parameters at the surface.

In this study we will use a linear (Newtonian) constitutive relationship between the stress and strain-rate tensors, since the present formulation of the mean-field equations cannot handle nonlinear (non-Newtonian) rheologies. For compressible, viscous fluids the deviatoric stress tensor τ_{ij} is related to the gradients of the velocity field through the dynamic viscosity η by

$$\tau_{ij} = \eta\left(\frac{\partial u_i}{\partial x_j} + \frac{\partial u_j}{\partial x_i} - \frac{2}{3}\delta_{ij}\,\nabla\cdot\underset{\sim}{u}\right) \tag{8}$$

As discussed above, the divergence of the velocity field does not vanish for compressible fluids and this term is related to the density profile by

$$\nabla\cdot\underset{\sim}{u} = -u_r\frac{\partial \ln\rho}{\partial r} \tag{9}$$

where u_r is the radial velocity component. We note that eqn. (9) is a general expression in which the characteristic scale-height of the density is given by the right hand side.

Temperature- and pressure-dependent viscosity is one of the unique physical aspects in mantle convection, because of the thermally activated process in mantle creep (e.g. Sammis et al., 1977). We have used a viscosity law with an Arrhenius type of dependence. This viscosity relationship is given by

$$\eta(T, p) = \eta_\infty \exp\left(\frac{Q^* + pV^*}{RT} - \frac{Q^*}{RT_\infty}\right) \tag{10}$$

where T is absolute temperature, Q* and V* are respectively the activation energy and volume, p is the hydrostatic pressure given by $\rho(r)\, g\, (r_0 - r)\, d$, R is the gas constant, and η_∞ is an interior viscosity for a given reference mantle temperature T_∞. In this work we will set $\eta_\infty = 10^{21}$ Pa s, a constraint from postglacial rebound (e.g. Yuen, Sabadini and Boschi, 1982) and $T_\infty = 2773$ K, a characteristic temperature in the lower mantle, below the solidus (e.g. Heinz and Jeanloz, 1987). At this point it is important to point out that in steady-state non-Newtonian variable viscosity convection, the effective activation enthalpy $H^* = Q^* + pV^*$ is approximately given by H^*/n, where n is the exponent of the stress power-law in non-Newtonian rheology (Karato, 1981; Christensen, 1984). However, this scaling relationship no longer holds in time-dependent non-Newtonian convection (Christensen and Yuen, 1988).

Mathematical Formulation

In this section we focus on the mathematical development of thermal convection with equation of state and rheology included. The equations governing thermal convection express mass balance, motion and energy balance. For the Earth's mantle the conservation of mass equation in spherical-shell coordinates with axisymmetry assumed reads

$$\frac{1}{r^2}\frac{\partial}{\partial r}\left(\rho\, r^2 u_r\right) + \frac{1}{r \sin\theta}\frac{\partial}{\partial \theta}\left(\rho u_\theta \sin\theta\right) = 0 \tag{11}$$

where u_r and u_θ are respectively the radial and tangential components of the axisymmetric velocity vectors and θ co-latitudinal angle measured from the polar axis. Eqn (11) is called the anelastic approximation (Ogura and Phillips, 1962) in which sound waves in the medium have been filtered out.

For highly viscous fluids, such as the Earth's mantle, the non-dimensional momentum equations in the radial and tangential directions are

$$-\frac{\partial \pi}{\partial r} + \left[\frac{1}{r^2}\frac{\partial}{\partial r}\left(r^2 \tau_{rr}\right) + \frac{1}{r\sin\theta}\frac{\partial}{\partial\theta}\left(\tau_{r\theta}\sin\theta\right)\right] - \frac{\tau_{\theta\theta} + \tau_{\phi\phi}}{r} + Ra_T\,\alpha(\rho)\rho(T - T_r) = 0 \tag{12}$$

$$-\frac{1}{r}\frac{\partial \pi}{\partial \theta} + \left[+\frac{1}{r^2}\frac{\partial}{\partial r}\left(r^2 \tau_{r\theta}\right) + \frac{1}{r\sin\theta}\frac{\partial}{\partial\theta}\left(\tau_{\theta\theta}\sin\theta\right) + \frac{\tau_{r\theta}}{r} - \frac{\cot\theta\,\tau_{\phi\phi}}{r}\right] = 0 \tag{13}$$

where T is the non-dimensional temperature, T_r is the dimensionless reference temperature, π is the fluid dynamical pressure, $\tau_{\phi\phi}$ is the normal stress along the longitudinal direction, given by the angle ϕ, and Ra_T is the Rayleigh number for basal heating given by

$$Ra_T = \alpha_0 \frac{\Delta T g \rho_0 d^3}{\kappa_0 \eta_\infty} \tag{14}$$

The thermal diffusivity κ_0 is based on the surface values and ΔT is the temperature difference across the convecting layer which is used to nondimensionalize the temperatures. We note that α in eqn. (12) may vary with density according to eqn. (2).

The elements τ_{ij} of the deviatoric stress tensor, which are used in axisymmetric spherical convection, are given by

$$\tau_{rr} = 2\eta\left(\frac{\partial u_r}{\partial r} - \frac{1}{3}\nabla \cdot \underset{\sim}{u}\right) \tag{15}$$

$$\tau_{\theta\theta} = 2\eta\left(\frac{1}{r}\frac{\partial u_\theta}{\partial \theta} + \frac{u_r}{r} - \frac{1}{3}\nabla \cdot \underset{\sim}{u}\right) \tag{16}$$

$$\tau_{\phi\phi} = 2\eta\left(\frac{u_r}{r} + \frac{u_\theta \cot\theta}{r} - \frac{1}{3}\nabla \cdot \underset{\sim}{u}\right) \tag{17}$$

$$\tau_{r\theta} = \eta\left(\frac{1}{r}\frac{\partial u_r}{\partial \theta} + r\frac{\partial}{\partial r}\left(\frac{u_\theta}{r}\right)\right) \tag{18}$$

The non-dimensionalized energy equation with mantle radioactivity present is given by

$$\frac{\partial T}{\partial t} = \frac{1}{\rho}\left(\frac{\partial k}{\partial r}\frac{\partial T}{\partial r} + k(\rho)\nabla^2 T + r + \frac{D_0}{Ra_T}\eta(T, p)\Phi\right) - u_r\frac{\partial T}{\partial r} - \frac{u_\theta}{r}\frac{\partial T}{\partial \theta} - D_0 u_r\, \alpha(\rho)\,(T + T_0) \tag{19}$$

where t is time nondimensionalized by d^2/κ_0. T_0 is the non-dimensional surface temperature, and D_0 is the dissipation number based on surface values. The magnitude of the radiogenic heating is given by the ratio $r = Ra_H/Ra_T$ where Ra_H is the internal-heating Rayleigh number defined by

$$Ra_H = \frac{\alpha_0 g \rho_0 d^5 H}{\kappa_0 \eta_\infty k_0} \tag{20}$$

The heat production rate per unit mass is denoted by H. The viscous dissipation function Φ is given by

$$\Phi = 2\left[\left(\frac{\partial u_r}{\partial r}\right)^2 + \left(\frac{1}{r}\frac{\partial u_\theta}{\partial \theta} + \frac{u_r}{r}\right)^2 + \left(\frac{u_r}{r} + \frac{u_\theta \cot\theta}{r}\right)^2\right]$$
$$+\left[r\frac{\partial}{\partial r}\left(\frac{u_\theta}{r}\right) + \frac{1}{r}\frac{\partial u_r}{\partial \theta}\right]^2$$
$$-\frac{2}{3}\left[\frac{1}{r^2}\frac{\partial}{\partial r}(r^2 u_r) + \frac{1}{r\sin\theta}\left(\frac{\partial}{\partial\theta}u_\theta \sin\theta\right)\right]^2 \quad (21)$$

We now make the single-mode, mean-field approximation on the governing eqns., (11), (12), (13) and (19) along with the ancillary eqns., (15) to (18) and (21). Basically single-mode mean-field analysis (e.g. Gough, Spiegel and Toomre, 1975) separates the dependent variables, velocity, dynamical pressure, and temperature into an horizontally averaged (mean) and fluctuating parts. This method has been employed as a reconnaissance tool for many diverse geological problems (e.g. Spera et al., 1986; Quareni et al., 1985). The horizontal average involves integrating the field variable over a spherical surface S at radius r. That is, for the mean temperature T(r,t) we have

$$T(r,t) = \frac{\int_S T(r,\theta,\phi,t)\,d\Omega}{\int_S d\Omega} \quad (22)$$

The horizontally averaged velocity is zero, since the motion is assumed to be periodic on the spherical surface. The single-mode representation assumes that the fluctuating temperature θ and velocity fields are concentrated at a particular degree n of the spherical harmonic expansion $Y_{nm}(\theta,\phi)$, m being zero in the axisymmetric configuration. This represents a simple but useful model in that the horizontally averaged structure is preserved. In this way we set

$$u_r(r,\theta,t) = y_{1n}(r,t)Y_n(\theta) \quad (23a)$$

$$u_\theta(r,\theta,t) = y_{2n}(r,t)Y_n(\theta) \quad (23b)$$

$$\sigma_{rr}(r,\theta,t) = -\pi + \tau_{rr} = \frac{y_{2n}(r,t)}{r}Y_n(\theta) \quad (24a)$$

$$\tau_{r\theta}(r,\theta,t) = \frac{y_{4n}(r,t)}{r}\frac{\partial}{\partial\theta}Y_n(\theta) \quad (24b)$$

where $Y_n(\theta) = (2n+1)^{1/2} P_n(\cos\theta)$ with $P_n(\cos\theta)$ the usual Legendre function of degree n. The fluctuating temperature field is given by

$$\Theta(r,t,\theta) = \Theta_n(r,t)Y_n(\theta) \quad (25)$$

The partial-differential equations can then be spectrally decomposed to a set of partial differential equations for each degree n. Henceforth, we will drop the subscript "n" in the radial-dependent functions.

Substituting eqns. (23a) and (23b) into eqn. (11), we may write the mass conservation equation as an ordinary differential equation (O.D.E.):

$$r\frac{d}{dr}y_1 = \left[-2 - \frac{d\ln\rho}{dr}\right]y_1 + Ly_2 \quad (26)$$

with $L = n(n+1)$. Similarly, we may write the shear stress as

$$y_4 = \eta\left(y_1 - y_2 + r\frac{d}{dr}y_2\right) \quad (27)$$

Next we transform the motion equations to O.D.E.'s by multiplying eqn. (12) by $Y_n(\theta)$ and integrating them from $\theta = 0$ to π. From the orthogonal properties of Legendre functions we may simplify the two motion equations to

$$r\frac{d}{dr}y_3 = 12\eta\left(1 + \frac{1}{3}\frac{d\ln\rho}{dr}\right)y_1 - 6L\eta y_2$$
$$+ y_3 + Ly_4 - Ra_T\rho\alpha(\rho)\Theta r^2 \quad (28)$$

$$r\frac{d}{dr}y_4 = -6\eta\left(1 + \frac{1}{3}\frac{d\ln\rho}{dr}\right)y_1 +$$
$$(4L-2)\eta y_2 - y_3 - 2y_4 \quad (29)$$

Eqns. (26) through (29) constitute a linear, fourth-order, nonhomogeneous O.D.E. system with the fluctuating temperature $\theta(r)$ providing the forcing to drive the flow.

The energy equation is decomposed into mean and fluctuating parts. Projecting out the angular dependence of the field variables as in the momentum equations, we arrive at the single-mode, mean-field energy equations, which read

$$\frac{\partial\Theta}{\partial t} = \frac{1}{\rho}\left[\frac{dk(\rho)}{dr}\frac{\partial\Theta}{\partial r} + k(\rho)\left(\frac{\partial^2}{\partial r^2} + \frac{2}{r}\frac{\partial}{\partial r} - \frac{L}{r^2}\right)\Theta\right]$$
$$-y_1\frac{\partial T}{\partial r} - D_o\alpha(\rho)y_1(T + T_o) \quad (30)$$

$$\frac{\partial T}{\partial t} = \frac{1}{\rho}\left[\frac{dk(\rho)}{dr}\frac{\partial T}{\partial r} + k(\rho)\left(\frac{\partial^2 T}{\partial r^2} + \frac{2}{r}\frac{\partial T}{\partial r}\right) + r + \frac{D_o}{Ra_T}\eta\Phi\right] - y_1\frac{\partial \Theta}{\partial r} - D_o\alpha(\rho)y_1\Theta - \frac{L\Theta y_2}{r} \qquad (31)$$

with the viscous dissipation term in the mean-field approximation given by

$$\Phi = 12\left(\frac{y_1}{r}\right)^2 + 2L(2L-1)\left(\frac{y_2}{r}\right)^2 - \frac{12\,Ly_1y_2}{r^2} + \frac{4}{3}\left(\frac{d\ln\rho}{dr}y_1\right)^2 + \frac{Ly_4^2}{r^2\eta^2} + \frac{8}{r}\frac{d\ln\rho}{dr}y_1^2 - 4\left(\frac{d\ln\rho}{dr}\right)\frac{Ly_1y_2}{r} \qquad (32)$$

Although we expect on theoretical grounds that compressible convection would be highly time-dependent, we will solve the mean-field equations in the steady-state in order to derive some physical understanding from the steady horizontally averaged profiles. Furthermore, time-averaged temperature profiles (Machetel and Yuen, 1987) of chaotic mantle convection show very similar features to horizontally averaged temperature profiles taken from mean-field calculations for incompressible convection. With the steady-state ansatz, the time-derivatives in eqns. (30) and (31) are set to zero. The entire O.D.E. system then consists of the fourth-order momentum system (eqns. (26) through (29)) and the fourth-order energy system (eqns. (30) and (31)).

In this study we focus only on whole-mantle convection. The effects of compressibility on two-layer convection have been studied elsewhere (Yuen, Zhang and Langenberger, 1988). We have imposed stress-free and zero radial velocity at the surface and the CMB, which is taken to be 2886 km in depth. The temperature at the surface is held at 273 K in the case of constant viscosity and it is set to 1100 K for variable viscosity. The fluctuating temperature vanishes at both the surface and at the CMB. The temperature at the CMB can be constrained by the melting point of iron (Böhler, 1986; Williams et al., 1987). We have imposed a temperature difference $\Delta T = 3500$ K across the mantle. This is an important parameter, especially for variable viscosity, as the results of Zhang and Yuen (1987) have shown. This coupled eighth-order, nonlinear O.D.E. system with the associated eight boundary conditions on the temperature and velocity fields is solved for each degree n as two fourth-order systems with the momentum equations leading. Underrelaxation was used to facilitate numerical convergence (Quareni et al., 1985). Up to 200 adaptively chosen points have been used along the radial direction.

Numerical Results

The importance of certain geophysical parameters, such as activation energy and volume, are not immediately obvious, until when effects on the thermal-mechanical state in the deep mantle are illustrated by way of dynamical calculations. It is the purpose of this section to expose the reader to the sensitivity of the mantle temperature, viscosity and the convective distortion of the CMB from variations of the rheological and equation of state parameters. We will also discuss the effects from using different functional dependences of the thermal conductivity.

We illustrate the action of varying the magnitude of the Grüneisen parameter γ in Fig. 2, in which a constant viscosity is taken with a dissipation number of $D_o = 0.6$ and a Raleigh number of 10^6. In this paper we will focus only on long wavelength cells, since from both seismology (Masters et al., 1982) and geoid anomalies (Crough and Jurdy, 1980) it has been demonstrated that the dominant power in these two geophysical signals lies with the degree two (n=2) mode. In panel (a) we observe that the mean temperature profiles are

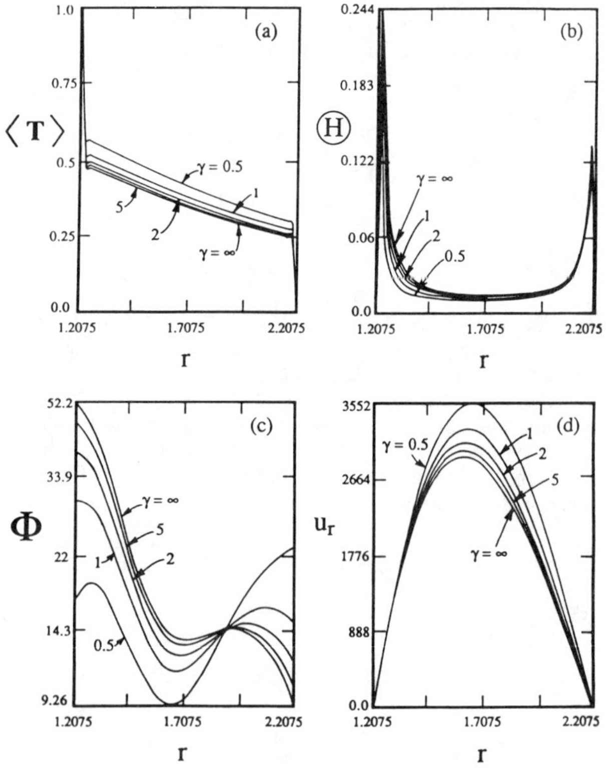

Fig. 2. Constant viscosity, compressible convection in spherical-shell. The Rayleigh number is 10^6, the mode used is n = 2 and the dissipation number D_o is 0.6. Panel (a) Mean temperature (b) fluctuating temperature with n = 2 angular dependence (c) dissipation function Φ', which is defined by $(D/\rho)(\eta/Ra_T)\Phi$ (d) radial velocity. Grüneisen parameters are given next to the curves. The temperature jump across the mantle is 3500 K. Radius has been non-dimensionalized with respect to the depth of the mantle.

characterized by the presence of two thin boundary layers joined by a gradual rise in temperature, which is adiabatic in character, i.e. dT/dr in the interior is equal to -DT. Temperatures rise as a consequence of decreasing γ or making the mantle material more compressible. This phenomenon is due to enhanced viscous heating from increased compressibility (Yuen et al., 1987). Lateral temperature variations θ induced by mantle convection contribute directly to the gravity anomaly signatures and deformations of the boundaries. They are also related to seismic anomalies from Birch's law relating seismic velocities to densities. We examine in panel (b) the influences of varying γ on the θ profiles. There are little differences among them. It is seen that greater concentration of lateral temperature variations is present at the CMB than at the top boundary layer.

As discussed above (see Fig. 1) one of the major sources of nonlinearity in compressible convection is viscous heating (see eqn. (32)). In panel (c) we plot Φ', which is defined by

$$\Phi'(r) = \frac{D_o\eta}{\rho Ra_T}\,\Phi(r) \tag{33}$$

We see that dissipative heating is concentrated in the deep part of the mantle for values of γ, greater than 1.0, characteristic of perovskite (e.g. Williams, Jeanloz and McMillan, 1987; Hemley, Jackson and Gordon, 1987). However, for lower values of γ, shear heating shifts to the upper mantle and two regions of intense heating in the upper and lower mantles result. As Φ' is divided by ρ, the magnitude of Φ' would be reduced by this factor, since density increases with smaller γ (see eqn. (5)). We display the radial velocity profiles in panel (d). Velocities increase with smaller γ, as viscous dissipation becomes more prominent. It is to be noted that a dimensionless velocity of 1000 corresponds to about 1 cm yr^{-1}. Thus these models deliver the proper magnitude of velocities, characteristic of surface plate motions.

We will now look at the effects of varying Q* in a purely temperature-dependent viscosity. We have in this study employed the same activation energy Q* and activation volume V* throughout the whole mantle, because of the much greater volume of the lower mantle. Furthermore, our purpose here is to illustrate the fundamental physics which would be obscured by the introduction of additional rheological parameters into the model. Panel (a) of Fig. 3 shows that the interior of the mantle becomes hotter with larger activation energies and that there is a sharp rise for Q* between 25 and 40 kcal/mole. This same phenomenon was also found for the cartesian geometry in which a heat-flux boundary condition was imposed at the bottom (Quareni and Yuen, 1988). These results would indicate that the bulk Q* for the mantle cannot be too high, say greater than 40 kcal/mole, but in the case of a non-Newtonian mantle, the actual Q* could conceivably be higher because the effective Q* may be reduced. The effects of varying the dissipation number D_o are shown in panel (b). The interior temperatures in the upper mantle diminish with D_o, while near the CMB the temperatures increase with D_o. Also displayed in panel (b) are calculations taken for Q* = 20 kcal/mole (dashed curves). It is seen that variations of Q* in the incompressible ($D_o = 0$) cases have much more pronounced effects than those

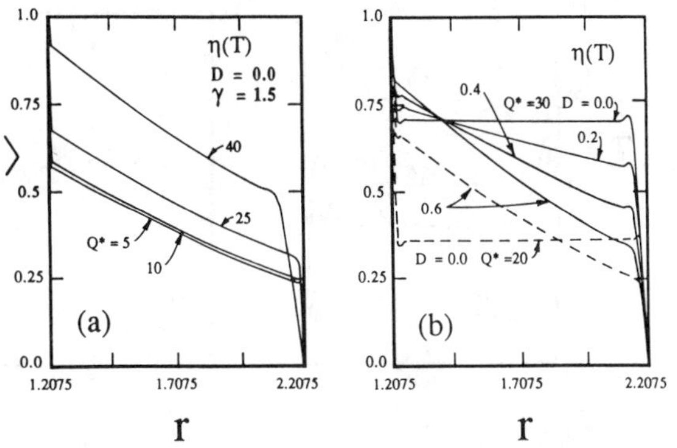

Fig. 3. Mean temperature as a function of radius for temperature-dependent viscosity. The Rayleigh number Ra_o, based on the interior viscosity η_∞, is set to 10^8 panel (a) Variations of activation energy. Values of Q* in kcal/mole are given next to the curves. D_o of 0.5 and γ of 1.5 have been employed. (b) Variations of the dissipation number for Q* = 30 kcal/mole (solid curves) and Q* = 20 kcal/mole (dashed curves). The temperature jump across the mantle is 3500 K. The Grüneisen parameter is 1.5.

with compressibility, because of the presence of the adiabatic temperature gradient in compressible models.

Fig. 4 shows the viscosity profiles for these cases with $\eta(T)$. Viscosities show a sharp drop near the surface due to the top thermal boundary layer. They all exhibit a monotonic decrease in the interior, terminating with a sharp drop of at most 0(10 Pa s) at the CMB. Viscosity reduction is found to increase with the activation energy. Since from geophysical inferences (Yuen et al., 1982; Hager et al., 1985) mantle viscosity increases at least slightly with depth, these results would then argue that there exist some amounts of pressure-dependence in the mantle flow laws in order to obtain a non-decreasing viscosity profiles when D_o is varied. Here we see that the viscosity gradients in the interior become more negative and the magnitudes of the lower mantle viscosity

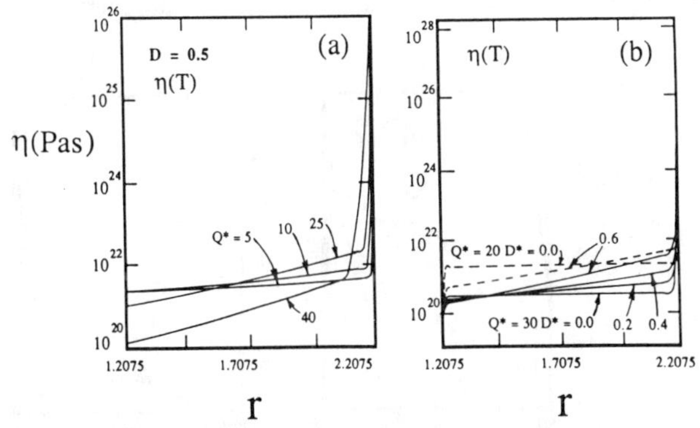

Fig. 4. Viscosity as a function of radius for temperature-dependent viscosity. These viscosity surves correspond to the temperature curves given in Fig. 3. All parameters are the same as in Fig. 3. Values of Q* and dissipation number are provided in the figure.

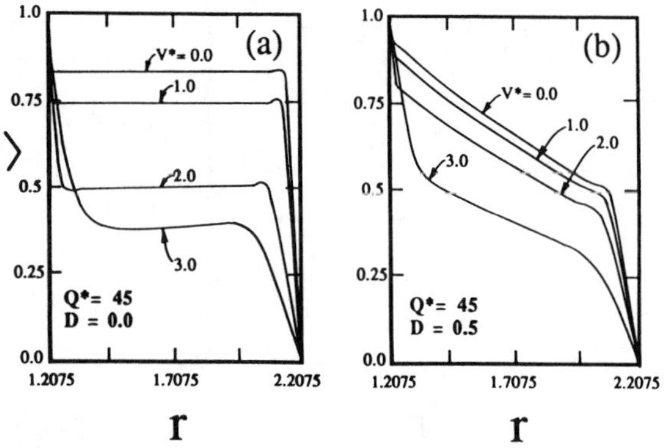

Fig. 5. Mean temperature profiles for temperature-and pressure-dependent viscosity. Ra_o is 10^8. Panel (a) Boussinesq ($D_o = 0$) case with $Q^* = 45$ kcal/mole, activation volumes V^* in cm^3/mole are given next to the curves (b) $D_o = 0.5$ and $Q^* = 45$ kcal/mole. Grüneisen parameter is 1.5.

increase, the larger the adiabatic gradients. These trends are useful for placing bounds on the dissipation number in the mantle.

In Figs. 5 and 6 we display respectively the results of the temperature and viscosity profiles for temperature- and pressure-dependent viscosity $\eta(T,p)$. Both incompressible (panel a) and compressible (panel b) cases are compared. The outstanding feature of these results is that the interior temperature is lowered with the introduction of pressure-dependence in the rheology. Boundary layers at the top and bottom are also broadened by $\eta(T, p)$. The effects of varying V^* are much greater for incompressible models, because of the absence of the adiabatic temperature gradient. The viscosity profiles in Fig. 6 demonstrate clearly their sensitivity to variations in the activation volume V^*. One can clearly rule out an activation volume greater than 3 cm^3/mole for Newtonian $\eta(T, p)$. The presence of the adiabatic temperature rise acts to moderate the viscosity increase. It is indeed quite possible for an isoviscous interior to be achieved self-consistently in compressible convection. This prospect has puzzled geodynamicists for a long time. There have been attempts in the past of using simplified temperature profiles from oceanic plate boundary-layer models (Karato, 1981), but they were ad hoc in character. Both the temperature and viscosity profiles would argue for a relatively small activation volume in the lower mantle, between 1 and 2 cm^3/mole, in order to satisfy geophysical inferences of deep mantle viscosity (c.g. Yuen and Sabadini, 1985). Recent work on the melting curves of perovskite (Heinz and Jeanloz, 1987) supports also the idea of a small activation volume in the lower mantle.

From the above figures we observe that the behavior of the variable viscosity convection solutions depends greatly on the values of Q^* and V^*. Figure 7 is a domain diagram on the (Q^*, V^*) plane where we give a summary of the various regimes of convective vigor and style. A Ra_T of 10^8, based on η_∞, has been used in these calculations. Other common parameters are $\gamma = 1.5$, $D_o = 0.5$, and $\Delta T = 3500$ K. Constant thermodynamical properties, have been employed in the construction of this diagram. For V^* greater than a certain magnitude, convective motion is not possible because of the strong pressure-dependence existing throughout the entire mantle. This region has been marked "no convection". A second region, in which the steady-state solution does not converge and weak convection of an oscillatory nature takes place, lies right below the convectively stagnant region. The third category, designated by the term "normal steady-state", extends from the origin ($Q^*=V^*=0$) and is situated right below the second region. The last sector on the lower right, called "hot convection", is characterized by an extremely hot interior, greater than $T = 0.75$. This domain diagram shows that there is a narrow band of Q^* and V^* for which satisfactory thermomechanical solutions can be obtained. Too large an activation energy without any pressure-dependence in the rheology will give rise to unbearably hot interiors. On the other hand, too large an activation energy, V^* greater than about 3 cm^3/mole, will stifle convection completely. It is to be emphasized that the boundaries delineating the different regions would shift with changes in the external control variables, such as Ra_T and D_o, and also with the introduction of variable thermodynamical properties. The construction of this domain diagram has required well over 100 separate

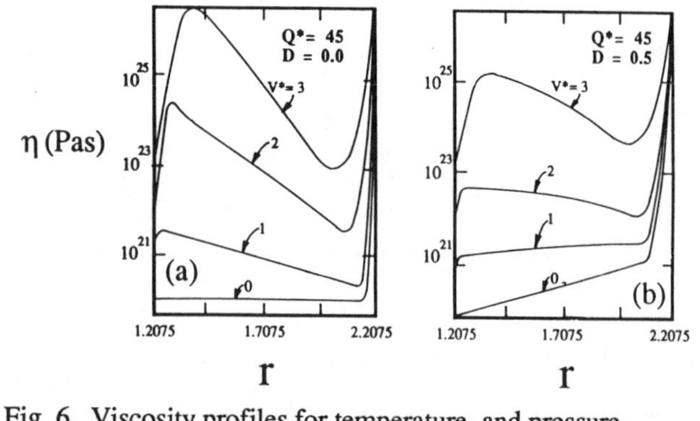

Fig. 6. Viscosity profiles for temperature- and pressure-dependent Newtonian rheology. All the parameters are the same as in Fig. 5.

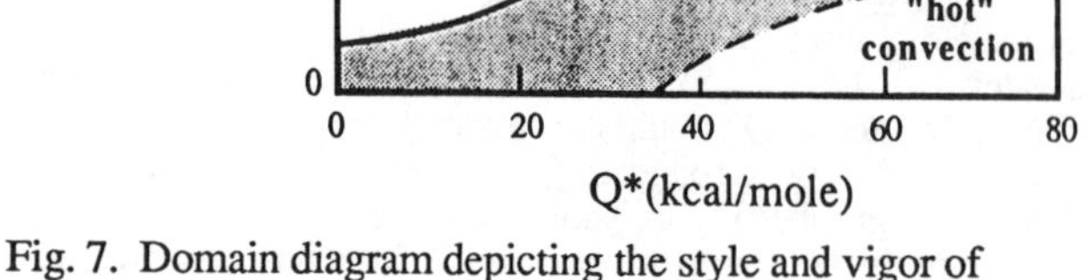

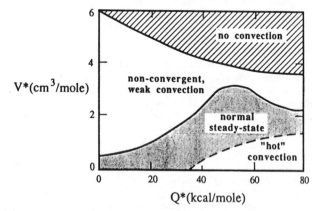

Fig. 7. Domain diagram depicting the style and vigor of temperature- and pressure-dependent viscosity compressible convection. The common parameters are $Ra_o = 10^8$, degree n = 2, $\gamma = 1.5$ and $D_o = 0.5$. The temperature parameters are ΔT = 3500 K and $T_o = 1100$ K.

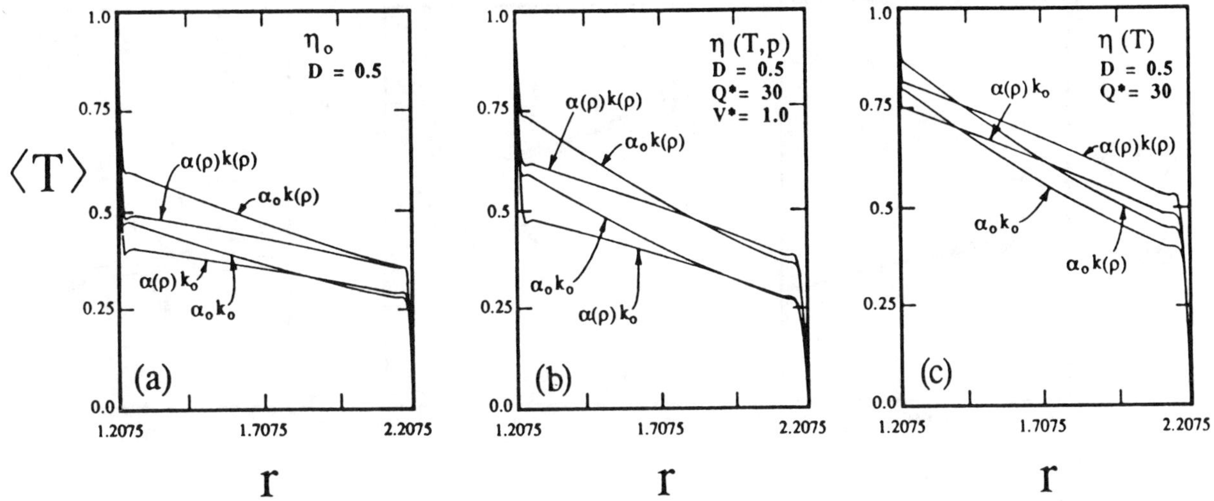

Fig. 8. Temperature profiles for variable thermodynamical properties. Panel (a) constant viscosity. The relevant input parameters are Ra_T = 10^6, D_0 = 0.5, Q* = 30 kcal/mole, V* = 1 cm^3/mole, and γ = 1.25 (c) temperature-dependent viscosity. Other parameters are the same as in panel (b). The symbol (α_0,k_0) denotes constant thermodynamical properties, (α_0, k(ρ)) denotes depth-dependent thermal conductivity only, ($\alpha(\rho)$, k_0) stands for depth-dependent thermal expansivity, and ($\alpha(\rho)$, k(ρ)) has both types of depth-dependences. Ra_0 for panels (b) and (c) is 10^8.

calculations, which would be an enormous task, if the full compressible convections were employed.

Next we study the effects of depth-dependent thermodynamic properties in the mantle on the thermal and viscosity structures. In what follows, we will employ the functional forms for the depth-dependences of $\alpha(\rho)$, $k(\rho)$ and $\gamma(\rho)$ derived by Anderson (1987) from lateral variations of seismic velocities in the lower mantle. It is worthwhile to point out here that, although there may be some uncertainties with the functional dependences in eqns. (2) to (4), the basic physics of incorporating these variable properties in compressible convection should be well demonstrated.

Fig. 8 shows the temperature profiles for constant viscosity (panel a), temperature- and pressure-dependent viscosity (panel b) and temperature-dependent viscosity (panel c). A temperature difference of 3500 K across the convecting layer is assumed. The Rayleigh numbers are 10^6 for constant viscosity and 10^8, based on η_∞, for $\eta(T)$ and $\eta(T, p)$. Profiles generated with constant thermodynamical properties and eqn. (5) are given the symbol (k_0, α_0). These curves are to be compared with those of (α_0, $k(\rho)$). We see that temperature profiles based on $k(\rho)$ alone are hotter and have a thicker D" layer with a much smaller temperature jump. On the other hand, those with $\alpha(\rho)$ alone have lower interior temperatures and a larger temperature rise across the D" layer. Profiles with both $\alpha(\rho)$ and $k(\rho)$ present have higher temperatures than the ones with constant thermal properties. The adiabatic temperature gradients for $\alpha(\rho)$ decrease with depth and remain linear for constant thermal expansivity. Inside the boundary layers the temperature gradients are not adiabatic but away from them the thermal profiles for the mean-field approximation lie on the adiabat, whose magnitude is determined by the local thermodynamical properties. For η(T,p) the temperature profile with both $\alpha(\rho)$ and $k(\rho)$ is several hundred degrees hotter than the one with constant thermodynamical properties. In general, effects of $\alpha(\rho)$ and $k(\rho)$ on mantle temperatures are not negligible.

The sensitivity to variations of the functional dependences of $\alpha(\rho)$, $k(\rho)$ and $\gamma(\rho)$ is examined next. Fig. 9 shows the temperature profiles for constant viscosity (panel a), $\eta(T)$ (panel b) and $\eta(T, p)$ (panel c), as we vary the power-law dependence of $\alpha(\rho) \sim 1/\rho^n$ from n = 0 to n = 4. Inspection of the figure reveals that the thermal profiles become more isothermal in the interior, hence more Boussinesq-like, as the power-law dependence becomes stronger. The effects of $k(\rho)$ on the thermal fields are illustrated next in Fig. 10. The influences of increasing the power-law dependence of $k(\rho)$ are stronger than for $\alpha(\rho)$, as we can observe from the greater differences in the temperature profiles, especially for η(T,p). Empirically, the Grüneisen parameter is found to decrease upon increasing compression with a power-law index between 1 and 2 (e.g. Anderson, 1987). In Fig. 11 we show the influences of $\gamma(\rho)$ on the thermal profiles associated with constant viscosity (panel a), $\eta(T)$ (panel b) and $\eta(T, p)$ (panel c). Here the effects from $\gamma(\rho)$ are clearly small, with $\eta(T, p)$ causing the biggest differences.

The expression for the thermal conductivity $k(\rho)$, given by eqn. (3), assumes that heat is conducted by the phonon scattering mechanism (Klemens, 1958). However, there can also exist radiative heat-transfer from electromagnetic waves, which may become substantial under lower mantle conditions (Kieffer, 1976; Shankland, 1979). Hence, the total thermal conductivity for the lower mantle is larger than the estimate based on only phonons. For constant opacity, the temperature

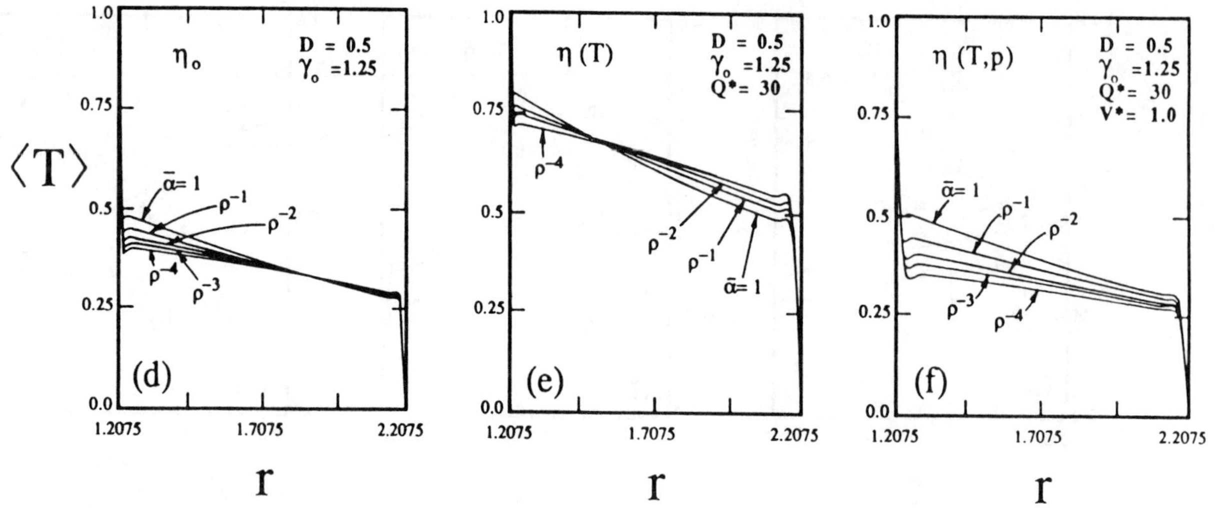

Fig. 9. Mean temperature profiles for various functional dependences of $\alpha(\rho)$. Panel (a) constant viscosity. Other parameters are the same as for panel (a) of Fig. 8. Panels (b) and (c) represent respectively η(T) and η(T,p). Functional dependences of $\alpha(\rho)$ are given adjacent to the curves.

dependence of the thermal conductivity due to electromagnetic radiation is proportional to T^4. But the absorption cross section of mantle minerals is such as to reduce greatly this strong dependence of the thermal conductivity on temperature. We will use k(T) of the form

$$k(T) = \left(\frac{T + T_o}{T_o}\right)^m \tag{35}$$

where m is a parameter used to characterize the degree of nonlinearity. Within the framework of the mean-field theory, k(T) goes into the equation for the mean-temperature (eqn. (31)). The divergence of the conductive flux, D_C, then becomes

$$D_C = k(T)\left(\frac{d^2T}{dr^2} + \frac{2}{r}\frac{dT}{dr}\right) + \frac{dk(T)}{dT}\left(\frac{dT}{dr}\right)^2 \tag{36}$$

Eqn. (36) replaces the first three terms on the right-hand side of eqn. (31). Substituting D_C into eqn. (31), we obtain now a strongly nonlinear O.D.E. for the mean temperature equation.

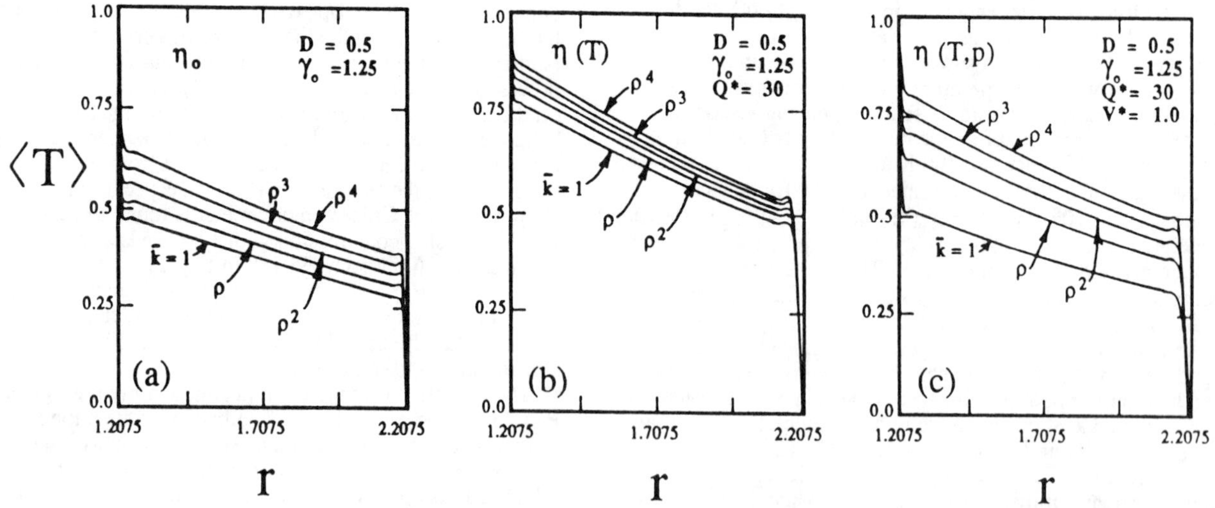

Fig. 10. Mean temperature profiles for different functional dependences of depth-dependent thermal conductivity. Panels (a), (b), (c) have the same parameters as the corresponding panels of Fig. 9. Functional dependences of $k(\rho)$ are given next to the curves.

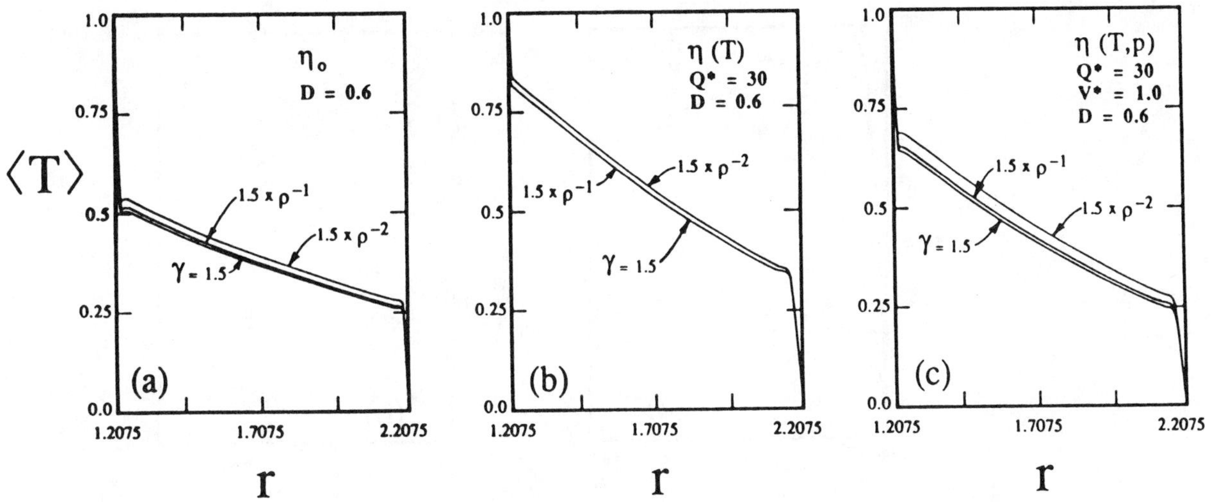

Fig. 11. Mean temperature curves for different functional dependences of the Grüneisen parameter. Panel (a) constant viscosity. Other parameters are $Ra_T = 10^6$, $\gamma_0 = 1.5$, $D_0 = 0.6$. (b) temperature-dependent viscosity. Other parameters are $Ra_0 = 10^8$, $\gamma_0 = 1.5$, $D_0 = 0.6$ and $Q^* = 30$ kcal/mole (c) temperature- and pressure-dependent viscosity. Other parameters are the same as in panel (b), but with $V^* = 1$ cm^3/mole. Functional dependences of $\gamma(\rho)$ are given next to the curves.

The nonlinear effects from k(T) have not been investigated before in mantle convection. We find that upper bounds on the power-law index m in eqn. (35) can be derived from looking at a suite of solutions with varying m. Fig. 12 shows that the interior temperatures rise rapidly with the nonlinearity index m. For $\eta(T, p)$, m certainly cannot be much greater than 2 in order to prevent melting in the lower mantle. This sensitivity of variable viscosity solutions to k(T) reflects the nonlinear feedback between temperature-dependent rheology and temperature-dependent thermal conductivity in the temperature equation. This feedback mechanism comes from the higher temperatures produced by k(T), which, in turn, reduce the temperature-dependent viscosity. It is to be noted that with the addition of mantle radioactivity, smaller upper bounds of m will be obtained, since internal heating would further raise the temperature and drives up the conductivity,

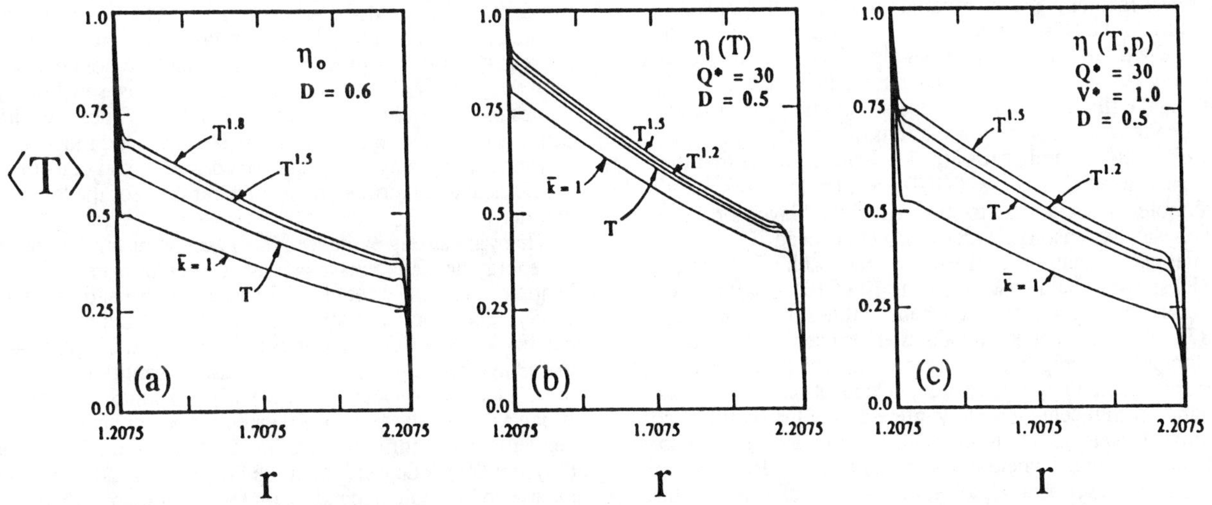

Fig. 12. Mean temperature profiles for various functional dependences of temperature-dependent thermal conductivity. Panel (a) constant viscosity (b) temperature-dependent viscosity (c) temperature- and pressure-dependent viscosity. The physical parameters in the three panels are the same as for the corresponding panels in Fig. 12. Functional dependences, given by eqn. (35), are given next to the curves.

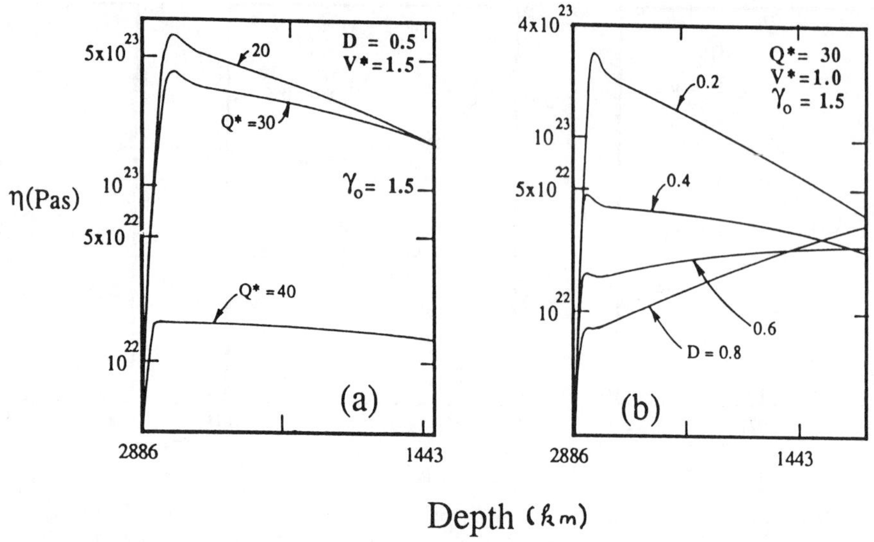

Fig. 13. Viscosity profiles of the lower mantle for temperature- and pressure-dependent rheology. Panel (a) $Ra_0 = 10^8$, $V^* = 1.5$ cm^3/mole, $\gamma_0 = 1.5$, $D_0 = 0.5$. Activation energies in kcal/mole are given next to the curves (b) $Ra_0 = 10^8$, $\gamma_0 = 1.5$, $Q^* = 30$ kcal/mole, and $V^* = 1.0$ cm^3/mole. Dissipation numbers D_0 are given next to curves.

leading to even higher interior temperatures. The central lesson in these calculations is that the temperature-dependence of lower mantle thermal conductivity is most probably weak, with m less than 2. Otherwise, it would be difficult to avoid some amounts of melting in the lower mantle, especially in view of the perovskite solidus (Heinz and Jeanloz, 1987).

Low viscosity channels exist both in the asthenosphere and at the CMB because of the competing effects between temperature and pressure in $\eta(p, T)$. The viscosity structure of the lower mantle plays an important role in the thermal structure of the D"-layer and also in the convective distortion of the CMB, an issue of particular significance to the thermal coupling between the core and mantle (Bloxham and Gubbins, 1987). We illustrate in Fig. 13 the sensitivity of the lower mantle viscosity profiles to variations of the activation energy and dissipation number for $\eta(p, T)$. Panel a shows that the viscosity drops abruptly to $0(10^{22}$ Pa s) for $Q^* = 40$ kcal/mole. Viscosity contrasts across the D" layer for these geophysically reasonable models do not exceed 10^2. Numerical simulations of thermal plumes with a Boussinesq model showed that a viscosity contrast of 10^3 to 10^4 is needed to initiate small-scale convective instabilities (Olson et al., 1987). There activation enthalpies H* exceeding 200 kcal/mole were employed in order to produce these small-scale features. Such a larger magnitude of activation enthalpy is at variance with recent laboratory inferences of Q* (Knittle and Jeanloz, 1987) and V* ((Heinz and Jeanloz, 1987) and also with the dynamical calculations presented here. But this incompressible model (Olson et al., 1987) is of a local nature. The mean-field models presented here give a global description of the thermal mechanical structure and, hence, can lead to much tighter constraints on the geophysically relevant parameters.

The low viscosity channel at the bottom becomes narrower for larger values of Q*. For small activation energies a high viscosity lid is developed right over the low viscosity zone (LVZ). For $V^* = 1.5$ cm^3/mole and $D_0 = 0.5$, the activation energy cannot be smaller than about 35 kcal/mole in the bulk of the lower mantle in order to satisfy viscosity constraints from postglacial rebound and geoid anomalies. The effects of varying D_0 are displayed in panel b. Too small a D_0 would give rise to geophysically unreasonable viscosity structure in the lower mantle. Even in these cases the viscosity contrast across the D" layer does not exceed 10^2. On the other hand, too large a D_0, like $D_0 = 0.8$, will produce a monotonically decreasing viscosity profile, which, again, are not consistent with the geophysical inferences. On this basis, satisfactory values of D_0 would lie between 0.3 and 0.6. These results demonstrate the capability of more sophisticated models, such as variable viscosity compressible convection, in constraining rheological and thermodynamical parameters of the lower mantle.

There are strong evidences from geochemistry that part of the energy needed to drive convection comes from radiogenic heating (e.g. Allegre et el., 1979; Wasserburg and DePaolo, 1979). The exact partitioning between basal and internal heating in mantle convection is an open question, although there have been attempts to estimate this quantity from parameterized convection studies (Spohn and Schubert, 1982). If one assumes the averaged concentrations of heat-producing elements in chondritic meteorites, then the ratio r from eqns. (14) and (20) is determined to lie between 20 and 60 for whole mantle convection, depending on the lower mantle thermal conductivity. In this study we will restrict our attention to low values of r, 5 and 10, for both $\eta(T)$ and $\eta(T, p)$. In Fig. 14 we show the influences on the lower mantle viscosity profiles from some amounts of internal heating. Panel a shows that with the introduction of internal heating viscosity decreases

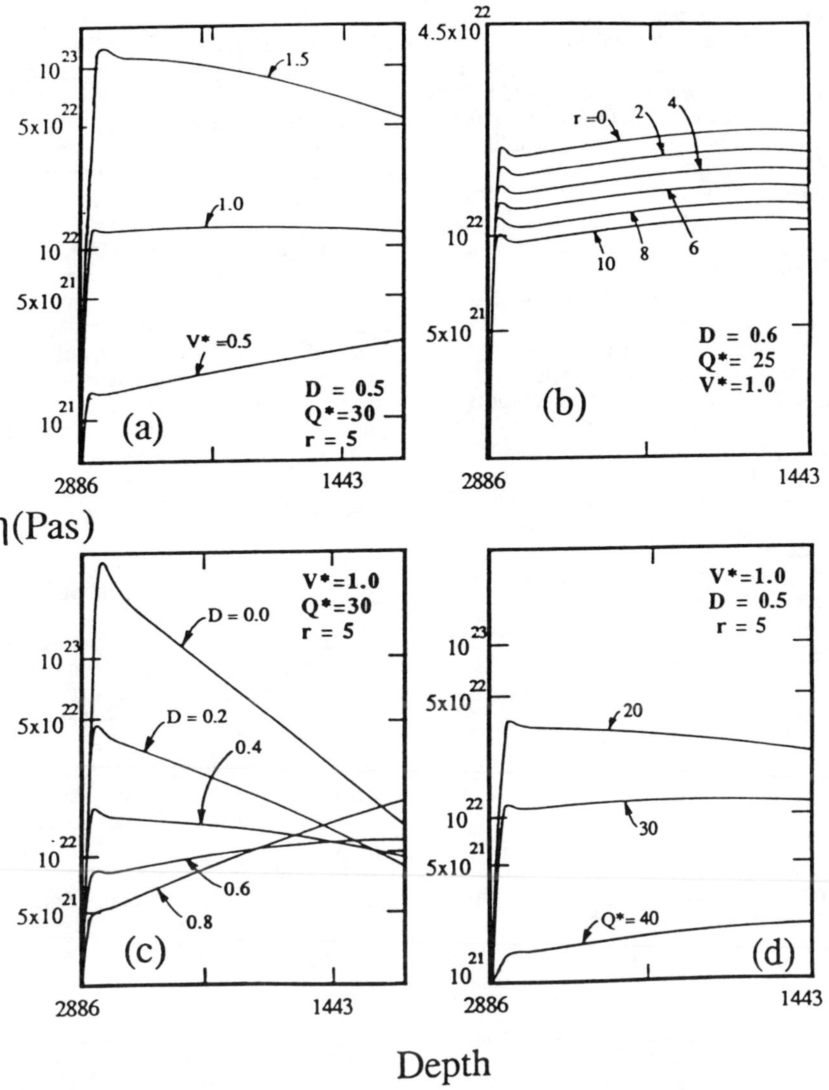

Fig. 14. Viscosity profiles of the lower mantle for temperature- and pressure-dependent rheology with internal heating. (a) Effects of varying activation volume. Other parameters are $Ra_0 = 10^8$, r = 5, D_0 = 0.5 and Q* = 30 kcal/mole. Values of activation volume (cm^3/mole) are given. (b) Effects of varying the amount of internal heating. Other parameters are $Ra_0 = 10^8$, D_0 = 0.6, Q* = 25 kcal/mole and V* = 1.0 cm^3/mole. The internal heating parameter $r = Ra_T/Ra_H$ is given next to the curves. (c) Effects of varying the dissipation number. Other parameters are $Ra_0 = 10^8$, r = 5, Q* = 30 kcal/mole and V* = 1 cm^3/mole. Dissipation numbers are given next to the curves. (d) Effects of varying the activation energy. Other parameters are $Ra_0 = 10^8$, r = 5, D_0 = 0.5 and V* = 1 cm^3/mole. Activation energies are next to the curves.

somewhat from the purely base-heated configuration (cf. panel a of Fig. 13). What is striking is the sharp decrease of lower mantle viscosity with slight changes of V*. In this case the lower mantle viscosity is nearly constant for V* = 1.0 cm^3/mole, but shows positive stratification for V* = 0.5 cm^3/mole. Panel b shows that variations in the viscosity are small for r between 0 and 10. In panel c we display the effects of varying the dissipation number for r = 5. The Boussinesq case (D_0 = 0) exhibits a sharp positive viscosity gradient, culminating in a viscosity lid above the LVZ. This lid is diminished by raising the dissipation number. For D_0 greater than 0.5, a negative viscosity results because of the rise in the temperature due to the steeper adiabat. Finally we show panel d the influences of increasing Q*. For D_0 = 0.5 viscosities for

Q* from 20 to 40 kcal/mole remain relatively constant in the lower mantle, although they differ by more than an order in magnitude. Viscosity contrast across the D"-layer decreases with increasing Q*. This trend is opposite to that produced by the viscosity function used by Olson et al. (1987), who have pinned the viscosity to a given value in the lower mantle above the CMB. Here the viscosity profile in the lower mantle is determined by the temperature imposed at the CMB. Because the temperature jump across the bottom thermal boundary layer decreases strongly with Q* (see Fig. 3), the viscosity contrast is reduced in spite of larger values of Q*.

The shape of the CMB, which is different from that predicted by the hydrostatic equilibrium theory, can yield information important to understanding the style of deep mantle convection and the thermal interaction between the core and mantle (Bloxham and Gubbins, 1987). Recent works from 3-D seismic topography indicate anomalous lateral structure in D" with topographic undulations of around 0 (10 km) (Morelli and Dziewonski, 1987; Creager and Jordan, 1986). The amount of CMB topography induced dynamically may be employed to put bounds on the rheological and thermodynamical parameters, as well as on whole mantle (Zhang and Yuen, 1987), as topography is governed by the viscosity structure in the D". To calculate the spectral amplitude of the CMB distortion ζ_{CMB}, we use the linearized formula, given in dimensional format by

$$\zeta_{CMB} = \frac{\sigma_{rr}}{\Delta\rho g} \qquad (37)$$

where $\Delta\rho = 5500$ kg/m^3 is the assumed density difference between the lower mantle and the core.

The point concerning the dependence of CMB topography on activation energy is emphasized in panel a of Fig. 15, where V* = 1 cm^3/mole is used and Q* spans between 10 and 70 kcal/mole. Extrapolating these results to Q* = 120 kcal/mole (Olson et al., 1987), one would find that, for D_o greater than 0.3, the long wavelength CMB deformation induced dynamically would be smaller than 100 meters! One observes that there is a sharp decrease of CMB topography for Q* around 35 kcal/mole. On the basis of the seismically inferred CMB topography, these results would argue for a relatively small Q* in the lower mantle, much smaller than Q* for olivine in the upper mantle, which is around 100 to 125 kcal/mole. This would, in fact, be consistent with the experimental work on Q* of perovskite by Knittle and Jeanloz (1987), in which small activation energy has been measured for the back transformation of the perovskite to the enstatite structure. The effects of activation volume on CMB topography of the degree two harmonic are summarized in panel b of Fig. 15. We observe a distinctly nonlinear increase of CMB distortion with V*. These results would imply a relatively small activation volume, V* less than 3 cm^3/mole in the lower mantle. Recent experiments on the melting curve of perovskite (Heinz and Jeanloz, 1987) give support to the idea of a low V* there.

Concluding Remarks

The central aim of this paper has been to present a unified description of the usage of both rheology and equation of state in mantle convection models. Our results have demonstrated quite clearly the importance of employing realistic thermodynamical and rheological parameters in the study of the thermal mechanical structure of the lower mantle, in particular, the D"-layer. We have so far been able to establish certain preliminary estimates on the rheological parameters of the lower mantle. Otherwise, the interior temperature would exceed the solidus. On the other hand, the lower mantle activation volume V* cannot be greater than 3 cm^3/mole. Otherwise, too large CMB deformation, exceeding 0(10 km), would be produced from thermal convection.

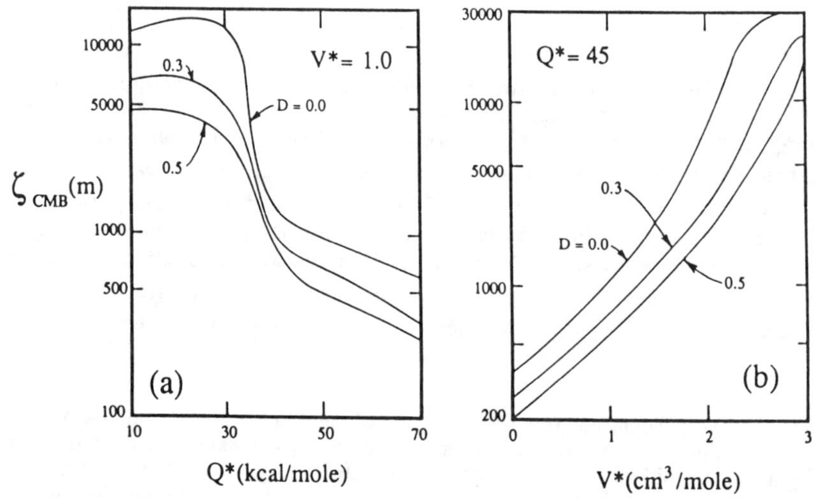

Fig. 15. CMB topography (maximum amplitude) as a function of activation energy (panel a) and of activation volume (panel b). Other common parameters are $Ra_o = 10^8$ and $\gamma = 1.5$. Dissipation numbers are given adjacent to the curves. Degree two harmonic has been employed.

We found that the temperature-dependence of thermal conductivity k(T) must be weak in the lower mantle. Or else the lower mantle temperatures will again exceed the solidus, due to nonlinear positive feedback between k(T) and η(T, p) in the temperature equation. The effects of the phonon contribution to the conductivity, k(ρ), also lead to higher mantle temperatures, whereas the decrease of thermal expansivity with depth in α(ρ) serves to decrease the interior temperature. Taken together, both k(ρ) and α(ρ) act to increase the lower mantle temperature and to decrease the temperature jump at D"-layer. Dynamical effects from the density variation of γ, γ(ρ), are found to be minor. Viscosity structures in the lower mantle are strongly influenced by the rheological parameters, the adiabatic gradient (Quareni, Yuen and Saari, 1986) and the amount of internal heating.

With the compressible model it is now possible to obtain an isoviscous lower mantle for a geophysically reasonable range of rheological and thermodynamical constants. In the past this aspect has eluded all of the variable viscosity, incompressible models (e.g. Christensen, 1984) and those based on a posteriori inclusion of the adiabatic gradient (Sammis et al., 1977). The topography at the CMB depends greatly on the local thermodynamical parameters of the lower mantle. From these results we have seen how each of the nonlinearities in mantle convection (see Fig. 1 again) can exert such a strong influence upon each other so that definite constraints can be placed on a range of plausible parameter values, such as activation volume, activation energy and the local adiabatic gradient. These zeroth order results, derived from the mean-field approximation, will certainly pose challenges for the next generation of compressible convection models. There is still a wealth of challenging fundamental questions in compressible convection that have yet to be adequately dealt with which will require the combined efforts of both dynamicists and mineral physicists.

Acknowledgements. This research has been supported by N.A.S.A. grant NAG 5-770, NAGW 1008 and N.S.F. grant EAR-8511200. We are thankful for discussions with Philippe Machetel, Ulli Hansen and Volker Steinbach. We are appreciative of the work done by Kari L. Rabie, Jill Borofka, Anne Boyd and Sue Selkirk in preparing this manuscript.

References

Allegre, C.J., Othman, D.B., Plove, M. and P. Richard, The Nd-Sr isotopic correlation in mantle materials and geodynamic consequences, Phys. Earth Planet. Inter., 19, 293-306, 1979.

Anderson, D.L., A seismic equation of state II. Shear properties and thermodynamics of the lower mantle, Phys. Earth Planet. Int., 45, 307-323, 1987.

Batchelor, G.K., An Introduction to Fluid Dynamics, Chapter 3, Cambridge Univ. Press, 1967.

Baumgardner, J. R., Three-dimensional treatment of convective flow in the Earth's mantle, J. Stat. Phys., 39, 501-511, 1985.

Birch, F., Elasticity and constitution of the Earth's interior, J. Geophys. Res., 57, 227-286, 1952.

Birch, F.,Finite strain isotherm and velocities for single-crystal and polycrystalline NaCl at high pressures and 300 °K, J. Geophys. Res., 83, 1257-1267, 1978.

Bloxham, J. and D. Gubbins, Thermal core-mantle interactions, Nature, 325, 511-513, 1987.

Böhler, R., The phase diagram of iron to 430 kbar, Geophys. Res. Lett., 13, 1153-1156, 1986.

Böhler, R., von Bargen, N., Hoffbauer, W. and E. Huang, dα/dP at high P and T from synchrotron measurements and systematics in αK, A.G.U. Fall meeting abstract, 1988.

Christensen, U.R., Convection with pressure- and temperature-dependent non-Newtonian rheology, Geophys. J.R. astr. Soc., 77, 343-384, 1984.

Christensen, U.R., Time-dependent convection in elongated Rayleigh-Bénard cells, Geophys. Res. Lett., 14, 220-223, 1987a.

Christensen, U.R. and D.A. Yuen, Time-dependent convection with non-Newtonian viscosity, J. Geophys. Res., in press, 1988.

Creager, K.C. and T.H. Jordan, Aspherical structure of the core-mantle boundary from PKP travel times, Geophys. Res. Lett., 13, 1497-1500, 1986.

Crough, S.T. and D.M. Jurdy, Subducted lithosphere, hotspots and the geoid, Earth Planet. Sci. Lett., 48, 15-22, 1980.

Dziewonski, A.M. and D.L. Anderson, Preliminary reference earth model, Phys. Earth Planet. Int., 25, 297-356, 1981.

Dziewonski, A.M., Mapping the lower mantle: Determination of lateral heterogeneities in P velocity up to degree and order 6, J. Geophys. Res., 89, 5929-5952, 1984.

Fleitout, L.M. and D.A. Yuen, Secondary convection and the growth of the oceanic lithosphere, Phys. Earth Planet. Inter., 36, 181-212, 1984.

Gough, D.O., Spiegel, E.A. and J. Toomre, Modal equations for cellular convection, J. Fluid Mech., 68, 695-719, 1975.

Hager, B.H., R.W. Clayton, M.A. Richards, R.P. Comer and A.M. Dziewonski, Lower mantle heterogeneity, dynamic topography and the geoid, Nature, 313, 541-545, 1985.

Heinz, D.L. and R. Jeanloz, Measurement of the melting curve of $Mg_{.9}Fe_{.1}SiO_3$ at lower mantle conditions and its geophysical implications, J. Geophys. Res., 92, 11437-11444, 1987.

Hemley, R.J., Jackson, M.D. and R.G. Gordon, Theoretical study of the structures, lattice dynamics, and equations of state of perovskite type $MgSiO_3$ and $CaSiO_3$, Phys. Chem. Miner., 14, 2-12, 1987.

Jarvis, G.T. and D.P. McKenzie, Convection in a compressible fluid with infinite Prandtl number, J. Fluid Mech., 96, 515-583, 1980.

Jarvis, G.T. and W.R. Peltier, Mantle convection as a boundary-layer phenomenon, Geophys. J.R. astr. Soc., 68, 385-424, 1982.

Jarvis, G.T. and W.R. Peltier, Lateral heterogeneities in the convecting mantle, J. Geophys. Res., 91, 435-451, 1986.

Karato, S., Rheology of the lower mantle, Phys. Earth Planet. Inter., 24, 1-14, 1981.

Kieffer, S.W., Thermodynamics and lattice vibrations of minerals, 1, Mineral heat capacities to simple lattice vibrational models, Rev. Geophys., 17, 1-19, 1979.

Klemens, P., Thermal conductivity and lattice vibration modes, in Solid State Physics, 7, ed. by F. Seitz and D. Turnbull, pp 1-98, Academic Press, New York, 1958.

Knittle, E., Jeanloz, R. and G.L. Smith, The thermal expansion of silicate perovskite and stratification of the earth's mantle, Nature, 319, 214-216, 1986.

Knittle, E. and R. Jeanloz, High-pressure metallization of FeO and implications for the earth's core, Geophys. Res. Lett., 13, 1541-1544, 1986.

Knittle, E. and R. Jeanloz, The activation energy of the back transformation of silicate perovskite to enstatite, in High-Pressure Research in Geophysics and Geochemistry, ed. by M. Manghnani and Y. Syono, pp 243-250, A.G.U. Washington, D.C., 1987.

Machetel, P. and D.A.Yuen, Chaotic axisymmetrical spherical convection and large-scale mantle circulation, Earth Planet. Sci. Lett., 86, 93-104, 1987.

Machetel, P. and D.A.Yuen, Penetrative convective flows induced by internal heating and mantle compressibility, submitted to J. Geophys. Res., 1988.

Masters, G., Jordan, T.H., Silver, P.G., and F. Gilbert, Aspherical earth structure from fundamental spheroidal mode data, Nature, 298, 609-613, 1982.

McKenzie, D.P., Roberts, J.M. and N.O. Weiss, Convection in the earth's mantle: toward a numerical simulation, J. Fluid Mech., 62, 465-538, 1974.

McQueen, R.G., Marsh, S.P., Taylor, J.W., Fritz, J.N. and W.J. Carter, The equation of state of solids from shock wave studies, in High Velocity Impact Phenomena, ed. by R. Kinslow, pp. 293-417, Academic Press, New York, 1970.

Morelli, A. and A.M. Dziewonski, Topography of the core-mantle boundary and lateral homogeneity of the liquid core, Nature, 325, 678-682, 1987.

Ogura, Y. and N.A.Phillips, Scale analysis of deep and shallow convection in the atmosphere, J. Atm. Sci., 19, 173-179, 1962.

Olson, P.L., and G.M. Corcos, A boundary layer model for mantle convection with surface plates, Geophys. J.R. astr. Soc., 62, 195-219, 1980.

Olson, P.L., G. Schubert and C. Anderson, Plume formation in the D"-layer and the roughness of the core-mantle boundary, Nature, 327, 409-415, 1987.

Quareni, F., D.A. Yuen, G. Sewell and U.R. Christensen, Mean-field convection solutions for strongly variable viscosity and high Rayleigh numbers: a comparison with two-dimensional solutions, J. Geophys. Res., 90, 12,633-12,644, 1985.

Quareni, F., Yuen, D.A. and M.R. Saari, Adiabaticity and viscosity in deep mantle convection, Geophys. Res. Lett., 13, 38-41, 1986.

Quareni, F. and D.A. Yuen, Mean-field methods in mantle convection, in Mathematical Geophysics, ed. by N. Vlaar et al., pp. 227-264, D. Reidel Publishing Co., Dordrecht, Netherlands, 1988.

Sammis, C.G., Smith, J.C., Scubert, G. and D.A.Yuen, Viscosity-depth profile in the Earth's mantle: effects of polymorphic phase transitions, J. Geophys. Res., 85, 3747-3761, 1977.

Shankland, T.J., Physical properties of minerals and melts, Rev. Geophys., 17, 792-802, 1979.

Spera, F.J., Yuen, D.A., Clark, S. and H.J. Hong, Double-diffusive convection in magma chambers: single or multiple layers? Geophys. Res. Lett., 13, 153-156, 1986.

Spohn, T. and G. Schubert, Modes of mantle convection and the removal of heat from the earth's interior, J. Geophys. Res., 87, 4682-4696, 1982.

Stacey, F.D., Brennan, B.J. and R.D. Irvine, Finite strain theories and comparisons with seismological data, Geophys. Surv., 4, 189-232, 1981.

Stevenson, D.J., Spohn, T. and G. Schubert, Magnetism and thermal evolution of the terrestrial planets, Icarus, 54, 466-489, 1983.

Tanimoto,T., The three-dimensional shear wave structure in the mantle by overtone waveform inversion - I. Radial seismogram inversion, Geophys. J.R. astr. Soc., 89, 713-740, 1987.

Turcotte, D.L. and E.R. Oxburgh, Finite amplitude convective cells and continental drift, J. Fluid Mech., 28, 29-42, 1967.

Vincent, A.P. and D.A. Yuen, Thermal attractor in chaotic convection with high Prandtl fluids, Phys. Rev. A., 38, 328-334, 1988.

Wasserburg. G.J. and D. DePaolo, Models of earth inferred from nodynium and strontium isotopic abundances, Proc. Natl. Acad. Sci., U.S.A., 76, 3594, 1979.

Williams, Q., R. Jeanloz, J. Bass, B. Svendsen and T.J. Ahrens, The melting curve of iron to 2.5 Mbar: First experimental constraint on the temperature at the Earth's center, Science, 236, 181-182, 1987.

Williams, Q., Jeanloz, R. and P. McMillan, Vibrational spectrum of $MgSiO_3$ perovskite: zero-pressure Raman and mid-infrared spectra to 27 GPa, J. Geophys. Res., 92, 8116-8128, 1987.

Williams, Q. and R. Jeanloz, Spectroscopic evidence for pressure-induced coordination changes in silicate glasses and melts, Science, 239, 902-905, 1988.

Woodhouse, J.H. and A.M. Dziewonski, Mapping the upper mantle: three-dimensional modelling of earth structure by inversion of seismic waveforms, J. Geophys. Res., 89, 5953-5986, 1984.

Young, C.J. and T. Lay, The core-mantle boundary, Annu. Rev. Earth and Planet. Sci., 15, 25-46, 1987.

Yuen, D.A., R. Sabadini and E.V. Boschi, Viscosity of the lower mantle as inferred from rotational data, J. Geophys. Res., 87, 10 745-10 762, 1982.

Yuen, D.A. and R. Sabadini, Viscosity stratification of the lower mantle as inferred from the J_2 observation, Ann.. Geophys., 3, 647-654, 1985.

Yuen, D.A., F. Quareni and H.-J. Hong, Effects from equation of state and rheology in dissipative heating in compressible mantle convection, Nature, 326, 67-69, 1987.

Yuen, D.A., Zhang, S. and S.E. Langenberger, Effects of compressibility on the temperature jump at the interface of layered, spherical-shell convection, Geophys. Res. Lett., 15, 447-450, 1988.

Zhang, S. and D.A. Yuen, Deformation of the core-mantle boundary induced by spherical-shell, compressible convection, Geophys. Res. Lett., 14, 899-902, 1987.

Zhao, W. and D.A. Yuen, The effects of adiabatic and viscous heatings on plumes, Geophys. Res. Lett., 14, 1223-1227, 1987.